Modern Birkhäuser Classics

Many of the original research and survey monographs, as well as textbooks, in pure and applied mathematics published by Birkhäuser in recent decades have been groundbreaking and have come to be regarded as foundational to the subject. Through the MBC Series, a select number of these modern classics, entirely uncorrected, are being re-released in paperback (and as eBooks) to ensure that these treasures remain accessible to new generations of students, scholars, and researchers.

Introduction to Commutative Algebra and Algebraic Geometry

Ernst Kunz

translated by Michael Ackerman
with a preface by David Mumford

Reprint of the 1985 Edition

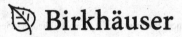 Birkhäuser

Ernst Kunz
Fakultät für Mathematik
Universität Regensburg
Regensburg, Germany

ISBN 978-1-4614-5986-6   ISBN 978-1-4614-5987-3 (eBook)
DOI 10.1007/978-1-4614-5987-3
Springer New York Heidelberg Dordrecht London

Library of Congress Control Number: 2012951658

Printed on acid-free paper

Springer is part of Springer Science+Business Media (www.birkhauser-science.com)

ERNST KUNZ

Introduction to Commutative Algebra and Algebraic Geometry

translated by Michael Ackerman

with a preface by David Mumford

Birkhäuser Boston • Basel • Berlin

Ernst Kunz
Fakultät für Mathematik
Universität Regensburg
8400 Regensburg
GERMANY

Originally published under the title
"Einführung in die kommutative Algebra und algebraische Geometrie"
© Friedr, Viehweg & Sohn Verlagsgesellschaft mbH, Braunschweig 1980

Library of Congress Cataloging-in-Publication Data
    Kunz, Ernst, 1933-
        Introduction to commutative algebra and
    algebraic geometry.
        Translation of: Einführung in die
    kommutative Algebra und algebraische
    Geometrie.
        Bibliography: p.
        Includes index.
        1. Comutative algebra.   2. Geometry,
    Algebraic.   I. Title
    QA 251.3.K8613   512'.24   81-18059
    ISBN 0-8176-3065-1   ISBN 3-7643-3065-1   AACR2

CIP-Kurztitelaufnahme der Deutschen Bibliothek
Kunz, Ernst:
Introduction to commutative algebra and
algebraic geometry/Ernst Kunz -- Boston;
Basel; Berlin: Birkhäuser, 1985
        Einheitssacht: Einführung in die kommutative
        Algebra und algebraische Geometrie (engl.)
        ISBN 0-8176-3605-1   ISBN 3-7643-3065-1

Printed on acid-free paper
© 1985 Birkhäuser Boston

Birkhäuser

ISBN 0-8176-3065-1
ISBN 0-8176-3065-1

Printed and bound by Edwards Brothers, Inc., Ann Arbor, Michigan.
Printed in the U.S.A.

9 8 7 6 5

Contents

Foreword

It has been estimated that, at the present stage of our knowledge, one could give a 200 semester course on commutative algebra and algebraic geometry without ever repeating himself. So any introduction to this subject must be highly selective.

I first want to indicate what point of view guided the selection of material for this book. This introduction arose from lectures for students who had taken a basic course in algebra and could therefore be presumed to have a knowledge of linear algebra, ring and field theory, and Galois theory. The present text shouldn't require much more.

In the lectures and in this text I have undertaken with the fewest possible auxiliary means to lead up to some recent results of commutative algebra and algebraic geometry concerning the representation of algebraic varieties as intersections of the least possible number of hypersurfaces and—a closely related problem—with the most economical generation of ideals in Noetherian rings.

The question of the equations needed to describe an algebraic variety was addressed by Kronecker in 1882. In the 1940s it was chiefly Perron who was interested in this question; his discussions with Severi made the problem known and contributed to sharpening the relevent concepts. Thanks to the general progress of commutative algebra many beautiful results in this circle of questions have been obtained, mainly after the solution of Serre's problem on projective modules. Because of their relatively elementary character they are especially suitable for an introduction to commutative algebra.

If one sets the goal of leading up to these results (and some still unsolved problems), one is led into dealing with a large part of the basic concepts of commutative algebra and algebraic geometry and to proving many facts which can then serve as a basic stock for a deeper study of these subjects. Through the close linking of ring-theoretic problems with those of algebraic geometry, the rôle of commutative algebra in algebraic geometry becomes clear, and conversely the algebraic inquiries are motivated by those of geometric origin.

Since the original question is classical, we begin with classical concepts of algebraic geometry: varieties in affine or projective space. This quite naturally presents us with an opportunity to lead up to the modern generalizations (spectra, schemes) and to exhibit their utility. If the detour is not too great, we shall also pass through neighboring subjects on the way to our main goal. Yet some elementary themes of commutative algebra have been entirely neglected, among them: flat modules, completions, derivations and differentials, Hilbert polynomials and multiplicity theory. From homological algebra we use only projective resolutions and the Snake Lemma. We do not try to derive the most general known form of a proposition if to do so would seem to harm the readability of the text or if the expense seems too great. The references at the end of each chapter and the many exercises, which often contain parts of recent publications, should help the reader to become more deeply informed.

The center of gravity of this book lies more in commutative algebra than in algebraic geometry. For a continued study of algebraic geometry I recommend one of the excellent works which have recently appeared and for which this text may serve as preparation.

I shall now indicate more precisely what knowledge this book assumes.

a) The most common facts of linear and multilinear algebra for modules over commutative rings.

b) The simplest basic concepts of set-theoretic topology.

c) The basic facts of the theory of rings and ideals, including factorial rings (unique factorization domains) and the Noether isomorphism theorems for rings and modules.

d) The theory of algebraic extensions of fields, including Galois theory, as well as the basic facts about transcendence degree and transcendence bases.

Most of what is needed should come up in any introductory course on algebra, so that the book can be read in connection with such a course.

In preparing the text I have been helped by the critical remarks and many good suggestions of H. Knebl, J. Koch, J. Rung, Dr. R. Sacher, and above all Dr. R. Waldi. I have much to thank them for, as well as the Regensburg students who industriously worked on the exercises. My special thanks also goes to Miss Eva Weber for her patience in typing the manuscript.

Ernst Kunz
Regensburg, November 1978

Preface

Dr. Klaus Peters of Birkhäuser Boston has suggested that I write a few words as a Preface to the English edition of Professor Kunz's book. This book will be particularly valuable to the American student because it covers material that is not available in any other textbooks or monographs. The subject of the book is not restricted to commutative algebra developed as a pure discipline for its own sake; nor is it aimed only at algebraic geometry where the intrinsic geometry of a general n-dimensional variety plays the central role. Instead this book is developed around the vital theme that certain areas of both subjects are best understood together. This link between the two subjects, forged in the nineteenth century, built further by Krull and Zariski, remains as active as ever. It deals primarily with polynomial rings and affine algebraic geometry and with elementary and natural questions such as: What are the minimal number of elements needed to generate certain modules over polynomial rings? Great progress has been made on these questions in the last decade. In this book, the reader will find at the same time a leisurely and clear exposition of the basic definitions and results in both algebra and geometry, as well as an exposition of the important recent progress due to Quillen–Suslin, Evans–Eisenbud, Szpiro, Mohan Kumar and others. The ample exericises are another excellent feature. Professor Kunz has filled a longstanding need for an introduction to commutative algebra and algebraic geometry that emphasizes the concrete elementary nature of the objects with which both subjects began.

David Mumford

Preface to the English Edition

The English text is—except for a few minor changes—a translation of the German edition of the book *Einführung in die Kommutative Algebra und Algebraische Geometrie.* Some errors found in the original text have been removed and several passages have been better formulated. In the references the reader's attention is drawn to new findings that are in direct correlation to the contents of the book; the references were expanded accordingly.

I would like to thank all of my colleagues whose criticisms contributed toward the improvement of the text, and naturally, of course, those mathematicians who expressed their recognition of the relevance of the book. My special thanks to the translator, Mr. Michael Ackermann, for his excellent work.

Ernst Kunz
Baton Rouge, November 1981

Terminology

Throughout the book the term ring is used for a commutative ring with identity. Every ring homomorphism $R \to S$ is supposed to map the unit element of R onto the unit element of S; in particular if S/R is an extension of S over R, both S and R have the same unit element. If we say that a subset S of a ring is multiplicatively closed we always assume that $1 \in S$. If M is a module over a ring R the unit element of R operates as identity on M ($1 \cdot m = m$ for all $m \in M$). $A^n(K)$ denotes the n-dimensional affine space over the field K ($n \in \mathbb{N}$), i. e. K^n with the usual affine structure. The affine subspaces of $A^n(K)$ are called "linear varieties." The same holds for the projective space $P^n(K)$.

If not otherwise specified, a corollary to a proposition will contain the same assumptions as the proposition itself. If a statement is quoted, it will be given by its number if the statement is contained in the same chapter. Otherwise the number of the chapter in which the statement is found will be given first (e. g. the theorem of Quillen and Suslin, Chap. IV, 3.14). References from the list of textbooks found at the end of the book are quoted by letter, research papers are quoted by numbers. Some papers which appeared after the publication of the German edition of the book will be referred to in the text or in the footnotes.

Introduction to Commutative Algebra
and Algebraic Geometry

Chapter I
Algebraic varieties

In this chapter affine algebraic varieties are introduced as the solution sets of systems of algebraic equations, and projective varieties are introduced as the solution sets in projective space of systems of algebraic equations involving only homogeneous polynomials. Hilbert's Nullstellensatz gives a necessary and sufficient condition for the solvability of a system of algebraic equations. The basic properties of varieties are discussed, and the relation to ideal theory is established. We then introduce the spectrum of a ring and the homogeneous spectrum of a graded ring and explain in what sense spectra generalize the concepts of affine and projective varieties.

1. Affine algebraic varieties

Let $\mathbb{A}^n(L)$ be n-dimensional affine space over a field L, $K \subset L$ a subfield.

Definition 1.1. A subset $V \subset \mathbb{A}^n(L)$ is called an *affine algebraic K-variety* if there are polynomials $f_1, \ldots, f_m \in K[X_1, \ldots, X_n]$ such that V is the solution set of the system of equations

$$f_i(X_1, \ldots, X_n) = 0 \qquad (i = 1, \ldots, m) \tag{1}$$

in $\mathbb{A}^n(L)$. (1) is called a *system of defining equations* of V, K a field of definition of V, and L the coordinate field.

A K-variety V is also a K'-variety for any subfield $K' \subset L$ that contains all the coefficients of a system of equations defining V (e.g. if $K \subset K'$). The concept of a K-variety is invariant under affine coordinate transformations

$$X_i = \sum_{k=1}^n a_{ik} Y_k + b_i \qquad (i = 1, \ldots, n) \tag{2}$$

if the coefficients a_{ik} and b_i are all in K.

We first consider some
Examples 1.2.

1. *Linear K-varieties.* These are the solution sets of systems of linear equations with coefficients in K. Their investigation is part of "linear algebra."

2. *K-Hypersurfaces.* These are defined by a single equation $f(X_1, \ldots, X_n) = 0$, where $f \in K[X_1, \ldots, X_n]$ is a nonconstant polynomial (cf. Figs. 3–5 and Exercise 2). For $n = 3$ hypersurfaces are also called simply "surfaces." By

Fig. 1

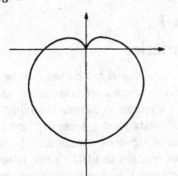

$$(X_1^2 + X_2^2 + 4X_2)^2$$
$$- 16(X_1^2 + X_2^2) = 0$$

Fig. 2

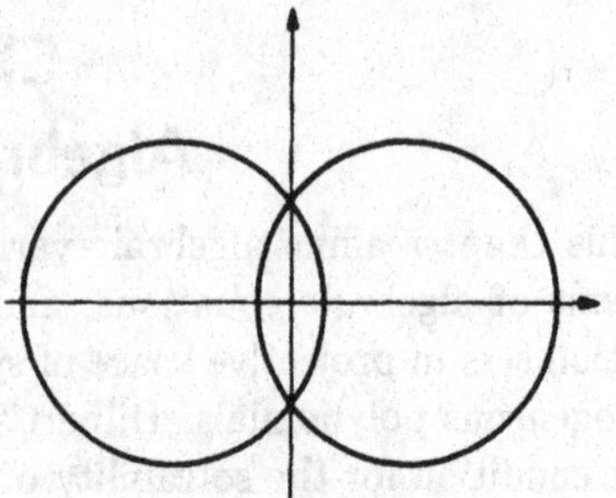

$$(X_1^2 - 9)^2 + (X_2^2 - 16)^2$$
$$+ 2(X_1^2 + 9)(X_2^2 - 16) = 0$$

Fig. 3

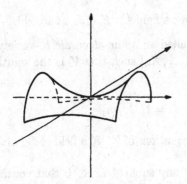

$$X_1^2 - X_2^2 - X_3 = 0$$

Fig. 4

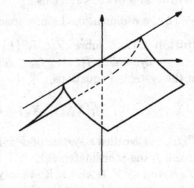

$$X_1^2 + X_3^3 = 0$$

Fig. 5

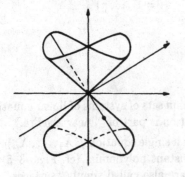

$$X_1^4 + (X_2^2 - X_1^2)X_3^2 = 0$$

Fig. 6

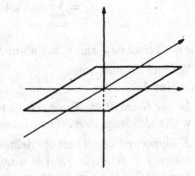

$$X_1 X_3 = 0, \qquad X_2 X_3 = 0$$

definition every affine variety is the intersection of finitely many hypersurfaces. Note that e.g. over the reals a "hypersurface" may be empty or may consist of a single point (cf. also Exercises 3 and 6). Later we shall always assume L algebraically closed; such phenomena cannot occur then.

Hypersurfaces of order 2 (quadrics) are described by equations

$$\sum_{i,k=1}^{n} a_{ik} X_i X_k + \sum_{i=1}^{n} b_i X_i + c = 0$$

(Fig. 3).

3. *Plane algebraic curves* are the hypersurfaces in $\mathbf{A}^2(L)$, i.e. the solution sets of equations $f(X_1, X_2) = 0$ with a nonconstant polynomial f in two variables (Figs. 1 and 2, Exercise 1). Such curves can be treated more simply than arbitrary varieties, and here one can often make more precise statements than in the general case. (Some textbooks that treat plane algebraic curves are: Fulton [L], Seidenberg [S], Semple-Kneebone [T], Walker [W] and Brieskorn-Knörrer [Z].)

4. *Cones.* If a variety V is defined by a system (1) with only homogeneous polynomials f_i, then it is called a K-cone with vertex at the origin. For each $x \in V$, $x \neq (0, \ldots, 0)$, the whole line through x and the origin also belongs to V (Fig. 5).

5. *Quasihomogeneous varieties.* A polynomial

$$f = \sum a_{\nu_1 \ldots \nu_n} X_1^{\nu_1} \ldots X_n^{\nu_n} \in K[X_1, \ldots, X_n]$$

is called quasihomogenous of type $\alpha = (\alpha_1, \ldots, \alpha_n) \in \mathbf{Z}^n$ and degree $d \in \mathbf{Z}$ if $a_{\nu_1 \ldots \nu_n} = 0$ for all $(\nu_1, \ldots, \nu_n)$ with $\sum_{i=1}^{n} \alpha_i \nu_i \neq d$. A variety is called quasihomogeneous if it is defined by a system (1) with only quasihomogeneous polynomials f_i of a fixed type α.

6. *Finite intersections and unions* of affine varieties are affine varieties (Fig. 6). It suffices to see this for two varieties. If one is defined by a system $f_i(X_1, \ldots, X_n) = 0$ $(i = 1, \ldots, m)$ and the other by a system $g_j(X_1, \ldots, X_n) = 0$ $(j = 1, \ldots, l)$, then to get the intersection one just puts the two systems together. To get the union one takes the system

$$f_i(X_1, \ldots, X_n) \cdot g_j(X_1, \ldots, X_n) = 0 \qquad (i = 1, \ldots, m; \, j = 1, \ldots, l).$$

7. *The product of two affine K-varieties.* Let $V \subset \mathbf{A}^n(L)$ be the solution set of a system $f_i(X_1, \ldots, X_n) = 0$ $(i = 1, \ldots, r)$ and $W \subset \mathbf{A}^m(L)$ the solution set of $g_j(Y_1, \ldots, Y_m) = 0$ $(j = 1, \ldots, s)$. Then the cartesian product $V \times W \subset \mathbf{A}^{n+m}(L)$ is described by the union of the two systems, the polynomials now considered as elements of $K[X_1, \ldots, X_n, Y_1, \ldots, Y_m]$.

8. *Affine algebraic groups.* For any matrix $A \in Gl(n, L)$ we can consider $(A, \det A^{-1})$ as a point of $\mathsf{A}^{n^2+1}(L)$. $Gl(n, L)$ is then identified with the hypersurface H:

$$\det(X_{ik})_{i,k=1,\ldots,n} \cdot T - 1 = 0,$$

where the X_{ik} are to be replaced by the coefficients of A and T by $\det A^{-1}$. Matrix multiplication defines a group operation on H:

$$H \times H \to H$$
$$(A, \det A^{-1}) \times (B, \det B^{-1}) \mapsto (A \cdot B, \det(AB)^{-1}).$$

Varieties which, like these, are provided with a group operation, where multiplication and inverse formation are, as with matrices, given by "algebraic relations," are called algebraic *groups*. Their theory is an independent branch of algebraic geometry (a textbook on this subject is Borel [I]).

9. *Rational points of algebraic varieties.* If $V \subset \mathsf{A}^n(L)$ is a variety and $R \subset L$ is a subring, then one is often interested in the question of whether there are points on V all of whose coordinates lie in R ("R-rational points"). For example, the Fermat Problem asks about the existence of nontrivial **Z**-rational points on the "Fermat variety"

$$X_1^n + X_2^n - X_3^n = 0 \qquad (n \geq 3).$$

(A reference for such difficult questions is Lang [Q].)

We now prove some facts about affine varieties, which easily follow from the definition.

Proposition 1.3.

a) If L has infinitely many elements and $n \geq 1$, then outside any K-hypersurface in $\mathsf{A}^n(L)$ there are infinitely many points of $\mathsf{A}^n(L)$. In particular, outside any K-variety $V \subset \mathsf{A}^n(L)$ with $V \neq \mathsf{A}^n(L)$ there are infinitely many points of $\mathsf{A}^n(L)$.

b) If L is algebraically closed and $n \geq 2$, then any K-hypersurface in $\mathsf{A}^n(L)$ contains infinitely many points.

Proof.

a) Let the hypersurface be given by a nonconstant polynomial

$$F \in K[X_1, \ldots, X_n].$$

We may assume that X_n, say, actually occurs in F; we then have a representation

$$F = \varphi_0 + \varphi_1 X_n + \cdots + \varphi_t X_n^t, \tag{3}$$

with $\varphi_i \in K[X_1, \ldots, X_{n-1}]$ $(i = 0, \ldots, t)$, $t > 0$, and $\varphi_t \neq 0$. By the induction hypothesis we may assume that there is an $(x_1, \ldots, x_{n-1}) \in L^{n-1}$

with $\varphi_t(x_1,\ldots,x_{n-1}) \neq 0$. $F(x_1,\ldots,x_{n-1},X_n)$ is then a nonvanishing polynomial in $L[X_n]$. It has only finitely many zeros, but L is infinite. Hence there are infinitely many $x_n \in L$ with $F(x_1,\ldots,x_{n-1},x_n) \neq 0$.

b) Let the hypersurface be given by a polynomial F of the form (3). Then there are infinitely many $(x_1,\ldots,x_{n-1}) \in L^{n-1}$ with $\varphi_t(x_1,\ldots,x_{n-1}) \neq 0$. Since L is algebraically closed, for each of these $(x_1,\ldots,x_{n-1})$ there is an $x_n \in L$ with $F(x_1,\ldots,x_{n-1},x_n) = 0$.

Definition 1.4. For a subset $V \subset \mathbb{A}^n(L)$ the set $\mathfrak{I}(V)$ of all $F \in K[X_1,\ldots,X_n]$ with $F(x) = 0$ for all $x \in V$ is called the *ideal* of V in $K[X_1,\ldots,X_n]$ (the "vanishing ideal").

For hypersurfaces we have

Proposition 1.5.

Let L be algebraically closed and $n \geq 1$. Let $H \subset \mathbb{A}^n(L)$ be a K-hypersurface defined by an equation $F = 0$, and let $F = c \cdot F_1^{\alpha_1} \cdot \ldots \cdot F_s^{\alpha_s}$ be a decomposition of F into a product of powers of pairwise unassociated irreducible polynomials F_i $(c \in K^\times)$. Then $\mathfrak{I}(H) = (F_1 \cdot \ldots \cdot F_s)$.

Proof. Of course $F_1 \cdot \ldots \cdot F_s \in \mathfrak{I}(H)$. It suffices to show that any $G \in \mathfrak{I}(H)$ is divisible by all the F_i $(i = 1,\ldots,s)$. Suppose that, for some $i \in [1,s]$, F_i is not a divisor of G. We can think of F_i as written in the form (3). F_i and G are then (according to Gauss) also relatively prime as elements of $K(X_1,\ldots,X_{n-1})[X_n]$. Hence there are polynomials $a_1, a_2 \in K[X_1,\ldots,X_n]$ and $d \in K[X_1,\ldots,X_{n-1}]$, $d \neq 0$, such that

$$d = a_1 F_i + a_2 G.$$

By 1.3a) there is an $(x_1,\ldots,x_{n-1}) \in L^{n-1}$ with

$$d(x_1,\ldots,x_{n-1}) \cdot \varphi_t(x_1,\ldots,x_{n-1}) \neq 0.$$

We choose $x_n \in L$ with $F_i(x_i,\ldots,x_{n-1},x_n) = 0$. Then $(x_1,\ldots,x_n) \in H$ and so $G(x_1,\ldots,x_n) = 0$. But this is a contradiction, since $d(x_1,\ldots,x_{n-1}) \neq 0$.

Between the K-varieties $V \subset \mathbb{A}^n(L)$ and the ideals of the polynomial ring $K[X_1,\ldots,X_n]$ there is a very close connection, which is the reason that ideal theory is of great significance for algebraic geometry.

We recall the following concepts of ideal theory in a commutative ring with unity.

Definition 1.6.

1. A *system of generators* of an ideal I is a family $\{a_\lambda\}_{\lambda \in \Lambda}$ of elements $a_\lambda \in I$ such that each $a \in I$ is a linear combination of the a_λ with coefficients in R. I is called *finitely generated* if I has a finite system of generators.
2. The *ideal generated by a family* $\{a_\lambda\}_{\lambda \in \Lambda}$ of elements $a_\lambda \in R$ is the set of all linear combinations of the a_λ with coefficients in R. In the future we shall write $(\{a_\lambda\}_{\lambda \in \Lambda})$ for this ideal. By definition the empty family generates the zero ideal.

3. The *sum* $\sum_{\lambda \in \Lambda} I_\lambda$ of a family $\{I_\lambda\}_{\lambda \in \Lambda}$ of ideals of a ring is the set of all sums $\sum_{\lambda \in \Lambda} a_\lambda$ with $a_\lambda \in I_\lambda$, $a_\lambda \neq 0$ for only finitely many λ.

4. The *product* $I_1 \cdot \ldots \cdot I_n$ of finitely many ideals $I_1, \ldots, I_n$ of a ring is the ideal generated by all the products $a_1 \cdot \ldots \cdot a_n$ with $a_j \in I_j (j = 1, \ldots, n)$. In particular, this defines the *n-th power* I^n of an ideal $I : I^n$ is the ideal generated by all the products $a_1 \cdot \ldots \cdot a_n$ $(a_i \in I)$.

5. The *radical* $\mathrm{Rad}(I)$ of an ideal I is the set of all $r \in R$ some power of which lies in I. It is easily shown that $\mathrm{Rad}(I)$ is indeed an ideal. $\mathrm{Rad}(0)$ is called the *nilradical* of R. It consists of all the nilpotent elements of R, so this set is an ideal of R. A ring R is called *reduced* if $\mathrm{Rad}(0) = (0)$. For any ring R, $R_{\mathrm{red}} := R/\mathrm{Rad}(0)$ is reduced. R_{red} is called the *reduced ring* belonging to R.

6. An ideal I of R is called a *prime* ideal if the following holds: If $a, b \in R$ and $a \cdot b \in I$, then $a \in I$ or $b \in I$. I is a prime ideal if and only if R/I is an integral domain. For an arbitrary ideal I we will call any prime ideal of R that contains I a *prime divisor* of I. A prime ideal $\mathfrak{P}$ is called a *minimal prime divisor* of I if $\mathfrak{P}' = \mathfrak{P}$ for any prime divisor $\mathfrak{P}'$ of I with $\mathfrak{P}' \subset \mathfrak{P}$. From the definition of a prime ideal it easily follows that: A prime ideal that contains the intersection (or the product) of two ideals contains one of the two ideals. Moreover, $\mathrm{Rad}(\mathfrak{P}) = \mathfrak{P}$ for any prime ideal $\mathfrak{P}$.

7. An ideal $I \neq R$ is called a *maximal ideal* of R if $I' = I$ for any ideal $I' \neq R$ with $I \subset I'$. An ideal I is maximal if and only if R/I is a field.

8. The *intersection* of a family $\{I_\lambda\}_{\lambda \in \Lambda}$ of ideals of a ring is an ideal. The same holds for the union if the following condition is satisfied: For all $\lambda_1, \lambda_2 \in \Lambda$ there is a $\lambda \in \Lambda$ with $I_{\lambda_1}, I_{\lambda_2} \subset I_\lambda$.

9. Let S/R be an extension of rings, $I \subset R$ an ideal. The *extension ideal* of I in S is the ideal generated by I in S. It is denoted by IS. More generally, if $\varphi : R \to S$ is a homomorphism of rings, IS denotes the ideal of S generated by $\varphi(I)$.

Definition 1.7. The *zero set* in $\mathsf{A}^n(L)$ of an ideal $I \subset K[X_1, \ldots, X_n]$ is the set of all common zeros in $\mathsf{A}^n(L)$ of the polynomials in I. We denote it by $\mathfrak{V}(I)$ (the "variety of I").

Once it is proved that any ideal $I \subset K[X_1, \ldots, X_n]$ has a finite system of generators $f_1, \ldots, f_m$ (§2), it will follow that $\mathfrak{V}(I)$ is a K-variety (with defining system of equations $f_i = 0$ $(i = 1, \ldots, m)$).

For the operations $\mathfrak{I}$ and $\mathfrak{V}$ the following rules hold.

Rules 1.8.

a) $\mathfrak{I}(\mathsf{A}^n(L)) = (0)$ if L is infinite; $\mathfrak{I}(\emptyset) = (1)$.

b) For any set $V \subset \mathsf{A}^n(L)$, $\mathfrak{I}(V) = \mathrm{Rad}(\mathfrak{I}(V))$.

c) For any variety $V \subset \mathsf{A}^n(L)$, $\mathfrak{V}(\mathfrak{I}(V)) = V$.

d) For two varieties V_1, V_2, we have $V_1 \subset V_2$ if and only if $\mathfrak{I}(V_1) \supset \mathfrak{I}(V_2)$, and $V_1 \subsetneq V_2$ if and only if $\mathfrak{I}(V_1) \supsetneq \mathfrak{I}(V_2)$.

e) For two varieties V_1, V_2, we have $\mathfrak{I}(V_1 \cup V_2) = \mathfrak{I}(V_1) \cap \mathfrak{I}(V_2)$ and $V_1 \cup V_2 = \mathfrak{V}(\mathfrak{I}(V_1) \cdot \mathfrak{I}(V_2))$.

f) For any family $\{V_\lambda\}_{\lambda \in \Lambda}$ of varieties V_λ,

$$\bigcap_{\lambda \in \Lambda} V_\lambda = \mathfrak{V}(\sum_{\lambda \in \Lambda} \mathfrak{I}(V_\lambda)).$$

Proof. a), b), e), and f) easily follow from the definitions.

c) Evidently $V \subset \mathfrak{V}(\mathfrak{I}(V))$. On the other hand, if V is the zero set of the polynomials $f_1, \ldots, f_m$, then $f_1, \ldots, f_m \in \mathfrak{I}(V)$ and hence $V = \mathfrak{V}(f_1, \ldots, f_m) \supset \mathfrak{V}(\mathfrak{I}(V))$.

d) From $\mathfrak{I}(V_1) \supset \mathfrak{I}(V_2)$ it follows by c) that $V_1 = \mathfrak{V}(\mathfrak{I}(V_1)) \subset \mathfrak{V}(\mathfrak{I}(V_2)) = V_2$. The remaining statements of d) are now clear.

In particular, the rules show that $V \mapsto \mathfrak{I}(V)$ is an injective, inclusion-reversing mapping of the set of all K-varieties $V \subset \mathsf{A}^n(L)$ into the set of all ideals I of $K[X_1, \ldots, X_n]$ with $\text{Rad}(I) = I$. Hilbert's Nullstellensatz (§3) will show that this mapping is also bijective if L is algebraically closed. Once it is shown that any ideal in $K[X_1, \ldots, X_n]$ is finitely generated, it will follow from 1.8f) that the intersection of an arbitrary family of K-varieties in $\mathsf{A}^n(L)$ is a K-variety.

Definition 1.9. A K-variety V is called *irreducible* if the following holds: If $V = V_1 \cup V_2$ with K-varieties V_1, V_2, then $V = V_1$ or $V = V_2$.

Fig. 6 shows an example of a reducible variety. The concept of irreducibility depends in general on the field of definition K; for example, the solution set in $\mathbb{C}$ of the equation $X^2 + 1 = 0$ is irreducible over $\mathbb{R}$ but not over $\mathbb{C}$.

Proposition 1.10. A K-variety $V \subset \mathsf{A}^n(L)$ is irreducible if and only if its ideal $\mathfrak{I}(V)$ is prime.

Proof. Let V be irreducible and let $f_1, f_2 \in K[X_1, \ldots, X_n]$ be polynomials with $f_1 \cdot f_2 \in \mathfrak{I}(V)$. For $H_i := \mathfrak{V}(f_i)$ $(i = 1, 2)$ we then have $V = (V \cap H_1) \cup (V \cap H_2)$ and so $V = V \cap H_1$ or $V = V \cap H_2$. From $V \subset H_1$ or $V \subset H_2$ it then follows that $f_1 \in \mathfrak{I}(V)$ or $f_2 \in \mathfrak{I}(V)$; i.e. $\mathfrak{I}(V)$ is prime.

Now let $\mathfrak{I}(V)$ be prime. Suppose there are K-varieties V_1, V_2 with $V = V_1 \cup V_2, V \neq V_i$ $(i = 1, 2)$. By 1.8 we have $\mathfrak{I}(V) = \mathfrak{I}(V_1) \cap \mathfrak{I}(V_2)$ and $\mathfrak{I}(V) \neq \mathfrak{I}(V_i)$ $(i = 1, 2)$. Then there are polynomials $f_i \in \mathfrak{I}(V_i), f_i \notin \mathfrak{I}(V)$ $(i = 1, 2)$. But, since $f_1 \cdot f_2 \in \mathfrak{I}(V_1) \cap \mathfrak{I}(V_2)$, we have reached a contradiction.

In the following statements let L be algebraically closed.

Corollary 1.11. A K-hypersurface $H \subset \mathsf{A}^n(L)$ is irreducible if and only if it is the zero set of an irreducible polynomial $F \in K[X_1, \ldots, X_n]$.

Namely, the principal ideal $\mathfrak{I}(H)$ (cf. 1.5) is prime if and only if it is generated by an irreducible polynomial.

Corollary 1.12. A K-hypersurface can be represented in the form

$$H = H_1 \cup \cdots \cup H_s \qquad (H_i \neq H_j \text{ for } i \neq j)$$

with irreducible K-hypersurfaces H_i. This representation is unique (up to its ordering).

Proof. Let $\Im(H) = (F_1 \cdot \ldots \cdot F_s)$ as in 1.5 and $H_i := \mathfrak{V}(F_i)$. Then $H = H_1 \cup \cdots \cup H_s$ ($H_i \neq H_j$ for $i \neq j$) and by 1.11 the H_i are irreducible hypersurfaces. If $H = H_1' \cup \cdots \cup H_t'$ is any such representation, where $\Im(H_j') = (G_j)$ with $G_j \in K[X_1, \ldots, X_n]$ $(j = 1, \ldots, t)$, then $\Im(H) = (F_1 \cdot \ldots \cdot F_s) = (G_1 \cdot \ldots \cdot G_t)$ and hence $F_1 \cdot \ldots \cdot F_s = a G_1 \cdot \ldots \cdot G_t$ with $a \in K^X$. By the theorem on unique factorization in $K[X_1, \ldots, X_n]$, we get $t = s$ and, with a suitable numbering, $H_i' = H_i$ $(i = 1, \ldots, s)$.

The considerations of the next section will show that, just as for hypersurfaces, there is a unique decomposition of an arbitrary variety into irreducible subvarieties. This is important because many questions about varieties can be reduced to questions about irreducible varieties, and these are often easier to answer.

Exercises

1. Sketch the algebraic curves in $\mathbf{R}^2$ given by the following equations (especially in the neighborhood of their "singularities," i.e. where both partial derivatives of the defining polynomial vanish):

$$X_1^3 - X_2^2 = 0, \qquad\qquad X_1^5 + X_1^4 + X_2^2 = 0,$$
$$X_1^3 + X_1^2 - X_2^2 = 0, \qquad\qquad X_1^6 - X_1^4 + X_2^2 = 0,$$
$$X_1^3 + X_1^2 + X_2^2 = 0, \qquad (X_1^2 + X_2^2)^3 - 4X_1^2 X_2^2 = 0,$$
$$X_1^4 - X_1^2 + X_2^2 = 0, \qquad\qquad X_1^n + X_2^n - 1 = 0.$$

   (It is often advantageous to consider the points of intersection of the curve with the lines $X_2 = tX_1$ in order to get a "parametric representation" of the curve.)

2. Describe the following algebraic surfaces in $\mathbf{R}^3$ by comparing their intersections with the planes $X = c$ for variable $c \in \mathbf{R}$:

$$X^2 - Y^2 Z = 0, \qquad (X^2 + Y^2)^3 - ZX^2 Y^2 = 0,$$
$$X^2 + Y^2 + XYZ = 0, \qquad\qquad X^3 + ZX^2 - Y^2 = 0.$$

3. If the field K is *not* algebraically closed, then any K-variety $V \subset \mathbf{A}^n(K)$ can be written as the zero set of a single polynomial in $K[X_1, \ldots, X_n]$. (Hint: It suffices to show that for any $m > 0$ there is a polynomial $\phi \in K[X_1, \ldots, X_m]$ whose only zero is $(0, \ldots, 0) \in \mathbf{A}^m(K)$. If V is defined by a system of equations (1), put $\phi(f_1, \ldots, f_m) = 0$.)

4. Let L/K be an extension of fields, $V \subset A^n(L)$ an L-variety. Then the set $V_K := V \cap A^n(K)$ of all K-rational points of V is a K-variety in $A^n(K)$.

5. Let L/K be a normal field extension. Two points $(x_1, \ldots, x_n)$ and $(y_1, \ldots, y_n)$ in $A^n(L)$ are called *conjugate* over K if there is a K-automorphism σ of L such that $(\sigma(x_1), \ldots, \sigma(x_n)) = (y_1, \ldots, y_n)$.

   a) Any K-variety $V \subset A^n(L)$ that contains x also contains all conjugates of x over K.

   b) If $V \subset A^n(L)$ is a finite set of points with the property that if V contains x then it also contains all conjugates of x over K, then V is a K-variety. (Hint: If $x = (x_1, \ldots, x_n)$, then $K[x_1, \ldots, x_n] \cong K[X_1, \ldots, X_n]/I$ for some ideal I generated by n elements.)

6. Let K be a finite field.

   a) For any $x \in K^n$ there is an $f \in K[X_1, \ldots, X_n]$ with $f(x) = 1$ and $f(y) = 0$ for all $y \in K^n \setminus \{x\}$.

   b) For any function $g : K^n \to K$ there is an $f \in K[X_1, \ldots, X_n]$ with $g(x) = f(x)$ for all $x \in K^n$.

   c) Any subset $V \subset K^n$ is the zero set of a suitable polynomial $f \in K[X_1, \ldots, X_n]$.

   (c) follows from b), but also from Exercise 3.)

7. Let K be a field. A system of equations

$$F(X_1, X_2) = 0, \qquad G(X_1, X_2) = 0$$

with two relatively prime polynomials $F, G \in K[X_1, X_2]$ has at most finitely many solutions in K^2. (Hint: Apply the argument in the proof of Proposition 1.5.)

8. Let $V \subset A^n(\mathbb{C})$ be an algebraic variety, $\mathbb{Z}^n \subset A^n(\mathbb{C})$ the set of all "lattice points", i.e. the set of points with integral coordinates. If $\mathbb{Z}^n \subset V$, then $V = A^n(\mathbb{C})$.

In the next two exercises the hypotheses and statements are not so sharply defined as hitherto. Through these exercises the reader should become confident that he can, for the most part, operate with polynomials over arbitrary rings "as usual."

9. Two polynomials $f, g \in \mathbb{Z}[X_1, \ldots, X_n]$ coincide if and only if their function values coincide when the variables are specialized to elements of arbitrary fields. A "polynomial formula" holds in every ring if and only if it holds in every field. In particular, the formulas of the theory of determinants over fields in which no "denominator" occurs also hold for determinants with coefficients in an arbitrary ring. If denominators occur (as in Cramer's Rule), a formula that holds in rings is gotten by multiplying by "the product of the denominators."

10. For a polynomial $F = \sum a_{\nu_1 \ldots \nu_n} X_1^{\nu_1} \ldots X_n^{\nu_n}$ with coefficients $a_{\nu_1} \ldots_{\nu_n}$ in a ring R, the formal partial derivative with respect to X_i is defined as

$$\frac{\partial F}{\partial X_i} := \sum \nu_i a_{\nu_1 \ldots \nu_n} X_1^{\nu_1} \ldots X_i^{\nu_i - 1} \ldots X_n^{\nu_n}.$$

Using the principle of Exercise 9 convince yourself that the formulas of differential calculus for polynomial functions (over the reals) also hold for polynomials with coefficients in arbitrary rings if the partial derivatives are taken formally. Some care is needed only because it can happen that $\nu_i a_{\nu_1 \ldots \nu_n} = 0$, although $\nu_i \neq 0$ and $a_{\nu_1 \ldots \nu_n} \neq 0$.

2. The Hilbert Basis Theorem. Decomposition of a variety into irreducible components

We first show that ideals in polynomial rings over fields are finitely generated and then derive some conseqences of this fact.

Definition 2.1. A ring† R is called *Noetherian* if any ideal of R has a finite system of generators.

Examples of Noetherian rings are the principal ideal rings, in particular all fields, as well as $\mathbf{Z}$ and $K[X]$, if K is a field. Any homomorphic image of a Noetherian ring is Noetherian.

In the sequel let R be a ring.

Proposition 2.2. The following statements are equivalent.

a) R is Noetherian.

b) The Ascending Chain Condition for ideals holds: Any ascending chain of ideals of R

$$I_1 \subset I_2 \subset \cdots \subset I_n \subset \cdots$$

becomes stationary.

c) The maximal condition for ideals holds: Any nonempty set of ideals of R contains a maximal element (with respect to inclusion).

Proof.

a)$\rightarrow$b). For a chain of ideals as in b), $I := \bigcup_{n=1}^{\infty} I_n$ is also an ideal of R. By hypothesis it is finitely generated: $I = (r_1, \ldots, r_m)$, $r_i \in R$. For sufficiently large n we then have $r_i \in I_n$ $(i = 1, \ldots, m)$, and it follows that $I_n = I_{n+1} = \cdots$.

b)$\rightarrow$c). Suppose there is a nonempty set M of ideals of R without a maximal element. For each ideal $I_1 \in M$ there is then an $I_2 \in M$ with $I_1 \subsetneq I_2$. In this way one can construct a chain of ideals that is not stationary.

---

† As stated at the beginning, by a "ring" we always mean a commutative ring with unity.

c)→b). Apply the maximal condition to the set of ideals in a chain of ideals.

b)→a). Suppose there is an ideal I of R that is not finitely generated. If $r_1, \ldots, r_m \in I$, then $(r_1, \ldots, r_m) \neq I$. Hence there is an $r_{m+1} \in I$, $r_{m+1} \notin (r_1, \ldots, r_m)$. Construct a chain of ideals

$$(r_1) \subsetneq (r_1, r_2) \subsetneq (r_1, r_2, r_3) \subsetneq \cdots,$$

contradicting hypothesis b).

The following theorem provides a large class of Noetherian rings.

Proposition 2.3. (Hilbert's Basis Theorem) If R is a Noetherian ring, so is $R[X]$.

What is probably the briefest possible proof is due to Heidrun Sarges [68]. We show that if $R[X]$ is not Noetherian, then neither is R. Let I be an ideal of $R[X]$ that is not finitely generated. Let $f_1 \in I$ be a polynomial of least degree. If f_k $(k \geq 1)$ has already been chosen, let f_{k+1} be a polynomial of least degree in $I \setminus (f_1, \ldots, f_k)$. Let n_k be the degree and $a_k \in R$ the leading coefficient of f_k $(k = 1, 2, \ldots)$. By the choice of f_k we have $n_1 \leq n_2 \leq \cdots$. Moreover, $(a_1) \subset (a_1, a_2) \subset \cdots$ is a chain of ideals that does not become stationary. For suppose $(a_1, \ldots, a_k) = (a_1, \ldots, a_{k+1})$. Then we have an equation $a_{k+1} = \sum_{i=1}^{k} b_i a_i$ $(b_i \in R)$ and $g := f_{k+1} - \sum_{i=1}^{k} b_i X^{n_{k+1} - n_i} f_i \in I \setminus (f_1, \ldots, f_k)$ is of lower degree than f_{k+1}, contradicting the choice of f_{k+1}.

Corollary 2.4. Let R be a Noetherian ring and S an extension ring of R that is finitely generated over R (in the ring sense). Then S is also Noetherian.

Proof. S is a homomorphic image of a polynomial ring $R[X_1, \ldots, X_n]$, so it suffices to show that the latter is Noetherian. But this follows from 2.3 by induction on n.

In particular, for a principal ideal ring R the polynomial ring $R[X_1, \ldots, X_n]$ and its homomorphic images are Noetherian, in particular $\mathbb{Z}[X_1, \ldots, X_n]$ and $K[X_1, \ldots, X_n]$ for any field K. The last fact has the following consequences for algebraic varieties.

Let L/K be a field extension.

Corollary 2.5. Every decreasing chain

$$V_1 \supset V_2 \supset \cdots \supset V_i \supset \cdots$$

of affine K-varieties $V_i \subset \mathbb{A}^n(L)$ is stationary.

This follows, because the corresponding chain $\mathfrak{I}(V_1) \subset \mathfrak{I}(V_2) \subset \cdots$ of ideals in $K[X_1, \ldots, X_n]$ becomes stationary (1.8d).

Corollary 2.6. For an ideal I of $K[X_1, \ldots, X_n]$, $\mathfrak{V}(I)$ is a K-variety in $\mathbb{A}^n(L)$.

For if $I = (f_1, \ldots, f_m)$, then $\mathfrak{V}(I)$ is the solution set of the system of equations $f_i = 0$ $(i = 1, \ldots, m)$ in $\mathbb{A}^n(L)$.

Corollary 2.7. If $\{V_\lambda\}_{\lambda \in \Lambda}$ is an arbitrary family of K-varieties in $\mathsf{A}^n(L)$, then $\bigcap_{\lambda \in \Lambda} V_\lambda$ is a K-variety.

By 1.8f), $\bigcap_{\lambda \in \Lambda} V_\lambda = \mathfrak{V}(\sum_{\lambda \in \Lambda} \mathfrak{I}(V_\lambda))$.

Since finite unions and arbitrary intersections of K-varieties in $\mathsf{A}^n(L)$ are K-varieties, the K-varieties form the closed sets of a topology on $\mathsf{A}^n(L)$, the *K-topology* or *Zariski topology* on $\mathsf{A}^n(L)$ with respect to K. If $V \subset \mathsf{A}^n(L)$ is a K-variety, then V carries the relative topology (the Zariski topology on V). Its closed sets are the *subvarieties* $W \subset V$, i.e. the K-varieties contained in V. This topology will often play a role later. We now apply the Zariski topology to show that any variety has a unique decomposition into irreducible components.

Definition 2.8. A topological space X is called *irreducible* if for any decomposition $X = A_1 \cup A_2$ with closed subsets $A_i \subset X (i = 1, 2)$ we have $X = A_1$ or $X = A_2$. A subset X' of a topological space X is called *irreducible* if X' is irreducible as a space with the induced topology.

It is clear that an algebraic K-variety is irreducible in the sense of Definition 1.9 if and only if it is irreducible as a topological space with the Zariski topology with respect to K.

Lemma 2.9. For a topological space X the following statements are equivalent.

a) X is irreducible.
b) If U_1, U_2 are open subsets of X, and if $U_i \neq \emptyset$ $(i = 1, 2)$, then $U_1 \cap U_2 \neq \emptyset$.
c) Any nonempty open subset of X is dense in X.

Proof. a)$\leftrightarrow$b) follows from Definition 2.8 by taking complements. b)$\leftrightarrow$c) is clear by the definition of density.

Corollary 2.10. For a subset X' of a topological space X the following statements are equivalent.

a) X' is irreducible.
b) If U_1, U_2 are open subsets of X with $U_i \cap X' \neq \emptyset$ $(i = 1, 2)$, then $U_1 \cap U_2 \cap X' \neq \emptyset$.
c) The closure $\overline{X}'$ of X' is irreducible.

Proof. a)$\leftrightarrow$b) is a consequence of 2.9; and b)$\leftrightarrow$c) follows from the fact that an open set meets X' if and only if it meets $\overline{X}'$.

Definition 2.11. An *irreducible component* of a topological space X is a maximal irreducible subset of X.

By 2.10 the irreducible components are closed, and so in the case of an algebraic variety are subvarieties.

Proposition 2.12.

a) Any irreducible subset of a topological space is contained in an irreducible component.
b) Any topological space is the union of its irreducible components.

Proof. Since every point $x \in X$ is irreducible, b) follows from a). a) is gotten using Zorn's Lemma. For an irreducible subset X' of a space X consider the set M of the irreducible subsets of X that contain X'. M is not empty, and for a linearly ordered family $\{X_\lambda\}_{\lambda \in \Lambda}$ of elements $X_\lambda \in M, Y := \bigcup_{\lambda \in \Lambda} X_\lambda$ is also an element of M: If U_1, U_2 are open sets with $U_i \cap Y \neq \emptyset$ $(i = 1, 2)$, then there are indices $\lambda_1, \lambda_2 \in \Lambda$ with $U_i \cap X_{\lambda_i} \neq \emptyset$ $(i = 1, 2)$. If, say $X_{\lambda_2} \subset X_{\lambda_1}$, then $U_1 \cap U_2 \cap X_{\lambda_1} \neq \emptyset$ by 2.10, so $U_1 \cap U_2 \cap Y \neq \emptyset$, and Y too is irreducible.

By Zorn's Lemma M has a maximal element. It is an irreducible component of X that contains X'.

Definition 2.13. A topological space X is called *Noetherian* if every descending chain $A_1 \supset A_2 \supset \cdots$ of closed subsets $A_i \subset X$ is stationary.

By 2.5 a K-variety V as a topological space with the Zariski topology is Noetherian. It is clear that a topological space is Noetherian if and only if in it ascending chains of open sets are stationary, or the maximal condition for open sets or the minimal condition for closed sets holds.

Proposition 2.14. A Noetherian topological space has only finitely many irreducible components. No component is contained in the union of the others.

Proof (by Noetherian recursion). Let X be a Noetherian topological space and M the set of all closed subsets of X that cannot be written as a finite union of irreducible subsets of X. Suppose M is not empty.

By the minimal condition there is a minimal element $Y \in M$. Y is not irreducible, so there are closed subsets Y_1, Y_2 of Y with $Y = Y_1 \cup Y_2, Y_i \neq Y$ $(i = 1, 2)$. Since $Y_i \notin M$, Y_i is a finite union of irreducible subsets of X, and therefore so is Y, a contradiction.

Since we now have $M = \emptyset$, any closed subset of X, and thus X itself, can be represented as a finite union of irreducible subsets. By 2.12a) we have $X = X_1 \cup \ldots \cup X_n$ with irreducible components X_i of X, $X_i \neq X_j$ for $i \neq j$.

If Y is an arbitrary irreducible component, it follows from the relation $Y = \bigcup_{i=1}^n (X_i \cap Y)$ that $Y = X_i \cap Y$ for suitable i, that is $Y = X_i$: therefore, all components occur among the X_i. Nor can we have $X_i \subset \bigcup_{j \neq i} X_j$, since otherwise $X_i = X_j$ for some $j \neq i$. This proves the proposition.

Corollary 2.15. Any K-variety $V \subset \mathbb{A}^n(L)$ has only finitely many irreducible components $V_1, \ldots, V_s$. We have $V = V_1 \cup \cdots \cup V_s$, and in this representation no V_i is superfluous.

It is easily seen that the irreducible components of a hypersurface H are just the irreducible hypersurfaces that were called the components of H in §1.

These observations on topological spaces are convenient in the general form given, for they can be applied again and again to cases that arise in algebraic geometry.

In the next section we need the Hilbert Basis Theorem for modules. We first recall some basic concepts about modules. We consider only R-modules M with $1 \cdot m = m$ for all $m \in M$.

1. A *system of generators* of M is a family $\{m_\lambda\}_{\lambda \in \Lambda}$ of elements $m_\lambda \in M$ such that any $m \in M$ can be represented as a linear combination of finitely many of the m_λ with coefficients in R. We then write $M = \langle \{m_\lambda\}_{\lambda \in \Lambda} \rangle$.

2. M is called *finitely generated* if it has a finite system of generators, *cyclic* (or *monogenic*) if it is generated by one element.

3. M is called *free* if it has a basis, i.e. a linearly independent system of generators. For example, $M = R^n$ is a free R-module. As is well known, it has a canonical basis. If M is free, then a linear mapping $l : M \to N$ into an R-module N is uniquely determined by the images of the elements of a basis. If one prescribes arbitrary images in N for the basis elements, then there is always a linear mapping that takes the prescribed values on the basis.

4. The *rank of a free R-module M* is by definition equal to the cardinality of a basis of M. One can show that this is independent of the basis chosen. If M has a (finite) basis $\{b_1, \ldots, b_n\}$ and if $m_i = \sum_{k=1}^{n} r_{ik} b_k$ $(i = 1, \ldots, n)$ are elements of M, then $\{m_1, \ldots, m_n\}$ is a basis of M if and only if $\det(r_{ik})$ is a unit of R. If M is free of rank n, then $M \cong R^n$.

Definition 2.16. An R-module M is called *Noetherian* if every submodule U of M is finitely generated.

Proposition 2.17. *If R is Noetherian and M is a finitely generated R-module, then M is Noetherian.*

Proof. Let $M = \langle m_1, \ldots, m_n \rangle$. Then there is a unique (surjective) linear mapping $\varphi : R^n \to M$ that maps the i-th canonical basis vector e_i to m_i $(i = 1, \ldots, n)$. It suffices to show that any submodule $U \subset R^n$ is finitely generated, since every submodule of M is a homomorphic image of such a module.

For the elements $u = (u_1, \ldots, u_n) \in U$ the first components u_1 form an ideal I in R. By hypothesis it is finitely generated: $I = (u_1^{(1)}, \ldots, u_1^{(k)})$. For $n = 1$ we have finished.

In the general case we consider elements $u^{(i)} \in U$ with first components $u_1^{(i)}$ $(i = 1, \ldots, k)$. For an arbitrary $u \in U$ let $u_1 = \sum_{i=1}^{k} r_i u_1^{(i)}$ $(r_i \in R)$. Then $u - \sum_{i=1}^{k} r_i u^{(i)}$ is of the form $(0, u_2^*, \ldots, u_n^*)$ and is thus an element of $U \cap R^{n-1}$ if R^{n-1} here denotes the submodule of R^n consisting of all the elements with first component 0. By the induction hypothesis, $U \cap R^{n-1}$ has a finite system of generators $\{v_1, \ldots, v_l\}$. Then $\{u^{(1)}, \ldots, u^{(k)}, v_1, \ldots, v_l\}$ is a system of generators of U.

Using the argument in this proof one can also easily show:

Remark 2.18. If R is a principal ideal domain, then every submodule $U \subset R^n$ has a basis (of length $\leq n$).

Namely, $I = (u_1^{(1)})$ is a principal ideal. If $u_1^{(1)} = 0$, then $U \subset R^{n-1}$, and, by the induction hypothesis, we have finished. If $u_1^{(1)} \neq 0$, then choose a $u^{(1)} \in U$ with first component $u_1^{(1)}$. It follows that $u^{(1)}$ together with a basis of $U \cap R^{n-1}$ forms a basis of U.

Exercises

1. Let K be an algebraically closed field. In K^2 consider the set C of all points (t^p, t^q) with fixed numbers $p, q \in \mathbb{Z}, p, q > 0$, where t varies over all of K. C is an algebraic variety. Determine its ideal $\mathfrak{I}(C) \subset K[X_1, X_2]$.

2. Let K be an infinite field, $V \subset \mathbb{A}^n(K)$ a finite set of points. Its ideal $\mathfrak{I}(V)$ in $K[X_1, \ldots, X_n]$ is generated by n polynomials. (Interpolation!)

3. Let L/K be an extension of fields, where L has infinitely many elements. For $f_1, \ldots, f_n \in K[T_1, \ldots, T_m]$ consider the closure V in $\mathbb{A}^n(L)$ (in the K-Zariski topology) of the set V_0 of all the points $(f_1(t_1, \ldots, t_m), \ldots, f_n(t_1, \ldots, t_m))$, where $(t_1, \ldots, t_m)$ varies over all of L^m. V is irreducible. (Hint: Consider the K-homomorphism

$$K[X_1, \ldots, X_n] \to K[T_1, \ldots, T_m] \quad (X_i \mapsto f_i(T_1, \ldots, T_m))$$

and apply Proposition 1.3a).) (In this situation one says that V is given by a "polynomial parametrization" with parameters $T_1, \ldots, T_m$.)

4. In the situation of Exercise 3 give an example of the case where V_0 need not be closed.

5. For L/K as in Exercise 3 show that any linear K-variety in $\mathbb{A}^n(L)$ is irreducible.

6. An irreducible real variety $V \subset \mathbb{R}^n$ is connected in the Zariski topology but need not be connected in the usual topology on $\mathbb{R}^n$. Give an example of this.

7. Let L/K be an extension of fields, $V \subset \mathbb{A}^n(K)$ a K-variety, $\overline{V} \subset \mathbb{A}^n(L)$ its closure in the L-topology on $\mathbb{A}^n(L)$.

   a) The ideal $\mathfrak{I}(\overline{V})$ of $\overline{V}$ in $L[X_1, \ldots, X_n]$ is the extension ideal

   $$\mathfrak{I}(V) \cdot L[X_1, \ldots, X_n]$$

   of the ideal $\mathfrak{I}(V)$ in $K[X_1, \ldots, X_n]$.

   b) $V = \overline{V} \cap \mathbb{A}^n(K)$.

   c) If $\overline{V} = V_1^* \cup \cdots \cup V_s^*$ is the decomposition of $\overline{V}$ into irreducible components (with respect to the L-topology), and $V_i := V_i^* \cap \mathbb{A}^n(K)$ $(i = 1, \ldots, s)$, then $V = V_1 \cup \cdots \cup V_s$ is the decomposition of V into irreducible components (with respect to the K-topology on $\mathbb{A}^n(K)$). Further, $V_i^* = \overline{V}_i$ is the closure of V_i in $\mathbb{A}^n(L)$. (In case $K = \mathbb{R}, L = \mathbb{C}, \overline{V}$ is called the "complexification" of the $\mathbb{R}$-variety $V \subset \mathbb{A}^n(\mathbb{R})$.)

8. Let $\{r_\lambda\}_{\lambda \in \Lambda}$ be a system of generators of a finitely generated module (or ideal) I. There are finitely many indices λ_i $(i = 1, \ldots, n)$ such that $\{r_{\lambda_1}, \ldots, r_{\lambda_n}\}$ is also a system of generators of I.

9. Let I be an ideal of the polynomial ring $K[X_1, \ldots, X_n]$ over a field K. A subfield $K' \subset K$ is called a *field of definition* of I if I has a system of generators consisting of elements of $K'[X_1, \ldots, X_n]$. Show that any ideal $I \subset K[X_1, \ldots, X_n]$ has a smallest field of definition K_0 (i.e. one that is contained in every field of definition of I). Hint: $K[X_1, \ldots, X_n]/I$ has a K-vector space basis consisting of the images of some of the monomials $X_1^{\alpha_1} \ldots X_n^{\alpha_n}$. All other monomials can be expressed modulo I as linear combinations of these with coefficients in K. One adjoins all these coefficients to the prime field of K and gets K_0.

10. Any surjective endomorphism φ of a Noetherian ring R is an automorphism. Hint: Consider $\mathrm{Ker}(\varphi^n)$ for all $n \in \mathsf{N}$.

3. Hilbert's Nullstellensatz

For an ideal I of the polynomial ring $K[X_1, \ldots, X_n]$ over a field K and an extension L/K, the zero set $\mathfrak{V}(I) \subset \mathsf{A}^n(L)$ may be empty. However, we do have the following theorem, which is fundamental to algebraic geometry.

Theorem 3.1. (Hilbert's Nullstellensatz) If L is algebraically closed and $I \neq K[X_1, \ldots, X_n]$, then $\mathfrak{V}(I)$ is not empty.

The theorem is equivalent to an assertion about field extensions:

Proposition 3.2. (Hilbert's Nullstellensatz, field-theoretic form) If A/K is an extension of fields and A arises from K through ring adjunction of finitely many elements, then A/K is algebraic.

From 3.2 follows 3.1. For I there is a maximal ideal M of $K[X_1, \ldots, X_n]$ with $M \supset I$. $A := K[X_1, \ldots, X_n]/M$ is then a field which arises from K through ring adjunction of the residue classes ξ_i of the X_i $(i = 1, \ldots, n)$. By 3.2 A/K is algebraic, so there is a K-homomorphism $\phi : A \to L$, since L is algebraically closed. Then $(\phi(\xi_1), \ldots, \phi(\xi_n)) \in L^n$ is a zero of M, and therefore of I too.

Conversely, 3.2 follows from 3.1. If the field A arises from K through ring adjunction of finitely many elements, then $A \cong K[X_1, \ldots, X_n]/M$ with a maximal ideal M. By 3.1 M has a zero $(\xi_1, \ldots, \xi_n) \in \overline{K}^n$, where $\overline{K}$ is the algebraic closure of K. One has a K-homomorphism $K[X_1, \ldots, X_n] \to \overline{K}, X_i \mapsto \xi_i$ with kernel M and thus a K-isomorphism $A \cong K[\xi_1, \ldots, \xi_n]$. Since the ξ_i are algebraic over K, A/K is also algebraic.

The proof of 3.2 (according to Artin–Tate [5] and Zariski [83]) rests on the next two lemmas.

Lemma 3.3. Let $R \subset S \subset T$ be rings, let R be Noetherian and $T = R[x_1, \ldots, x_n]$ with $x_1, \ldots, x_n \in T$. Assume T is finitely generated as an S-module. Then S is also finitely generated as a ring over R.

Proof. We consider a system of generators $\{w_1, \ldots, w_m\}$ of the S-module T that contains the x_i $(i = 1, \ldots, n)$ and its "multiplication table"

$$w_i w_k = \sum_{l=1}^{m} a_l^{ik} w_l \qquad (i, k = 1, \ldots, m; a_l^{ik} \in S).$$

For $S' := R[\{a_l^{ik}\}_{i,k,l=1,\ldots,m}]$ we then have $T = S'w_1 + \cdots + S'w_m$, since along with the a_l^{ik} all the products of powers of the x_i lie in $S'w_1 + \cdots + S'w_m$, therefore so does all of T, since $T = R[x_1, \ldots, x_n]$.

Since R is Noetherian, S' is also Noetherian by Hilbert's Basis Theorem for rings; and T is a Noetherian S'-module by Hilbert's Basis Theorem for modules. Since $S' \subset S \subset T$, it follows that S is finitely generated as an S'-module, and therefore also as a ring over R.

Lemma 3.4. Let $S = K(Z_1, \ldots, Z_t)$ be a rational function field over a field K, where $t > 0$. Then S is not finitely generable as a ring over K.

Proof. Suppose $\{x_1, \ldots, x_m\}$ is a ring generating system of S/K, where $x_i = \frac{f_i(Z_1, \ldots, Z_t)}{g_i(Z_1, \ldots, Z_t)}$ $(i = 1, \ldots, m)$ with polynomials f_i, g_i. Because $S = K[x_1, \ldots, x_m]$, every element of S can be represented as a quotient of two polynomials in $K[Z_1, \ldots, Z_t]$, where the denominator contains at most the irreducible polynomials that divide one of the g_i. For a prime polynomial p that divides none of the g_i, by the theorem on unique factorization in $K(Z_1, \ldots, Z_t)$ we see that $\frac{1}{p}$ can have no such representation. Since $K[Z_1, \ldots, Z_t]$ has infinitely many pairwise unassociated prime polynomials, but only finitely many prime polynomials divide the g_i, there is such a p, and we have reached a contradiction.

The proof of 3.2 is now gotten at once. If A/K is transcendental and $\{Z_1, \ldots, Z_t\}, t > 0$, is a transcendence basis, then by 3.3 $S := K(Z_1, \ldots, Z_t)$ is finitely generated as a ring over K, which is impossible by 3.4. (For other proofs of the Nullstellensatz see Ch. II, §2, and Ch. II, §3, Exercise 1.)

Corollary 3.5. Let L/K be an extension of fields, where L is algebraically closed. A system of algebraic equations

$$f_i = 0 \qquad (i = 1, \ldots, m)$$

with polynomials $f_1, \ldots, f_m \in K[X_1, \ldots, X_n]$ has a solution in L^n if and only if $(f_1, \ldots, f_m) \neq K[X_1, \ldots, X_n]$. (This statement is more of theoretical interest than of practical use.)

Corollary 3.6. Let K be a field and M a maximal ideal of the polynomial ring $K[X_1, \ldots, X_n]$. Then the following are true.

a) $K[X_1, \ldots, X_n]/M$ is a finite field extension of K.
b) For every extension field L of K the ideal M has at most finitely many zeros in L^n.
c) If K is algebraically closed, then there are elements $\xi_1, \ldots, \xi_n \in K$ such that $M = (X_1 - \xi_1, \ldots, X_n - \xi_n)$.

a) is a consequence of 3.2. b) If L/K is a field extension, then by a) there are at most finitely many K-homomorphisms $K[X_1,\ldots,X_n]/M \to L$, hence M has at most finitely many zeros in L^n. c) If K is algebraically closed, then M has a zero $(\xi_1,\ldots,\xi_n) \in K^n$ by 3.1. The polynomials not belonging to M then do not have $(\xi_1,\ldots,\xi_n)$ as a zero; otherwise, every polynomial would have this zero. It follows that $(X_1 - \xi_1,\ldots,X_n - \xi_n) \subset M$. But $(X_1 - \xi_1,\ldots,X_n - \xi_n)$ is a maximal ideal, hence $(X_1 - \xi_1,\ldots,X_n - \xi_n) = M$.

The Nullstellensatz can be sharpened as follows.

Proposition 3.7. Let L/K be an extension of fields, where L is algebraically closed. The assignment $V \mapsto \Im(V)$ defines a bijection of the set of all K-varieties $V \subset \mathbb{A}^n(L)$ onto the set of all ideals I of $K[X_1,\ldots,X_n]$ with $\mathrm{Rad}(I) = I$. For any ideal I of $K[X_1,\ldots,X_n]$,

$$\mathrm{Rad}(I) = \Im(\mathfrak{V}(I)).$$

Proof. From the rules 1.8 we have already found that the assignment $V \mapsto \Im(V)$ is injective. That it is also surjective will follow once the second assertion of 3.7 is proved.

For any ideal I of $K[X_1,\ldots,X_n]$ we have $\mathrm{Rad}(I) \subset \Im(\mathfrak{V}(I))$. Now let $F \in \Im(\mathfrak{V}(I)), F \neq 0$. We will show that $F \in \mathrm{Rad}(I)$ with the aid of "Rabinowitsch's argument":

In the polynomial ring $K[X_1,\ldots,X_n,T]$ with one more indeterminate T, we form the ideal J generated by I and $F \cdot T - 1$. If $(x_1,\ldots,x_n,t) \in L^{n+1}$ is a zero of J, then $(x_1,\ldots,x_n) \in \mathfrak{V}(I)$, so $F(x_1,\ldots,x_n) \cdot t - 1 = -1$. But since $(x_1,\ldots,x_n,t)$ is also a zero of $F \cdot T - 1$, this is a contradiction. Since J has no zeros, $J = K[X_1,\ldots,X_n,T]$ by 3.1. We then have an equation

$$1 = \sum_{i=1}^{s} R_i F_i + S(FT - 1)$$

with $R_i, S \in K[X_1,\ldots,X_n,T]$ and $F_i \in I$ $(i = 1,\ldots,s)$. Let

$$\varphi : K[X_1,\ldots,X_n,T] \to K(X_1,\ldots,X_n)$$

be the K-homomorphism with $\varphi(X_i) = X_i(i = 1,\ldots,n), \varphi(T) = \frac{1}{F}$. Then

$$1 = \sum_{i=1}^{s} \varphi(R_i) \cdot F_i, \quad \varphi(R_i) = \frac{A_i}{F^{\rho_i}} \text{ with } A_i \in K[X_1,\ldots,X_n], \quad \rho_i \in \mathbb{N}.$$

If $\rho := \mathrm{Max}_{i=1,\ldots,s}\{\rho_i\}$, we then have $F^\rho \in (F_1,\ldots,F_s)$, so $F \in \mathrm{Rad}(I)$.

Corollary 3.8.

a) For two ideals $I_1, I_2 \subset K[X_1,\ldots,X_n]$, $\mathfrak{V}(I_1) = \mathfrak{V}(I_2)$ if and only if

$$\mathrm{Rad}(I_1) = \mathrm{Rad}(I_2).$$

b) Two systems of algebraic equations

$$F_i = 0 \quad (i = 1, \ldots, m) \quad \text{and} \quad G_j = 0 \quad (j = 1, \ldots, l)$$

have the same solution sets in $A^n(L)$ if and only if: For all $i \in [1, m]$ there is a $\rho_i \in \mathbb{N}$ with $F_i^{\rho_i} \in (G_1, \ldots, G_l)$ and for all $j \in [1, l]$ there is a $\sigma_j \in \mathbb{N}$ with $G_j^{\sigma_j} \in (F_1, \ldots, F_m)$.

Corollary 3.9.

a) Let K'/K be an extension of fields with $K' \subset L$, and let $\mathfrak{I}(V)$ be the ideal of a K-variety $V \subset A^n(L)$ in $K[X_1, \ldots, X_n]$. Then

$$\mathrm{Rad}(\mathfrak{I}(V) \cdot K'[X_1, \ldots, X_n])$$

is the ideal of V in $K'[X_1, \ldots, X_n]$.

b) Let $V \subset A^n(L)$ and $W \subset A^m(L)$ be two K-varieties, $\mathfrak{I}(V) \subset K[X_1, \ldots, X_n]$ and $\mathfrak{I}(W) \subset K[Y_1, \ldots, Y_m]$ the corresponding ideals. Then to the product variety $V \times W \subset A^{n+m}(L)$ corresponds the ideal

$$\mathfrak{I}(V \times W) = \mathrm{Rad}(\mathfrak{I}(V), \mathfrak{I}(W))$$

in $K[X_1, \ldots, X_n, Y_1, \ldots, Y_m]$.

These assertions immediately follow from the formulas in 3.7. The reader should compare 3.9a) with the statement of Exercise 7a) in §2.

In the following let L/K be an extension of fields, where L is algebraically closed.

Definition 3.10.

a) A K-algebra which (as a ring) is finitely generated over K is called an *affine K-algebra*.

b) For a K-variety $V \subset A^n(L)$

$$K[V] := K[X_1, \ldots, X_n]/\mathfrak{I}(V)$$

is called the *coordinate ring* of V (or the *affine K-algebra* of V).

For $V = A^n(L)$ we have $K[V] = K[X_1, \ldots, X_n]$. The relation given in 3.7 between the K-varieties in $A^n(L)$ and the ideals of $K[X_1, \ldots, X_n]$ can be generalized. One considers on the one hand the K-subvarieties of a fixed variety V and on the other hand the ideals of $K[V]$. In deriving this relation we use the following ring-theoretic facts, which will often be tacitly used later on.

Let R be a ring, $\mathfrak{a}$ an ideal of R, $\epsilon : R \to R/\mathfrak{a}$ the canonical epimorphism.

1. The mapping that assigns to any ideal I of $R/\mathfrak{a}$ its inverse image $\epsilon^{-1}(I)$ in R is an inclusion-preserving bijection of the set of all ideals of $R/\mathfrak{a}$ onto the set of all ideals of R that contain $\mathfrak{a}$.
2. ϵ induces an isomorphism $R/\epsilon^{-1}(I) \cong \epsilon(R)/I$.
3. I is a prime (maximal) ideal of $R/\mathfrak{a}$ if and only if $\epsilon^{-1}(I)$ is a prime (maximal) ideal of R.
4. $\mathrm{Rad}(I) = I$ if and only if $\mathrm{Rad}(\epsilon^{-1}(I)) = \epsilon^{-1}(I)$.

1. and 2. form the contents of one of the Noether Isomorphism theorems of ring theory: 3. and 4. easily follow therefrom.

The elements φ of the coordinate ring $K[V]$ of a K-variety $V \subset A^n(L)$ can be considered as functions $\varphi : V \to L$. If $\varphi = F + \mathfrak{I}(V)$ with $F \in K[X_1, \ldots, X_n]$ and $x = (\xi_1, \ldots, \xi_n)$, put

$$\varphi(x) = F(\xi_1, \ldots, \xi_n).$$

This is independent of the choice of the representative F of the residue class φ, so that one gets a well-defined function on V. For example, if $x_i := X_i + \mathfrak{I}(V)$, then x_i is the i-th coordinate function: it assigns to each $(\xi_1, \ldots, \xi_n) \in V$ its i-th coordinate ξ_i.

For a subset I (in particular, an ideal) of $K[V]$ we can now speak of the zero set $\mathfrak{V}_V(I)$ of I on V (a K-subvariety of V) and for a subset $W \subset V$ (in particular, a K-subvariety of V) we can speak of the vanishing ideal $\mathfrak{I}_V(W)$ in $K[V]$.

It immediately follows that $\mathfrak{I}_V(W) = \mathfrak{I}(W)/\mathfrak{I}(V)$. Moreover, we have rules analogous to 1.8 for the operations $\mathfrak{V}_V$ and $\mathfrak{I}_V$. From 3.7 and 1.10 now follows the more general version of the Nullstellensatz.

Proposition 3.11. Let $V \subset A^n(L)$ be a K-variety. The mapping $W \mapsto \mathfrak{I}_V(W)$ which to each K-variety $W \subset V$ assigns its ideal $\mathfrak{I}_V(W)$ in $K[V]$ is an inclusion-reversing bijection of the set of all K-subvarieties of V onto the set of all ideals I of $K[V]$ with $\mathrm{Rad}(I) = I$. For each ideal I of $K[V]$ we have

$$\mathrm{Rad}(I) = \mathfrak{I}_V(\mathfrak{V}_V(I)).$$

A K-subvariety $W \subset V$ is irreducible if and only if $\mathfrak{I}_V(W)$ is a prime ideal of $K[V]$.

With these propositions the question, for example, of what irreducible subvarieties V has, is reduced to the search for the prime ideals of $K[V]$.

We conclude this section with some general statements about the coordinate rings of affine varieties.

Rules 3.12.

a) The coordinate ring $K[V]$ of a variety V is a reduced affine K-algebra. V is irreducible if and only if $K[V]$ is an integral domain.

b) Under the hypotheses of 3.9a)

$$K'[V] \cong (K' \otimes_K K[V])_{\mathrm{red}}.$$

c) Under the hypotheses of 3.9b)

$$K[V \times W] \cong (K[V] \otimes_K K[W])_{\mathrm{red}}.$$

d) If W is a K-subvariety of the K-variety V, then we have a canonical isomorphism

$$K[W] \cong K[V]/\mathfrak{I}_V(W),$$

which is induced by restricting the functions in $K[V]$ to W.

e) If U and W are two subvarieties of V and if $I \subset K[W]$ is the image of $\mathfrak{I}_V(U)$ under the canonical epimorphism $K[V] \to K[W]$, then

$$\mathfrak{V}_W(I) = U \cap W.$$

f) Every reduced affine K-algebra A is K-isomorphic to the coordinate ring of a K-variety V (in a suitable affine space $\mathbb{A}^n(L)$).

Proof.

a) Since $\mathfrak{I}(V) = \operatorname{Rad}(\mathfrak{I}(V))$, $K[V]$ is reduced. $K[V]$ is an integral domain if and only if $\mathfrak{I}(V)$ is a prime ideal, i.e. if V is irreducible.

b) By 3.9a)

$$\begin{aligned} K'[V] &= K'[X_1, \ldots, X_n]/\operatorname{Rad}(\mathfrak{I}(V) \cdot K'[X_1, \ldots, X_n]) \\ &\cong (K'[X_1, \ldots, X_n]/\mathfrak{I}(V) \cdot K'[X_1, \ldots, X_n])_{\text{red}} \\ &\cong (K' \otimes_K K[V])_{\text{red}}. \end{aligned}$$

c) By 3.9b)

$$\begin{aligned} K[V \times W] &= K[X_1, \ldots, X_n, Y_1, \ldots, Y_m]/\operatorname{Rad}(\mathfrak{I}(V), \mathfrak{I}(W)) \\ &\cong (K[X_1, \ldots, X_n, Y_1, \ldots, Y_m]/(\mathfrak{I}(V), \mathfrak{I}(W)))_{\text{red}} \\ &\cong (K[V] \otimes_K K[W])_{\text{red}}. \end{aligned}$$

d) We have

$$\begin{aligned} K[W] &= K[X_1, \ldots, X_n]/\mathfrak{I}(W) \cong K[X_1, \ldots, X_n]/\mathfrak{I}(V)/\mathfrak{I}(W)/\mathfrak{I}(V) \\ &= K[V]/\mathfrak{I}_V(W). \end{aligned}$$

If the elements of $K[V]$ and $K[W]$ are considered as functions, one sees that the functions in $K[W]$ are the restrictions of the functions in $K[V]$, where two functions on V have the same restriction to W if and only if they are congruent modulo $\mathfrak{I}_V(W)$.

e) We have $I = \mathfrak{I}_V(U) + \mathfrak{I}_V(W)/\mathfrak{I}_V(W)$ and $\mathfrak{V}_V(\mathfrak{I}_V(U) + \mathfrak{I}_V(W)) = U \cap W$. The assertion follows from this.

f) We have $A = K[x_1, \ldots, x_n]$ with $x_i \in A$ $(i = 1, \ldots, n)$, so that $A \cong K[X_1, \ldots, X_n]/I$ with an ideal $I \subset K[X_1, \ldots, X_n]$. Since A is assumed reduced, it follows that $\operatorname{Rad}(I) = I$. If V is the zero set of I in $\mathbb{A}^n(L)$, then $A \cong K[V]$.

Exercises

1. a) Any maximal ideal in the polynomial ring $K[X_1, \ldots, X_n]$ over a field K is generated by n elements.

   b) Let R be a ring, $M \subset R[X_1, \ldots, X_n]$ a maximal ideal for which $M \cap R$ is a maximal ideal of R generated by p elements. Then M is generated by $p + n$ elements.

2. Let A be an affine algebra over a field K, M a maximal ideal of A, and $B \subset A$ a K-subalgebra. Then $M \cap B$ is a maximal ideal of B.

3. For any ideal $I \neq A$ of an affine algebra A, $\mathrm{Rad}(I)$ is the intersection of all the maximal ideals of A that contain I.

4. Let A be an affine algebra over the field K, d its dimension as a vector space over K. A has at most d maximal ideals.

5. In the polynomial ring $K[X_1, X_2]$ over an algebraically closed field K, there are only the following prime ideals.

   a) The zero ideal (0) and the unit ideal (1).

   b) The principal ideals (f), where f is an irreducible polynomial.

   c) The maximal ideals $(X_1 - \xi_1, X_2 - \xi_2)$, where $(\xi_1, \xi_2) \in K^2$.
     (Apply §1, Exercise 7 and 3.7.)

6. Let L/K be an extension of fields, where L is algebraically closed. For a point $x = (\xi_1, \dots, \xi_n)$ of an affine K-variety $V \subset \mathsf{A}^n(L)$ let $\varphi_x : K[V] \to L$ be the mapping assigning to any function $f \in K[V]$ its value $f(x) \in L$. The mapping $x \mapsto \varphi_x$ is a bijection of V onto the set $\mathrm{Hom}_K(K[V], L)$ of K-algebra homomorphisms of $K[V]$ into L.

7. Let K be an arbitrary field, S the set of all polynomials in $K[X_1, \dots, X_n]$ that have no zeros in $\mathsf{A}^n(K)$, and I an ideal in $K[X_1, \dots, X_n]$ with $I \cap S = \emptyset$. Then I has a zero in $\mathsf{A}^n(K)$. (In the proof of this *generalization of the Nullstellensatz* one can use the hint to §1, Exercise 3.)

4. The spectrum of a ring

In the last section the close relation between algebraic geometry and ring theory became clear. In this section we describe a generalization of the concept of affine variety that starts from an arbitrary commutative ring with 1. This generalization has proved very significant in modern algebraic geometry. The formal analogy to the concept of affine variety will soon be evident.

Definition 4.1. For a ring R we write:

 a) $\mathrm{Spec}(R)$ for the set of all prime ideals $\mathfrak{p}$ of R, $\mathfrak{p} \neq R$;

 b) $J(R)$ for the set of all prime ideals in $\mathrm{Spec}(R)$ that can be written as the intersection of maximal ideals;

 c) $\mathrm{Max}(R)$ for the set of all maximal ideals of R.

$\mathrm{Spec}(R)$ is called the *spectrum*, $J(R)$ the *J-spectrum*, and $\mathrm{Max}(R)$ the *maximal spectrum* of R.

Obviously $\mathrm{Max}(R) \subset J(R) \subset \mathrm{Spec}(R)$. If X is one of these sets and I an ideal of R, then
$$\mathfrak{V}(I) := \{\mathfrak{p} \in X \mid \mathfrak{p} \supset I\}$$
is called the *zero set* of I in X. A subset $A \subset X$ is called closed if there is an ideal I of R such that $A = \mathfrak{V}(I)$.

For two closed subsets $A_k = \mathfrak{V}(I_k)$ $(k = 1, 2)$ of X, $A_1 \cup A_2 = \mathfrak{V}(I_1 \cap I_2)$ is also closed, and for a family $A_\lambda = \mathfrak{V}(I_\lambda)$ of closed subsets $(\lambda \in \Lambda)$ $\bigcap_{\lambda \in \Lambda} A_\lambda = \mathfrak{V}(\sum_{\lambda \in \Lambda} I_\lambda)$ is also closed in X. The sets $\mathfrak{V}(I)$, as I runs over all the ideals of R, are the closed sets of a topology on X, the *Zariski topology* on X.

Obviously, $\mathrm{Max}(R)$ and $J(R)$ carry the relative topology of the Zariski topology on $\mathrm{Spec}(R)$.

For an arbitrary set $A \subset X$,

$$\mathfrak{I}(A) := \bigcap_{\mathfrak{p} \in A} \mathfrak{p}$$

is called the *ideal of A in R* (the vanishing ideal).

We have

Rule 4.2. For any subset A of X, $\mathfrak{V}(\mathfrak{I}(A)) = \overline{A}$ is the closure of A in X.

Proof. From the definition of $\mathfrak{I}(A)$ it immediately follows that $A \subset \mathfrak{V}(\mathfrak{I}(A))$, so $\overline{A} \subset \mathfrak{V}(\mathfrak{I}(A))$. Conversely, if $\mathfrak{V}(I)$ is a closed subset of X containing A, I an ideal of R, then $\mathfrak{p} \supset I$ for all $\mathfrak{p} \in A$, so that $I \subset \bigcap_{\mathfrak{p} \in A} \mathfrak{p} = \mathfrak{I}(A)$ and hence $\mathfrak{V}(I) \supset \mathfrak{V}(\mathfrak{I}(A))$, whence $\mathfrak{V}(\mathfrak{I}(A)) = \overline{A}$ follows.

As an analogue to Hilbert's Nullstellensatz we have

Proposition 4.3. Let $X = \mathrm{Spec}(R)$. For any ideal I of R, $\mathfrak{I}(\mathfrak{V}(I)) = \mathrm{Rad}(I)$. The closed subsets of X correspond bijectively to the ideals of R that equal their radicals; this correspondence reverses inclusions.

In the proof of this proposition we use

Lemma 4.4. (Krull) Let I be an ideal of R, S a multiplicatively closed subset of R with $I \cap S = \emptyset$. Then the set M of all ideals J of R with $I \subset J$ and $J \cap S = \emptyset$ has a maximal element. It is a prime ideal of R.

Proof. If $\{J_\lambda\}_{\lambda \in \Lambda}$ is a totally ordered (with respect to inclusion) family of ideals of M, then $J := \bigcup_{\lambda \in \Lambda} J_\lambda$ is an ideal of R with $I \subset J$ and $J \cap S = \emptyset$. By Zorn's Lemma M has a maximal element $\mathfrak{p}$.

Suppose that for two elements $a_1, a_2 \in R \setminus \mathfrak{p}$ we have $a_1 \cdot a_2 \in \mathfrak{p}$. Since $a_i \notin \mathfrak{p}$, we have $(Ra_i + \mathfrak{p}) \cap S \neq \emptyset$ $(i = 1, 2)$. Hence there are elements $r_i \in R$, $p_i \in \mathfrak{p}$ such that

$$r_i a_i + p_i \in S \qquad (i = 1, 2).$$

But then $(r_1 a_1 + p_1) \cdot (r_2 a_2 + p_2) = r_1 r_2 a_1 a_2 + r_1 a_1 p_2 + r_2 a_2 p_1 + p_1 p_2 \in \mathfrak{p} \cap S$, contradicting $\mathfrak{p} \cap S = \emptyset$. Thus $\mathfrak{p}$ is a prime ideal.

Applying this lemma to $S = \{1\}$ yields the well-known fact that any ideal I of a ring R with $I \neq R$ is contained in a maximal ideal. Further, we have

Corollary 4.5. For any ideal I of a ring R with $I \neq R$,

$$\mathrm{Rad}(I) = \bigcap_{\substack{\mathfrak{p} \supset I \\ \mathfrak{p} \in \mathrm{Spec}(R)}} \mathfrak{p}.$$

In particular, $\bigcap_{\mathfrak{p}\in\mathrm{Spec}(R)} \mathfrak{p} = \mathrm{Rad}(O)$ is the set of all nilpotent elements of R.

Proof. Passing to R/I, we see that it suffices to prove the second statement. It is clear that $\bigcap_{\mathfrak{p}\in\mathrm{Spec}(R)} \mathfrak{p}$ contains all the nilpotent elements of R. If $x \in \bigcap_{\mathfrak{p}\in\mathrm{Spec}(R)} \mathfrak{p}$, consider $S := \{x^n \mid n \in \mathbb{N}\}$. If x is not nilpotent, then $(0) \cap S \neq \emptyset$, and by 4.4 there is a prime ideal $\mathfrak{p}$ with $\mathfrak{p} \cap S = \emptyset$, contradicting the hypothesis $x \in \mathfrak{p}$.

Corollary 4.5 also proves Proposition 4.3. We have $\mathfrak{I}(\mathfrak{V}(I)) = \bigcap_{\mathfrak{p}\in\mathfrak{V}(I)} \mathfrak{p} = \bigcap_{\mathfrak{p}\supset I} \mathfrak{p} = \mathrm{Rad}(I)$. The other assertion of the proposition follows with the aid of 4.2. (Of course, 4.3 does not provide a new proof of Hilbert's Nullstellensatz.)

As an analogue to 1.10 we have

Proposition 4.6. Let R be a ring, X its spectrum, J-spectrum, or maximal spectrum. A closed subset $A \subset X$ is irreducible if and only if $\mathfrak{I}(A)$ is a prime ideal.

Proof.
  a) Let A be irreducible and $f \cdot g \in \mathfrak{I}(A)$ with $f, g \in R$. For any $\mathfrak{p} \in A$, $f \in \mathfrak{p}$ or $g \in \mathfrak{p}$ and hence $A = (A \cap \mathfrak{V}(f)) \cup (A \cap \mathfrak{V}(g))$. Therefore, $A \subset \mathfrak{V}(f)$ or $A \subset \mathfrak{V}(g)$, i.e. $f \in \mathfrak{I}(A)$ or $g \in \mathfrak{I}(A)$.
  b) Let $\mathfrak{I}(A)$ be a prime ideal and $A = A_1 \cup A_2$ with closed sets $A_1, A_2 \subset A$. Then $\mathfrak{I}(A_i) \supset \mathfrak{I}(A)$ and on the other hand $\mathfrak{I}(A) = \mathfrak{I}(A_1 \cup A_2) = \mathfrak{I}(A_1) \cap \mathfrak{I}(A_2)$; therefore, $\mathfrak{I}(A_1) \subset \mathfrak{I}(A)$ or $\mathfrak{I}(A_2) \subset \mathfrak{I}(A)$. It follows that $\mathfrak{I}(A_1) = \mathfrak{I}(A)$ or $\mathfrak{I}(A_2) = \mathfrak{I}(A)$; and by 4.2, $A_1 = A$ or $A_2 = A$.

We now want to investigate the relations between an affine variety and the spectrum of its coordinate ring. Let L/K be an extension of fields, where L is algebraically closed, and $V \subset \mathbb{A}^n(L)$ a K-variety.

For any $x \in V$, the set $\mathfrak{p}_x := \mathfrak{I}_V(\{x\})$ of all functions $\varphi \in K[V]$ with $\varphi(x) = 0$ is a prime ideal $\neq K[V]$. We thus have a mapping

$$\varphi : V \to \mathrm{Spec}(K[V]),$$
$$x \mapsto \mathfrak{p}_x.$$

φ is continuous, for if $A = \mathfrak{V}(I)$ is a closed set of $\mathrm{Spec}(K[V])$, then $\varphi^{-1}(A) = \{x \in V \mid \mathfrak{p}_x \supset I\}$ is the zero set of I in V and so is also a closed set. However, in general φ is neither injective nor surjective.

For $x = (x_1, \ldots, x_n) \in V$, $\mathfrak{p}_x$ is the kernel of the K-epimorphism $K[V] \to K[x_1, \ldots, x_n]$ with $\varphi \mapsto \varphi(x)$ for all $\varphi \in K[V]$. If $y = (y_1, \ldots, y_n) \in V$ is another point, then $\mathfrak{p}_x = \mathfrak{p}_y$ if and only if there is a K-isomorphism $K[x_1, \ldots, x_n] \cong K[y_1, \ldots, y_n]$ mapping x_i to y_i $(i = 1, \ldots, n)$. In this case we say that x and y are *conjugate points* over K.

For many purposes of algebraic geometry conjugate points can be identified. This is effected by passing from V to $\mathrm{Spec}(K[V])$.

The points $x = (x_1, \ldots, x_n) \in V$ whose coordinates x_i are algebraic over K are called the *K-algebraic points* of V. By Hilbert's Nullstellensatz, for every

$\mathfrak{m} \in \text{Max}(K[V])$ there is a K-algebraic point $x \in V$ with $\mathfrak{m} = \mathfrak{p}_x$. Therefore, $\varphi^{-1}\text{Max}(K[V])$ is the set of K-algebraic points of V, and φ maps this set *onto* $\text{Max}(K[V])$.

In particular, if K is algebraically closed, then φ induces a homeomorphism of the space V_K of all K-rational points of V onto $\text{Max}(K[V])$.

In the general case we shall show that $\mathfrak{p} \in \text{Spec}(K[V])$ lies in $\text{Im}(\varphi)$ if and only if the corresponding subvariety $\mathfrak{V}_V(\mathfrak{p}) \subset V$ has a "generic point."

Definition 4.7. Let A be a closed set of a topological space X. $x \in A$ is called a *generic point* of A if $A = \overline{\{x\}}$, the closure of $\{x\}$ in X.

If A has a generic point, then A is irreducible, for since $\{x\}$ is irreducible, so is $\overline{\{x\}}$ (2.10).

For $\mathfrak{p} \in \text{Spec}(K[V])$ there is an $x \in V$ with $\mathfrak{p} = \mathfrak{p}_x = \mathfrak{I}_V(\{x\})$ if and only if $\mathfrak{V}_V(\mathfrak{p}) = \mathfrak{V}_V(\mathfrak{I}_V(\{x\})) = \overline{\{x\}}$, i.e. when x is a generic point of $\mathfrak{V}_V(\mathfrak{p})$.

Not every (nonempty) irreducible variety V need have a generic point; for example, if $L = K$, then $\overline{\{x\}} = \{x\}$ for any point $x \in V$. However, if the transcendence degree of L over K is at least n, then any nonempty irreducible K-variety $V \subset \mathbf{A}^n(L)$ has a generic point. In this case it is easily shown that $K[V]$ can be embedded in L by a K-homomorphism; the images $x_i \in L$ of the coordinate functions then provide a generic point $x = (x_1, \ldots, x_n)$ of V. For example, any nonempty irreducible affine $\mathbf{Q}$-variety in $\mathbf{A}^n(\mathbf{C})$ (n arbitrary) has a generic point, since $\mathbf{C}$ has infinite transcendence degree over $\mathbf{Q}$.

One technique of classical algebraic geometry (see $[X]$) was, when studying K-varieties, immediately to admit points with coordinates in a "universal field" L over K (i.e. an algebraically closed extension field of infinite transcendence degree over K) in order always to have at hand generic points. In this case the mapping φ above is always surjective. The spectrum provides a substitute for this technique.

Proposition 4.8. Let X be the spectrum or J-spectrum of a ring R. Any nonempty irreducible closed subset $A \subset X$ has a unique generic point $\mathfrak{p}$, namely $\mathfrak{p} := \mathfrak{I}(A)$.

Proof. If $\mathfrak{p}$ is a generic point of A, then by 4.2 we have the formula $A = \overline{\{\mathfrak{p}\}} = \mathfrak{V}(\mathfrak{I}(\{\mathfrak{p}\})) = \mathfrak{V}(\mathfrak{p})$, so $\mathfrak{I}(A) = \mathfrak{I}(\mathfrak{V}(\mathfrak{p}))$. But from the definition of $\mathfrak{I}$ and $\mathfrak{V}$ it follows at once that $\mathfrak{I}(\mathfrak{V}(\mathfrak{p})) = \mathfrak{p}$, so $\mathfrak{p} = \mathfrak{I}(A)$.

In general, by 4.6 $\mathfrak{I}(A) = \bigcap_{\mathfrak{p} \in A} \mathfrak{p}$ is a prime ideal of R. In case $X = J(R)$, the $\mathfrak{p} \in A$ are the intersections of maximal ideals, so that $\mathfrak{I}(A) \in J(R)$. Applying the rule $\overline{\{x\}} = \mathfrak{V}(\mathfrak{I}(\{x\}))$ to $x = \mathfrak{I}(A)$, we see that $\overline{\{\mathfrak{I}(A)\}} = \mathfrak{V}(\mathfrak{I}(A)) = A$, i.e. $\mathfrak{I}(A)$ is indeed a generic point of A.

Like any topological space the spectra introduced above have decompositions into irreducible components. By means of 4.3 the statements on topological spaces in §2 can be translated into the language of rings, thus yielding propositions about rings.

Proposition 4.9. Let R be a ring, I an ideal of R. The minimal prime divisors of I correspond bijectively to the irreducible components of the subset $\mathfrak{V}(I) \subset$

Spec(R). In particular, any ideal in a ring R has minimal prime divisors, and any $\mathfrak{p} \in \mathrm{Spec}(R)$ contains a minimal prime ideal of R.

a) If Spec(R) is a Noetherian topological space (for example, if R is a Noetherian ring), then I has only finitely many minimal prime divisors and R has only finitely many minimal prime ideals.

b) If $J(R)$ is Noetherian, then the set of prime divisors of I that are contained in $J(R)$ has only finitely many elements.

Proof. The minimal prime divisors of I are just the generic points of the irreducible components of the set $\mathfrak{V}(I) \subset J(R)$.

If a ring has only finitely many minimal prime ideals, we can say something about its zero divisors.

Proposition 4.10. Let $R \neq \{0\}$ be a ring with only finitely many minimal prime ideals $\mathfrak{p}_1, \ldots, \mathfrak{p}_s$. Then $\mathrm{Rad}(0) = \bigcap_{i=1}^{s} \mathfrak{p}_i$. Further, $\bigcup_{i=1}^{s} \mathfrak{p}_i$ consists of zero divisors alone. If R is reduced, then $\bigcup_{i=1}^{s} \mathfrak{p}_i$ is the set of all zero divisors of R.

Proof. The first statement follows from 4.5 and 4.9; the second remains to be proved only for $s > 1$. If $r \in \mathfrak{p}_j$ for some $j \in [1, s]$, choose $t \in \bigcap_{i \neq j} \mathfrak{p}_i$, $t \notin \mathfrak{p}_j$. There is such a t, since otherwise we would have $\bigcap_{i \neq j} \mathfrak{p}_i \subset \mathfrak{p}_j$, and then $\mathfrak{p}_i \subset \mathfrak{p}_j$ for some $i \neq j$, contradicting the assumption that $\mathfrak{p}_j$ is minimal. From $rt \in \bigcap_{i=1}^{s} \mathfrak{p}_i$, it follows that $(rt)^\rho = 0$ for a suitable $\rho \in \mathbb{N}$. Since $t \notin \mathfrak{p}_j$, we have $t^\rho \neq 0$. Hence there is a $\sigma \in \mathbb{N}$ with $r^\sigma t^\rho \neq 0$, $r^{\sigma+1} t^\rho = 0$; that is, r is a zero divisor of R.

If R is reduced, then $\bigcap_{i=1}^{s} \mathfrak{p}_i = (0)$. If $r \in R$ is a zero divisor, then there is a $t \in R \setminus \{0\}$ with $rt = 0$. There is also a $j \in [1, s]$ with $t \notin \mathfrak{p}_j$. From $rt = 0 \in \mathfrak{p}_j$ it follows that $r \in \mathfrak{p}_j$.

We now want to compare the spectra of different rings with one another.

Let $\alpha : R \to S$ be a ring homomorphism. For any $\mathfrak{p} \in \mathrm{Spec}(S)$, $\alpha^{-1}(\mathfrak{p}) \in \mathrm{Spec}(R)$. Hence α induces a mapping

$$\mathrm{Spec}(\alpha) : \mathrm{Spec}(S) \to \mathrm{Spec}(R), \qquad \mathrm{Spec}(\alpha)(\mathfrak{p}) = \alpha^{-1}(\mathfrak{p}).$$

Remark 4.11. Spec(α) is continuous.

Indeed, if $A = \mathfrak{V}(I)$ is a closed subset of Spec(R) with an ideal I of R, then

$$\mathrm{Spec}(\alpha)^{-1}(A) = \{\mathfrak{p} \in \mathrm{Spec}(S) \mid \alpha^{-1}(\mathfrak{p}) \supset I\}$$
$$= \{\mathfrak{p} \in \mathrm{Spec}(S) \mid \mathfrak{p} \supset \alpha(I) \cdot S\} = \mathfrak{V}(\alpha(I) \cdot S),$$

where $\alpha(I) \cdot S$ is the ideal generated by $\alpha(I)$ in S. The inverse image of a closed subset of Spec(R) is therefore closed in Spec(S), i.e. Spec(α) is continuous.

Proposition 4.12. If α is surjective with kernel I, then Spec(α) induces a homeomorphism of Spec(S) onto $\mathfrak{V}(I) \subset \mathrm{Spec}(R)$. Spec($\alpha$) is a homeomorphism of Spec(S) with Spec(R) if and only if I consists of nilpotent elements alone.

Proof. That Spec(α) is a bijection of Spec(S) onto $\mathfrak{V}(I)$ is a reformulation of the statement that the prime ideals $\mathfrak{p}$ of S and their inverse images $\alpha^{-1}(\mathfrak{p})$

in R correspond bijectively, and these are precisely the prime ideals of R that contain I. If $A = \mathfrak{V}(J)$ is a closed subset of $\operatorname{Spec}(S)$, then $\operatorname{Spec}(\alpha)(A) = \mathfrak{V}(\alpha^{-1}(J))$ is a closed subset of $\mathfrak{V}(I)$, and so the bijection above is indeed a homeomorphism.

By 4.5 $\mathfrak{V}(I) = \operatorname{Spec}(R)$ if and only if $\operatorname{Rad}(I) = \operatorname{Rad}(0)$, i.e. if I consists of nilpotent elements alone.

By 4.12 we find, in particular, that $\operatorname{Spec}(R_{\mathrm{red}}) \to \operatorname{Spec}(R)$ is a homeomorphism. We see that the pair $(R, \operatorname{Spec}(R))$ contains more information than the topological space $\operatorname{Spec}(R)$ alone. In modern algebraic geometry one considers the pairs $(R, \operatorname{Spec}(R))$ as the objects to be studied in affine geometry; for the moment we want to call them "affine schemes" (for the precise concept of an affine scheme see [M] or [N] (cf. also Ch. III, §4, Exercise 1); yet some ideas related to this concept can be clarified in our simplified development).

By a closed subscheme of an affine scheme $(R, \operatorname{Spec}(R))$ we understand a pair $(R/I, \operatorname{Spec}(R/I))$, where I is an ideal of R. By 4.12 $\operatorname{Spec}(R/I)$ is identified with the closed subset $\mathfrak{V}(I)$ of $\operatorname{Spec}(R)$, but this set is endowed with the ring R/I. In contrast to classical algebraic geometry (Proposition 3.7), the closed subschemes of $(R, \operatorname{Spec}(R))$ and the ideals of R are in one-to-one correspondence.

For two closed subschemes $(R/I_k, \operatorname{Spec}(R/I_k))$ $(k = 1, 2)$, $(R/I_1 \cap I_2, \operatorname{Spec}(R/I_1 \cap I_2))$ is called their *union* and $(R/I_1 + I_2, \operatorname{Spec}(R/I_1 + I_2))$ is called their *intersection*. Since

$$\mathfrak{V}(I_1 \cap I_2) = \mathfrak{V}(I_1) \cup \mathfrak{V}(I_2) \quad \text{and} \quad \mathfrak{V}(I_1 + I_2) = \mathfrak{V}(I_1) \cap \mathfrak{V}(I_2),$$

these are actually set-theoretic unions and intersections.

The concept of a scheme opens the possibility of speaking of "varieties that are counted several times" or "multiple varieties." We suggest the meaning of this through some examples.

Let $R := K[X, Y]$ be the polynomial ring over an algebraically closed field K. $\operatorname{Spec}(R/(X^2))$ is identified topologically with the line $X = 0$ in K^2 (the Y-axis). On the other hand, the affine scheme $(R/(X^2), \operatorname{Spec}(R/(X^2)))$ can be considered as "the Y-axis taken twice" (we say $X^2 = 0$ defines a "double line"). $(R/(X), \operatorname{Spec}(R/(X)))$ is a proper closed subscheme of $(R/(X^2), \operatorname{Spec}(R/(X^2)))$ with the same "underlying topological space."

The affine scheme $(R/(X^2, XY), \operatorname{Spec}(R/(X^2, XY)))$ is, in view of the relation $(X^2, XY) = (X) \cap (X^2, Y)$, the union of the two affine schemes

$$(R/(X), \operatorname{Spec}(R/(X))) \quad \text{(the Y-axis)}$$
$$(R/(X^2, Y), \operatorname{Spec}(R/(X^2, Y))) \quad \text{(the origin counted twice)}.$$

It can be interpreted as the Y-axis on which the origin is counted twice, the other points once (Fig. 7).

The necessity of considering multiple varieties is seen when one wants to investigate more closely the intersections of algebraic varieties. We illustrate this with examples.

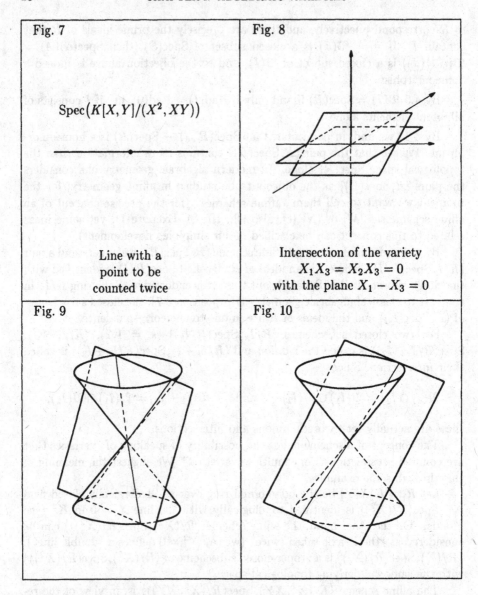

Fig. 7

$$\mathrm{Spec}(K[X,Y]/(X^2,XY))$$

Line with a
point to be
counted twice

Fig. 8

Intersection of the variety
$X_1 X_3 = X_2 X_3 = 0$
with the plane $X_1 - X_3 = 0$

Fig. 9

Fig. 10

The variety $V : X_1 X_3 = X_2 X_3 = 0$ is cut by the plane $X_1 - X_3 = 0$ along the line $X_1 = X_3 = 0$ (Fig. 8). Because

$$K[X_1, X_2, X_3]/(X_1 X_3, X_2 X_3, X_1 - X_3) \cong K[X_1, X_2]/(X_1^2, X_1 X_2),$$

the intersection of the corresponding schemes is the line with its zero point counted twice, corresponding to the fact that the origin lies on two distinct irreducible components of V. If the cone $X_1^2 + X_2^2 - X_3^2 = 0$ is cut by the plane $X_1 = 0$, then, because

$$(X_1^2 + X_2^2 - X_3^2, X_1) = (X_1, X_2 + X_3) \cap (X_1, X_2 - X_3),$$

the resulting intersection scheme is (if Char $K \neq 2$) the union of two "lines counted once." (This is always the case if the intersection plane is not "tangent" to the cone (Fig. 9).) On the other hand, because $(X_1^2 + X_2^2 - X_3^2, X_1 - X_3) = (X_2^2, X_1 - X_3)$, its intersection with the plane $X_1 - X_3 = 0$ is the "doubly counted" line $X_1 - X_3 = X_2 = 0$ (Fig. 10). In general, a plane through the vertex of the cone (over an algebraically closed coordinate field) always cuts it in two lines, if they are counted "with multiplicity."

In the multiplicity theory of schemes the suggestions above are made precise. A modern treatise on intersection theory is Fulton [AA].

Exercises

1. Let R be a ring. $\mathrm{Spec}(R)$ is irreducible if and only if R_{red} is an integral domain.

2. An element e of a ring R is called *idempotent* if $e^2 = e$.

   a) $e \in R$ is idempotent if and only if $1 - e$ is.

   b) If $1 = e_1 + e_2$ with nonunits $e_1, e_2 \in R$ and if $e_1 \cdot e_2$ is nilpotent, then there is an idempotent element $e \in R$ with $e \neq 0, e \neq 1$.

   c) $\mathrm{Spec}(R)$ is a connected topological space if and only if R contains no idempotent element $\neq 0, 1$.

3. Indicate a ring with infinitely many minimal prime ideals.

4. Let L/K be an extension of fields, $V \subset \mathbb{A}^n(L)$ a K-variety, and $W \subset V$ an irreducible subvariety that has a generic point x. $\varphi \in K[V]$ vanishes on all of W if and only if $\varphi(x) = 0$.

5. Under the hypotheses of Exercise 4, $(x_1, \ldots, x_n) \in \mathbb{A}^n(L)$ is a generic point of V if and only if the K-homomorphism $K[V] \to K[x_1, \ldots, x_n]$ $(\varphi \mapsto \varphi(x_1, \ldots, x_n))$ is an isomorphism. If this is the case, then $(y_1, \ldots, y_n) \in \mathbb{A}^n(L)$ belongs to V if and only if there is a K-homomorphism

$$\alpha : K[x_1, \ldots, x_n] \to K[y_1, \ldots, y_n]$$

   with $\alpha(x_i) = y_i$ $(i = 1, \ldots, n)$. (In this case we say that $(y_1, \ldots, y_n)$ is a *specialization* of $(x_1, \ldots, x_n)$.)

6. For an affine algebra A over a field K, $J(A) = \mathrm{Spec}(A)$. If B is any K-algebra and $\alpha : B \to A$ is a K-homomorphism, then $\mathrm{Spec}(\alpha)$ $(\mathrm{Max}(A)) \subset \mathrm{Max}(B)$.

5. Projective varieties and the homogeneous spectrum

The n-dimensional projective space $\mathbb{P}^n(L)$ over a field L is the set of all lines through the origin in L^{n+1}. A point $x \in \mathbb{P}^n(L)$ can be represented by an $(n+1)$-tuple $(x_0, \ldots, x_n) \neq (0, \ldots, 0)$ in L^{n+1}, and $(x_0', \ldots, x_n') \in L^{n+1}$ defines the same point if and only if there is a $\lambda \in L^\times$ with $(x_0, \ldots, x_n) = \lambda(x_0', \ldots, x_n')$.

An $(n+1)$-tuple representing x is called a *system of homogeneous coordinates* of x. We write $x = \langle x_0, \ldots, x_n \rangle$. A *projective coordinate transformation* is a mapping $\mathbf{P}^n(L) \to \mathbf{P}^n(L)$ that is given by a matrix $A \in Gl(n+1, L)$ through the equation

$$(Y_0, \ldots, Y_n) = (X_0, \ldots, X_n) \cdot A \tag{1}.$$

If K is a subfield of L and $F \in K[X_0, \ldots, X_n]$, then $x \in \mathbf{P}^n(L)$ is called a *zero* of F if $F(x_0, \ldots, x_n) = 0$ for *every* system $(x_0, \ldots, x_n)$ of homogeneous coordinates of x. If F is homogeneous, it suffices that this condition be fulfilled for *one* system $(x_0, \ldots, x_n)$ with $x = \langle x_0, \ldots, x_n \rangle$. In general, if $F = F_0 + \cdots + F_d$ is the decomposition of F into homogeneous polynomials F_k of degree k and if L is an infinite field, then $F(x) = 0$ if and only if $F_k(x_0, \ldots, x_n) = 0$ for all k and one $(x_0, \ldots, x_n)$ with $x = \langle x_0, \ldots, x_n \rangle$. Indeed,

$$F(\lambda(x_0, \ldots, x_n)) = F_0(x_0, \ldots, x_n) + \lambda F_1(x_0, \ldots, x_n) + \cdots + \lambda^d F_d(x_0, \ldots, x_n),$$

so the right side vanishes for infinitely many $\lambda \in L$ if and only if $F_k(x_0, \ldots, x_n) = 0$ for $k = 0, \ldots, d$.

Projective varieties are the solution sets of systems of algebraic equations that contain only homogeneous polynomials.

Definition 5.1. $V \subset \mathbf{P}^n(L)$ is called a *projective algebraic K-variety* if there are homogeneous polynomials $F_1, \ldots, F_m \in K[X_0, \ldots, X_n]$ such that V is the set of all common zeros of the F_i in $\mathbf{P}^n(L)$.

This concept is invariant under coordinate transformations (1) as long as A has coefficients in K. The solution sets of systems of homogeneous linear equations are called *linear varieties*. If the system of equations has rank $n - d$, one gets a d-dimensional linear variety, in particular for $d = 1$ a *projective line*. A *projective hypersurface* is given by an equation $F = 0$ with a nonconstant homogeneous polynomial F. If it is linear, one speaks of a *projective hyperplane*.

An advantage of working in projective space is that there we have stronger intersection theorems than in affine space, often allowing us to avoid troublesome case distinctions. Of this we give two examples, which will later turn out to be special cases of a more general proposition (Ch. V, 3.10).

Proposition 5.2. Let L be algebraically closed, $n \geq 2$.

a) A linear variety of dimension $d \geq 1$ and a hypersurface always intersect.

b) Any two projective hypersurfaces intersect.

Proof.

a) It suffices to prove the assertion for $d = 1$. After a coordinate transformation we may assume that the line is given by the system $X_2 = \cdots = X_n = 0$. If the hyperplane is defined by $F = 0$, then $F(X_0, X_1, 0, \ldots, 0)$ is a homogeneous polynomial in X_0, X_1 of positive degree (or it is 0). Since L is algebraically closed, there are elements $x_0, x_1 \in L$ not both zero such that $F(x_0, x_1, 0, \ldots, 0) = 0$. Then $x := \langle x_0, x_1, 0, \ldots, 0 \rangle$ is a point of intersection.

b) Let the two hypersurfaces be given by the equations $F = 0$ and $G = 0$ with nonconstant homogeneous polynomials $F, G \in L[X_0, \ldots, X_n]$. We may assume that F and G are irreducible and not associated to each other. (The irreducible factors of homogeneous polynomials are also homogeneous.) Through a suitable choice of coordinate system we can arrange that $\langle 0, \ldots, 0, 1 \rangle$ lies on neither of the two hypersurfaces. If we then consider F and G as polynomials in X_n with coefficients in $L[X_0, \ldots, X_{n-1}]$, their highest coefficient lies in L.

F and G are also relatively prime as elements of $L(X_0, \ldots, X_{n-1})[X_n]$. We then have an equation

$$1 = R \cdot F + S \cdot G \qquad (R, S \in L(X_0, \ldots, X_{n-1})[X_n]).$$

After multiplying by the product of the denominators of all the coefficients of R and S in $L(X_0, \ldots, X_{n-1})$, we get an equation

$$N = A \cdot F + B \cdot G \qquad (A, B \in L[X_0, \ldots, X_n], N \in L[X_0, \ldots, X_{n-1}]).$$

Since F and G are homogeneous and nonconstant, we can (after decomposing N, A, B into homogeneous components) also find such an equation with homogeneous polynomials N, A, B, where N is nonconstant. After dividing A by G (with remainder), we may also assume that $\deg_{X_n} A < \deg_{X_n} G$. Finally, we may assume that no irreducible divisor of N divides either of the polynomials A and B; for if it divides one, it divides both, and then the fraction is not in lowest terms.

If φ is an irreducible factor of N, then there is a zero $(x_0, \ldots, x_{n-1}) \in L^n$, $(x_0, \ldots, x_{n-1}) \neq (0, \ldots, 0)$, of φ that is not a zero of all the coefficients of A, if A is considered as a polynomial in X_n with coefficients in $L[X_0, \ldots, X_{n-1}]$; otherwise these coefficients would belong to $\mathfrak{I}(\mathfrak{V}(\varphi)) = (\varphi)$, and so all of them would be divisible by φ. From the equation

$$0 = A(x_0, \ldots, x_{n-1}, X_n) \cdot F(x_0, \ldots, x_{n-1}, X_n)$$
$$+ B(x_0, \ldots, x_{n-1}, X_n) \cdot G(x_0, \ldots, x_{n-1}, X_n),$$

which holds in $L[X_n]$, we find, because $\deg_{X_n} A < \deg_{X_n} G$, that the polynomials $F(x_0, \ldots, x_{n-1}, X_n)$ and $G(x_0, \ldots, x_{n-1}, X_n)$ have a nonconstant divisor in $L[X_n]$, hence a common zero $x_n \in L$, q. e. d.

We now want, just as in the affine case, to establish the connection between projective varieties and ideal theory. To this end we first introduce some general ring-theoretic concepts which generalize notions like the degree of a polynomial and the decomposition of a polynomial into homogeneous components.

Definition 5.3. A *grading* of a ring G is a family $\{G_k\}_{k \in \mathbf{Z}}$ of subgroups G_k of the additive group of G such that

a) $G = \bigoplus_{k \in \mathbf{Z}} G_k$,
b) $G_i \cdot G_j \subset G_{i+j}$ for all $i, j \in \mathbf{Z}$.

G is called a *graded ring* if it is provided with a grading $\{G_k\}_{k\in\mathbf{Z}}$. If $G_k = 0$ for $k < 0$, G is called *positively graded*. The elements of G_k are the *homogeneous elements of degree* k of G. If, by a), $g \in G$ is written as $g = \sum_{k\in\mathbf{Z}} g_k$ $(g_k \in G_k)$, g_k is called the *homogeneous component of degree* k of g.

Examples.

1. Any polynomial ring $G = R[X_1,\ldots,X_n]$ over a ring R is positively graded. The homogeneous elements of degree k are the homogeneous polynomials of degree k:

$$\sum_{\nu_1+\cdots+\nu_n=k} \rho_{\nu_1\ldots\nu_n} X_1^{\nu_1} \cdots X_n^{\nu_n}.$$

We call this grading the *canonical grading* of the polynomial ring.

2. The polynomial ring can be endowed with different gradings. Let a point $(\alpha_1,\ldots,\alpha_n) \in \mathbf{Z}^n$ be given. Now let G_k be the set of the polynomials

$$\sum_{\alpha_1\nu_1+\cdots+\alpha_n\nu_n=k} \rho_{\nu_1\ldots\nu_n} X_1^{\nu_1} \cdots X_n^{\nu_n}$$

("quasihomogeneous" polynomials of type $(\alpha_1,\ldots,\alpha_n)$ and degree k).

If $\{G_k\}_{k\in\mathbf{Z}}$ is a grading on a ring G, then by 5.3b) G_0 is a subring of G with $1 \in G_0$. Indeed, if $1 = \sum_{k\in\mathbf{Z}} e_k$ is the decomposition of 1 into homogeneous components, then for any homogeneous element $g \in G$ we have the equation $g = 1 \cdot g = \sum_{k\in\mathbf{Z}} e_k \cdot g$, and comparing coefficients yields $g = e_0 \cdot g$. But then $g = e_0 \cdot g$ for all $g \in G$; that is, $e_0 = 1$.

Definition 5.4. An ideal of a graded ring is called *homogeneous* if it can be generated by homogeneous elements.

Lemma 5.5. For an ideal I of a graded ring G the following statements are equivalent.

a) I is homogeneous.

b) For any $a \in I$ the homogeneous components a_k of a also belong to I $(k \in \mathbf{Z})$.

c) G/I is a graded ring with the grading $\{(G/I)_k\}_{k\in\mathbf{Z}}$, where

$$(G/I)_k := G_k + I/I$$

for all $k \in \mathbf{Z}$.

Proof.

a)$\to$b). Let $\{b_\lambda\}_{\lambda\in\Lambda}$ be a system of generators of I consisting of homogeneous elements b_λ of degree d_λ. Further, let $a = \sum_{i=1}^m r_{\lambda_i} b_{\lambda_i}$ be an element of I and let $r_{\lambda_i} = \sum_{k\in\mathbf{Z}} r_{\lambda_i}^{(k)}$ be the decomposition of r_{λ_i} into homogeneous components. Then $a = \sum_{k\in\mathbf{Z}} a_k$ with $a_k := r_{\lambda_1}^{(k-d_{\lambda_1})} \cdot b_{\lambda_1} + \cdots + r_{\lambda_m}^{(k-d_{\lambda_m})} \cdot b_{\lambda_m}$, where a_k is of degree k. Obviously, $a_k \in I$ for all $k \in \mathbf{Z}$.

b)$\to$a). If a generating system of I is given, the homogeneous components of all the elements of the generating system also form a generating system of I.

b)→c). It is clear that $G/I = \sum_{k\in\mathbf{Z}}(G/I)_k$. Hence it suffices to show that the representation of an element of G/I as the sum of elements in the $(G/I)_k$ is unique. Suppose $\sum_{k\in\mathbf{Z}} \bar{g}_k = 0$ with $\bar{g}_k \in (G/I)_k$, that is $\bar{g}_k = g_k + I$ for some $g_k \in G_k$. Then $\sum_{k\in\mathbf{Z}} g_k \in I$ and so $g_k \in I$; therefore, $\bar{g}_k = 0$ for all $k \in \mathbf{Z}$.

c)→b). Let $a = \sum_{k\in\mathbf{Z}} a_k$ be an element of I, $a_k \in G_k$ for all $k \in \mathbf{Z}$. Then $\sum_{k\in\mathbf{Z}} \bar{a}_k = 0$ in G/I, if $\bar{a}_k$ is the residue class of a_k. It follows that $a_k \in I$ for all $k \in \mathbf{Z}$.

In the situation of 5.5 G/I will always be considered a graded ring with the grading given by 5.5c). If I is finitely generated, it also has a finite system of generators consisting of homogeneous elements alone. If I and J are homogeneous ideals of G, then so are their sum, product, and intersection. Further, the image of J in G/I is a homogeneous ideal of G/I, and the inverse image of a homogeneous ideal in G/I in G is also homogeneous.

Under the hypotheses made in 5.1, just as in the affine case for a nonempty projective K-variety $V \subset \mathbf{P}^n(L)$ we define the vanishing ideal $\mathfrak{I}(V) \subset K[X_0,\ldots,X_n]$ as the set of all polynomials that vanish at all points of V. Further, we put $\mathfrak{I}(\emptyset) := (X_0,\ldots,X_n)$. If L is an infinite field, then $\mathfrak{I}(V)$ is always a homogeneous ideal with $\mathrm{Rad}(\mathfrak{I}(V)) = \mathfrak{I}(V)$. $K[V] := K[X_0,\ldots,X_n]/\mathfrak{I}(V)$ is called the *homogeneous* (or projective) *coordinate ring* of V. It is a positively graded, reduced Noetherian K-algebra.

For any homogeneous ideal $I \subset K[X_0,\ldots,X_n]$ the zero set $\mathfrak{V}(I)$ is defined as the set of all common zeros of all polynomials in I. Since I is generated by finitely many homogeneous polynomials, $\mathfrak{V}(I)$ is a projective K-variety.

For the formation of the zero set and the vanishing ideal in the projective case there are rules analogous to those that hold in the affine case, and they can be proved just as easily. In particular:

Lemma 5.6. Finite unions and arbitrary intersections of projective K-varieties in $\mathbf{P}^n(L)$ are projective K-varieties. The projective K-varieties in $\mathbf{P}^n(L)$ are the closed sets of a topology on $\mathbf{P}^n(L)$ (the K-Zariski topology); $\mathbf{P}^n(L)$ (and hence also any projective K-variety) is, when endowed with the K-Zariski topology, a Noetherian topological space.

Lemma 5.7. Any projective K-variety $V \subset \mathbf{P}^n(L)$ has a unique decomposition into irreducible components. If L is an infinite field, V is irreducible if and only if $\mathfrak{I}(V)$ is a prime ideal of $K[X_0,\ldots,X_n]$.

Definition 5.8. The affine cone $\tilde{V}$ of a projective K-variety $V \subset \mathbf{P}^n(L)$ is the set of all $(x_0,\ldots,x_n) \in \mathbf{A}^{n+1}(L)$ that occur as homogeneous coordinate systems of a point of V, together with the point $(0,\ldots,0) \in \mathbf{A}^{n+1}(L)$.

If $F_i = 0$ $(i = 1,\ldots,m)$ is a system of equations defining V by homogeneous polynomials, then $\tilde{V}$ is just the solution set of this system in $\mathbf{A}^{n+1}(L)$. The assignment $V \mapsto \tilde{V}$ is a bijection of the set of the projective K-varieties onto the set of all nonempty affine K-cones in $\mathbf{A}^{n+1}(L)$ (in the sense of Example 1.2, 4). If $V \neq \phi$ we have $\mathfrak{I}(V) = \mathfrak{I}(\tilde{V})$, the (affine) vanishing ideal of the cone $\tilde{V}$. If L is algebraically closed, then the affine cones in $A^{n+1}(L)$ (with vertex at the origin) correspond uniquely, by Hilbert's Nullstellensatz, to the homogeneous ideals $I \subsetneq K[X_0, \ldots, X_n]$ with $\mathrm{Rad}(I) = I$. From this follows

Proposition 5.9. (Projective Nullstellensatz) Let L be algebraically closed. The assignment $V \mapsto \mathfrak{I}(V)$ is a bijection of the set of all K-varieties $V \subset \mathbf{P}^n(L)$ onto the set of all homogeneous ideals $I \subset (X_0, \ldots, X_n)$ of $K[X_0, \ldots, X_n]$ with $\mathrm{Rad}(I) = I$. The inverse mapping is given by the formation of the zero set. For any homogeneous ideal $I \neq K[X_0, \ldots, X_n]$ we have $\mathfrak{I}(\mathfrak{V}(I)) = \mathrm{Rad}(I)$. This bijection is inclusion-reversing and the empty variety is assigned the ideal $(X_0, \ldots, X_n)$. The irreducible K-varieties correspond bijectively to the homogeneous prime ideals $\neq K[X_0, \ldots, X_n]$.

Corollary 5.10. A system of equations $F_i = 0$ $(i = 1, \ldots, m)$ with nonconstant homogeneous polynomials $F_i \in K[X_0, \ldots, X_n]$ has a nontrivial solution in L^{n+1}, where L is an algebraically closed extension field of K, if and only if

$$\mathrm{Rad}(F_1, \ldots, F_m) \neq (X_0, \ldots, X_n).$$

From 5.9 it follows that for a homogeneous ideal I of the polynomial ring $\mathrm{Rad}(I)$ is also homogeneous, which can also easily be proved directly. More generally we have:

Proposition 5.11. Let $G = \bigoplus_{i \in \mathbf{Z}} G_i$ be a graded ring, I a homogeneous ideal of G. Then all minimal prime divisors of I are homogeneous.

This follows immediately from the

Lemma 5.12. If $\mathfrak{P}$ is a prime ideal of G and $\mathfrak{P}^*$ is the ideal generated by the homogeneous elements of $\mathfrak{P}$, then $\mathfrak{P}^*$ is also a prime ideal.

Indeed, if $\mathfrak{P}$ is a minimal prime divisor of I, then $I \subset \mathfrak{P}^* \subset \mathfrak{P}$ and from the lemma it follows that $\mathfrak{P} = \mathfrak{P}^*$ is homogeneous.

To prove the lemma we may assume that G is positively graded; in the general case the proof is analogous. Let $a, b \in G$ with $a \cdot b \in \mathfrak{P}^*$ be given. Suppose that $a = a_0 + \cdots + a_n, b = b_0 + \cdots + b_m$, where the a_i, b_i are homogeneous of degree i. Suppose that $a \notin \mathfrak{P}^*, b \notin \mathfrak{P}^*$. Then there is a greatest index i with $a_i \notin \mathfrak{P}^*$ and a greatest index j with $b_j \notin \mathfrak{P}^*$. The homogeneous component of degree $i + j$ of the element $a \cdot b$ is $\sum_{\alpha+\beta=i+j} a_\alpha b_\beta$. Since $\mathfrak{P}^*$ is homogeneous, $\sum_{\alpha+\beta=i+j} a_\alpha b_\beta \in \mathfrak{P}^*$. Since all summands except for $a_i b_j$ belong to $\mathfrak{P}^*$, we also have $a_i b_j \in \mathfrak{P}^* \subset \mathfrak{P}$ and so $a_i \in \mathfrak{P}$ or $b_j \in \mathfrak{P}$, and therefore $a_i \in \mathfrak{P}^*$ or $b_j \in \mathfrak{P}^*$, a contradiction.

Corollary 5.13. Under the hypotheses of 5.11, $\mathrm{Rad}(I)$ is the intersection of the homogeneous prime ideals of G that contain I and is therefore itself homogeneous. In particular, $\mathrm{Rad}(0)$ is homogeneous and G_{red} is a graded ring.

Corollary 5.14. Let L be an algebraically closed field. The irreducible components of a cone in $A^{n+1}(L)$ are also cones (with the same vertex).

Proof. To the irreducible components of V belong the minimal prime divisors of $\mathfrak{I}(V)$. Because $\mathfrak{I}(V)$ is homogeneous, by 5.11 the minimal prime divisors are also homogeneous. (This corollary holds more generally for arbitrary infinite fields L; see Exercise 2.)

The coordinate ring $K[V]$ of a projective K-variety $V \subset P^n(L)$ can also be considered as the coordinate ring of the affine cone $\tilde{V}$ of V. From 3.10 it follows immediately that (if L is algebraically closed) the projective subvarieties $W \subset V$ and the homogeneous ideals $I \neq K[V]$ with $\mathrm{Rad}(I) = I$ correspond bijectively, the irreducible subvarieties $W \subset V$ corresponding to the homogeneous prime ideals in $K[V]$.

We now briefly present the projective analogue of the spectrum of a ring. We start from a positively graded ring $G = \bigoplus_{i \in \mathbb{N}} G_i$. In such a ring, $G_+ := \bigoplus_{i>0} G_i$ is a homogeneous ideal.

Definition 5.15. $\mathrm{Proj}(G)$ denotes the set of all homogeneous prime ideals $\mathfrak{P}$ of G with $G_+ \not\subset \mathfrak{P}$. These are also called the *relevant prime ideals* of G. $\mathrm{Proj}(G)$ is called the *homogeneous spectrum* of G.

According to the projective Nullstellensatz, in the case $G = K[V]$ for a projective K-variety V the relevant prime ideals of G correspond bijectively to the nonempty irreducible subvarieties of V.

Since $\mathrm{Proj}(G) \subset \mathrm{Spec}(G)$, we can endow $\mathrm{Proj}(G)$ with the relative topology of $\mathrm{Spec}(G)$. A subset $A \subset \mathrm{Proj}(G)$ is then closed if and only if there is an ideal I of G such that A is the set of all relevant prime ideals of G that contain I. As for the spectrum, it can be shown that any nonempty irreducible closed subset of $\mathrm{Proj}(G)$ has precisely one generic point. If $\mathrm{Proj}(G)$ is Noetherian, then it has only finitely many irreducible components and so there are only finitely many relevant minimal prime ideals in G.

Along with passing to the affine cone of a projective variety, an important role is played by another connection between affine and projective algebraic geometry—the projective closure of an affine variety.

For an extension of fields L/K we first consider the embedding of $A^n(L)$ into $P^n(L)$ that to any $(x_1, \ldots, x_n) \in A^n(L)$ assigns the point $\langle 1, x_1, \ldots, x_n \rangle \in P^n(L)$. This identifies $A^n(L)$ with the complement of the hyperplane $X_0 = 0$, which is called the *hyperplane at infinity*, its points the *points at infinity*. (These concepts depend on the coordinate system. With a suitable choice of coordinates any hyperplane can be the hyperplane at infinity.)

Definition 5.16. For any affine K-variety $V \subset A^n(L)$ the closure $\overline{V} \subset P^n(L)$ of V in the K-topology on $P^n(L)$ is called the *projective closure* of V. The points of $\overline{V} \setminus V$ are called the *points at infinity* of V.

By definition $\overline{V}$ is the smallest projective K-variety in $P^n(L)$ that contains

V. If V is given by a system of equations

$$F_i(X_1, \ldots, X_n) = 0 \qquad (i = 1, \ldots, m),$$

then $\overline{V}$ is contained in the solution set V^* in $\mathbf{P}^n(L)$ of the system of homogeneous equations

$$F_i^*(Y_0, \ldots, Y_n) = 0 \qquad (i = 1, \ldots, m),$$

where

$$F_i^*(Y_0, \ldots, Y_n) := Y_0^{\deg F_i} F_i\left(\frac{Y_1}{Y_0}, \ldots, \frac{Y_n}{Y_0}\right)$$

is the "homogenization" of F_i. Since $V^* \cap \mathbf{A}^n(L) = V$, we also have $\overline{V} \cap \mathbf{A}^n(L) = V$. Now let V be an arbitrary projective K-variety given by the system of equations

$$F_i(Y_0, \ldots, Y_n) = 0 \qquad (F_i \ \text{homogeneous}, i = 1, \ldots, m);$$

then $V_a := V \cap \mathbf{A}^n(L)$ is an affine K-variety with system of equations

$$F_i(1, X_1, \ldots, X_n) = 0 \qquad (i = 1, \ldots, m).$$

(We say that $F_i(1, X_1, \ldots, X_n)$ arises from F_i through "dehomogenization" with respect to Y_0.)

It turns out that the K-topology on $\mathbf{A}^n(L)$ is the relative topology of the K-topology on $\mathbf{P}^n(L)$.

Proposition 5.17.

a) The mapping $V \mapsto \overline{V}$ that assigns to each K-variety $V \subset \mathbf{A}^n(L)$ its projective closure $\overline{V} \subset \mathbf{P}^n(L)$ is a bijection of the set of nonempty affine K-varieties in $\mathbf{A}^n(L)$ onto the set of projective K-varieties in $\mathbf{P}^n(L)$ no irreducible component of which lies entirely in the hyperplane at infinity.

b) V is irreducible if and only if $\overline{V}$ is.

c) If $V = V_1 \cup \cdots \cup V_s$ is the decomposition of V into irreducible components, then $\overline{V} = \overline{V}_1 \cup \cdots \cup \overline{V}_s$, where $\overline{V}_i$ is the projective closure of V_i, is the decomposition of $\overline{V}$ into irreducible varieties.

Proof. b) follows from the fact that a subset of a topological space is irreducible if and only if its closure is (2.10).

Conversely, if $V^* \subset \mathbf{P}^n(L)$ is an irreducible projective K-variety that is not entirely contained in the hyperplane at infinity, then V_a^* is a nonempty open set in V^*. Since such a set is dense in V^* (2.9c), we have $V^* = \overline{V_a^*}$.

Thus we have a one-to-one correspondence between the irreducible affine K-varieties in $\mathbf{A}^n(L)$ and the irreducible projective K-varieties in $\mathbf{P}^n(L)$ that do not lie on the hyperplane at infinity. The other assertions of the proposition now follow immediately.

Exercises

1. Let G be a positively graded ring, with $G_0 = K$ a field with infinitely many elements. Suppose every $n + 1$ elements of G are algebraically dependent over K.

   a) If $F_1, \ldots, F_{n+1} \in G$ are homogeneous elements of the same degree and one puts $\widetilde{F}_i := F_i - \lambda_i F_{n+1}$ ($i = 1, \ldots, n$), then for a suitable choice of the $\lambda_i \in K$ there is an equation

   $$F_{n+1}^N = \sum_{i=0}^{N-1} \varphi_i(\widetilde{F}_1, \ldots, \widetilde{F}_n) F_{n+1}^i,$$

   where each φ_i is a homogeneous polynomial of degree $N - i$ ($i = 0, \ldots, N - 1$).

   b) For each finitely generated homogeneous ideal $I \subset G$ there are homogeneous elements $F_1, \ldots, F_n \in I$ with

   $$\mathrm{Rad}(I) = \mathrm{Rad}(F_1, \ldots, F_n).$$

   c) If L is an arbitrary extension field of K, then any projective K-variety $V \subset \mathbf{P}^n(L)$ is the intersection of $n + 1$ hypersurfaces (a theorem of Kronecker, which will be sharpened in Ch. V).

2. Let L/K be an extension of fields, where L is infinite, $V \subset \mathbf{A}^n(L)$ an irreducible K-variety, $V \neq \{(0, \ldots, 0)\}$. Further, let

   $$V^* = \bigcup_{x \in V \setminus \{(0,\ldots,0)\}} g_x,$$

   where g_x denotes the line through x and $(0, \ldots, 0)$.

   a) $\mathfrak{J}(V^*)$ is the ideal generated by all the homogeneous elements of $\mathfrak{J}(V)$.

   b) The closure $\overline{V^*}$ of V^* in the K-topology is a cone and is irreducible.

   c) The irreducible components of a K-cone in $\mathbf{A}^n(L)$ are cones (with the same vertex).

3. Let K be a field, and let $F \in K[X_1, \ldots, X_n]$ be quasihomogeneous of type $(\alpha_1, \ldots, \alpha_n) \in \mathbf{Z}^n$ and degree d. Then the Euler relation $\sum_{i=1}^n \alpha_i X_i \frac{\partial F}{\partial X_i} = d \cdot F$ holds. Conversely, if this relation holds for a polynomial F and if $\mathrm{Char}(K) = 0$, then F is quasihomogeneous.

4. Let $G = \bigoplus_{i \in \mathbf{N}} G_i$ be a positively graded ring, $G_+ := \bigoplus_{i > 0} G_i$. The following statements are equivalent.

   a) G is a Noetherian ring.

   b) G_0 is Noetherian and G_+ is a finitely generated ideal of G.

   c) G_0 is Noetherian and G is finitely generated as a G_0-algebra.

   d) The ring $G^{(n)} := \bigoplus_{i \equiv 0 \bmod n} G_i$ is Noetherian for all $n \in \mathbf{N}_+$.

5. With the notation of Exercise 4 we have:

    a) For any $\mathfrak{P} \in \text{Proj}(G)$, $\mathfrak{P} \cap G^{(n)} \in \text{Proj}(G^{(n)})$.

    b) The mapping $\text{Proj}(G) \to \text{Proj}(G^{(n)})$, $\mathfrak{P} \mapsto \mathfrak{P} \cap G^{(n)}$ is a homeomorphism.

6. Let I be an ideal of the polynomial ring $K[X_1, \ldots, X_n]$ over a field K. Let I^* denote the ideal of $K[Y_0, \ldots, Y_n]$ generated by all the homogenizations

$$F^* := Y_0^{\deg F} F\left(\frac{Y_1}{Y_0}, \ldots, \frac{Y_n}{Y_0}\right)$$

of the $F \in I$.

    a) If $G \in K[Y_0, \ldots, Y_n]$ is homogeneous, Y_0 is not a divisor of G, and $F = G(1, X_1, \ldots, X_n)$ is the dehomogenization of G with respect to Y_0, then $G = F^*$.

    b) If $I = \mathfrak{I}(V)$ for a K-variety $V \subset \mathbf{A}^n(L)$ where L is an algebraically closed extension field of K, then I^* is the ideal of the projective closure $\overline{V}$ of V.

    c) I is a prime ideal $(I = \text{Rad}(I))$ if and only if I^* is a prime ideal $(I^* = \text{Rad}(I^*))$.

References

The main theorems of this chapter, the Basis Theorem and the Nullstellensatz, were proved by Hilbert in [38] and [39], whose other results will play a role in our later work. There are several different proofs of the Nullstellensatz; treatments different from that of the text can be found in Krull [47] and Goldman [26]. For another proof of the Nullstellensatz for homogeneous ideals see: P. Cartier and J. Tate, A simple proof of the main theorem of elimination theory in algebraic geometry, *Enseign. Math.* **24**, 311–317 (1978). The treatment presented in the text of the decomposition of algebraic varieties into components goes back to Bourbaki [B]. Formerly its ideal-theoretic consequences were proved directly (cf. Emmy Noether's fundamental work [59]).

The J-spectrum of a ring has mainly technical significance. Its usefulness was first demonstrated by Swan [76]. General facts about algebraic varieties with coordinate fields that are not algebraically closed are contained in Silhol's paper [74]. More precise information on real algebraic varieties than is given in the text can be obtained from Whitney's article [82]. For special questions about complex varieties (e.g. in connection with the complex topology) consult Mumford [P].

The language of schemes was developed by Grothendieck (cf. [M]) and built up into a powerful system for formulating algebraic geometry; in this framework many problem—even of classical algebraic geometry—have been solved. Together with the seminar notes [30] this work comprises several thousand pages.

On the history of algebraic geometry much can be learned from Dieudonné's book [K].

Chapter II
Dimension

We now turn to the problem of measuring the "size" of algebraic varieties by assigning them a "dimension." We first introduce a very general concept of dimension which, we eventually see, provides a "natural" measure of the size of a variety and which in special cases agrees with usual concepts of dimension.

When in the rest of this book we speak of a K-variety, the coordinate field L of the variety is always an *algebraically closed* extension field of the field of definition K.

1. The Krull dimension of topological spaces and rings

Let X be a topological space, $Y \subset X$ a closed irreducible subset.

Definition 1.1. If $X \neq \emptyset$, the *Krull dimension* $\dim X$ (or combinatorial dimension) of X is the supremum of the lengths n of all chains

$$X_0 \subset X_1 \subset \cdots \subset X_n \qquad (X_{i+1} \neq X_i) \tag{1}$$

of nonempty closed irreducible subsets X_i of X. If $Y \neq \emptyset$, then the *codimension* $\operatorname{codim}_X Y$ of Y in X is defined as the supremum of the lengths of all chains (1) with $X_0 = Y$. The codimension of an arbitrary nonempty closed subset A of X is the infimum of the codimensions of the irreducible components of A. The empty topological space is assigned Krull dimension -1, and the empty subset of X is assigned codimension ∞.

These dimension concepts are mainly applied to algebraic varieties and the spectra of rings, considered as topological spaces with the Zariski topology. It is in no way clear at first that the Krull dimension of a variety is always finite. However, it is certainly independent of the coordinates, since it is defined in terms of the Zariski topology, which is independent of the coordinates.

In this section we collect properties of these concepts which easily follow from their definitons.

Rules 1.2.

a) If $\{X_\lambda\}_{\lambda \in \Lambda}$ is the family of irreducible components of X, then $\dim X = \operatorname{Sup}_{\lambda \in \Lambda} \{\dim X_\lambda\}$. Indeed, for any chain (1) X_n is contained in one of the irreducible components of X (I.2.12a)).

b) If $X = A_1 \cup \cdots \cup A_n$ with closed subsets A_i, then

$$\dim X = \operatorname{Sup}_{i=1,\dots,n} \{\dim A_i\}.$$

39

Again, for any chain (1) the irreducible set X_n is contained in one of the A_i.

c) $\dim Y + \operatorname{codim}_X Y \le \dim X$ if $Y \ne \emptyset$.

Join a chain (1) ending with Y to one beginning with Y.

d) If X is irreducible and $\dim X < \infty$, then $\dim Y < \dim X$ if and only if $Y \ne X$.

Definition 1.3. The *Krull dimension* $\dim R$ of a ring R is the dimension of $\operatorname{Spec}(R)$; therefore, for $R \ne \{0\}$ it is the supremum of the lengths n of all prime ideal chains

$$\mathfrak{p}_0 \subset \mathfrak{p}_1 \subset \cdots \subset \mathfrak{p}_n \qquad (\mathfrak{p}_{i+1} \ne \mathfrak{p}_i) \tag{2}$$

in $\operatorname{Spec}(R)$. The *height* $h(\mathfrak{p})$ of $\mathfrak{p} \in \operatorname{Spec}(R)$ is the supremum of the lengths of all chains (2) with $\mathfrak{p} = \mathfrak{p}_n$. For an arbitrary ideal $I \ne R$ the height $h(I)$ is defined as the infimum of the heights of the prime divisors of I. Further, we also call $\dim(I) := \dim R/I$ the *dimension* (or coheight) of the ideal I.

On the basis of the connection between the closed irreducible subsets of $\operatorname{Spec}(R)$ and the prime ideals of R, we find that $\dim \mathfrak{p} = \dim(\mathfrak{V}(\mathfrak{p}))$ and $h(\mathfrak{p}) = \operatorname{codim}_{\operatorname{Spec}(R)} \mathfrak{V}(\mathfrak{p})$ for all $\mathfrak{p} \in \operatorname{Spec}(R)$. Further, $\dim R = \dim R_{\mathrm{red}}$, since $\operatorname{Spec}(R)$ and $\operatorname{Spec}(R_{\mathrm{red}})$ are homeomorphic.

$\dim J(R)$ is called the *J-dimension* of R. Since every nonempty irreducible closed subset of $J(R)$ has precisely one generic point, we see that J-$\dim R$ (if $R \ne \{0\}$) equals the supremum of the lengths of all chains (2) with $\mathfrak{p}_i \in J(R)$ for $i = 0, \ldots, n$.

For a positively graded ring G let g-$\dim(G) := \dim(\operatorname{Proj}(G))$. g-$\dim(G)$ is (if $\operatorname{Proj}(G) \ne \emptyset$) equal to the supremum of the lengths of all chains (2) consisting of only relevant prime ideals $\mathfrak{p}_i$ of $G(i = 0, \ldots, n)$.

For a K-variety $V \subset \mathbf{A}^n(L)$, because the chains (1) in V correspond bijectively to the chains (2) in $K[V]$, we have

$$\dim V = \dim K[V]. \tag{3}$$

(Later it will turn out that $\dim V$ does not depend on the choice of the field of definition K (3.11a)), on account of which we can avoid speaking of the K-dimension of V.)

Likewise, for a projective K-variety $V \subset \mathbf{P}^n(L)$ we have

$$\dim V = g\text{-}\dim K[V]. \tag{4}$$

If $\overline{V} \subset \mathbf{P}^n(L)$ is the projective closure of an affine K-variety $V \subset \mathbf{A}^n(L)$, then

$$\dim \overline{V} \ge \dim V.$$

This follows immediately from I.5.17. As we shall see in (4.1), in reality the equality sign holds.

Through (3) and (4) the study of the dimension of varieties is reduced to the study of chains of prime ideals in finitely generated algebras over fields.

Examples 1.4.

a) In the polynomial ring $K[X_1, \ldots, X_n]$ over a field K,

$$(0) \subset (X_1) \subset (X_1, X_2) \subset \cdots \subset (X_1, \ldots, X_n)$$

is a chain of prime ideals (2) of length n ; therefore, $\dim K[X_1, \ldots, X_n] \geq n$. Later we shall prove that in $K[X_1, \ldots, X_n]$ every prime ideal chain (2) has length $\leq n$ (3.4). It then follows that $\dim A^n(L) = n$, and so affine and projective varieties are of finite dimension.

b) In a factorial ring R the prime ideals of height 1 are precisely the principal ideals generated by prime elements.

Every $\mathfrak{p} \in \mathrm{Spec}(R)$ with $h(\mathfrak{p}) = 1$ contains an $r \in R, r \neq 0$, and therefore also contains a prime divisor π of r. It follows that $\mathfrak{p} = (\pi)$, since (π) is also a prime ideal. Conversely, if a prime element π of R is given and if $\mathfrak{p} \subset (\pi)$ for some $\mathfrak{p} \in \mathrm{Spec}(R), \mathfrak{p} \neq (0)$, then $\mathfrak{p}$ contains a prime element π'. It is divisible by π and is therefore associated to π. It follows that $\mathfrak{p} = (\pi)$ and so $h(\pi) = 1$.

In particular, it follows that principal ideal domains that are not fields have Krull dimension 1. In particular, $\dim \mathbb{Z} = 1$ and $\dim K[X] = 1$.

Applying b) to the factorial ring $K[X_1, \ldots, X_n]$, we see that hypersurfaces (in both affine and projective space) are of codimension 1 (in the ambient affine or projective space).

c) A ring $R \neq \{0\}$ has dimension 0 if and only if $\mathrm{Spec}(R) = \mathrm{Max}(R)$. An integral domain R has dimension 0 if and only if it is a field.

d) For a 0-dimensional ring R with Noetherian spectrum (in particular, a Noetherian ring), $\mathrm{Spec}(R)$ has only finitely many elements (which are both maximal and minimal prime ideals of R).

By I.4.9 such a ring R has only finitely many minimal prime ideals.

The 0-dimensional reduced rings with Noetherian spectrum can be completely determined.

Proposition 1.5. For a reduced ring R with only finitely many minimal prime ideals the following statements are equivalent.

a) $\dim R = 0$.

b) R is isomorphic to a direct product of finitely many fields.

This proposition follows as a consequence of the so-called Chinese Remainder Theorem, which we will now prove.

Definition 1.6. Two ideals I_1, I_2 of a ring R are called *relatively prime* (or *comaximal*) if they are $\neq R$ but $I_1 + I_2 = R$.

Proposition 1.7. (Chinese Remainder Theorem) Let $I_1, \ldots, I_n$ ($n > 1$) be pairwise relatively prime ideals of a ring R. Then the canonical ring homomorphism

$$\varphi : R \to R/I_1 \times \cdots \times R/I_n$$
$$r \mapsto (r + I_1, \ldots, r + I_n)$$

is an epimorphism with kernel $\bigcap_{k=1}^{n} I_k$.

Proof. The statement about the kernel follows immediately from the definition of φ and the definition of the direct product of rings.

We prove the surjectivity of φ by induction on n. Let $n = 2$ and $(r_1 + I_1, r_2 + I_2) \in R/I_1 \times R/I_2$ be given. We have an equation $1 = a_1 + a_2$ with $a_k \in I_k$ $(k = 1, 2)$ and so $a_k \equiv 1 \bmod I_l$ $(l \neq k)$. If we put $r := r_2 a_1 + r_1 a_2$, then $r \equiv r_k \bmod I_k$ $(k = 1, 2)$ which proves that φ is surjective for $n = 2$.

Now let $n > 2$ and suppose the proposition has been proved for fewer than n pairwise relatively prime ideals. For given $(r_1 + I_1, \ldots, r_n + I_n) \in R/I_1 \times \cdots \times R/I_n$ there is then an $r' \in R$ with $r' = r_k \bmod I_k$ for $k = 1, \ldots, n - 1$.

We prove that $I_1 \cap \cdots \cap I_{n-1}$ is relatively prime to I_n. There are equations $1 = a_1 + a_3 = a_2 + a_3'$ with $a_k \in I_k$ $(k = 1, 2, 3)$, $a_3' \in I_3$, and so $1 = a_1 a_2 + (a_2 + a_3') a_3 + a_1 a_3' \in (I_1 \cap I_2) + I_3$. Therefore, $I_1 \cap I_2$ and I_3 are relatively prime. By induction it immediately follows that $I_1 \cap \cdots \cap I_{n-1}$ is relatively prime to I_n.

Since the proposition has already been proved for $n = 2$, for every $(r' + I_1 \cap \cdots \cap I_{n-1}, r_n + I_n) \in R/I_1 \cap \cdots \cap I_{n-1} \times R/I_n$ there is an $r \in R$ with $r \equiv r' \bmod (I_1 \cap \cdots \cap I_{n-1}), r \equiv r_n \bmod I_n$. Then $r \equiv r_k \bmod I_k (k = 1, \ldots, n)$, which proves the proposition.

Proof of 1.5. Let $\mathfrak{p}_1, \ldots, \mathfrak{p}_n$ be the minimal prime ideals of R. If $\dim R = 0$, then these are all maximal and therefore also pairwise relatively prime. Since R is reduced, $\bigcap_{k=1}^{n} \mathfrak{p}_k = (0)$. By 1.7 $R \cong R/\mathfrak{p}_1 \times \cdots \times R/\mathfrak{p}_n$, a direct product of fields.

Conversely, if $R \cong K_1 \times \cdots \times K_n$ with fields K_i $(i = 1, \ldots, n)$, then for any ideal I of R its projection in K_i is the zero ideal or K_i itself. The only elements of the spectrum of $K_1 \times \cdots \times K_n$ are therefore the ideals

$$\mathfrak{p}_i := K_1 \times \cdots \times K_{i-1} \times (0) \times K_{i+1} \times \cdots \times K_n.$$

These are both minimal and maximal.

For a later application we prove one more statement about the maximal spectrum and the J-spectrum of a ring.

Proposition 1.8. For any ring R:

a) $\mathrm{Max}(R)$ is Noetherian if and only if $J(R)$ is;

b) $\dim \mathrm{Max}(R) = \dim J(R) \leq \dim \mathrm{Spec}(R)$.

Proof. If $A \subset \mathrm{Max}(R)$ is closed, let $\overline{A}$ be the closure of A in $J(R)$. For any closed set $B \subset J(R)$ let $B^* := B \cap \mathrm{Max}(R)$. We shall show that $\overline{A}^* = A$ and $\overline{B^*} = B$. The closed subsets of $\mathrm{Max}(R)$ and $J(R)$ then correspond bijectively, whereby inclusions are preserved and irreducible sets are mapped to irreducible ones. The assertions of the proposition then follow at once.

$\overline{A}^* = A$ follows immediately from the definition of the Zariski topology on $\mathrm{Max}(R)$ and $J(R)$. For any closed set $B \subset J(R)$ we have $\mathfrak{I}(B) = \bigcap_{\mathfrak{p} \in B} \mathfrak{p}$. Since every $\mathfrak{p} \in J(R)$ is an intersection of maximal ideals of R, it follows that $\mathfrak{I}(B) = \bigcap_{\mathfrak{m} \in B^*} \mathfrak{m} = \mathfrak{I}(B^*)$. By I.4.2, $\overline{B^*} = \mathfrak{V}(\mathfrak{I}(B^*)) = \mathfrak{V}(\mathfrak{I}(B)) = B$.

Definition 1.9. A ring $R \neq \{0\}$ is called *local (semilocal)* if $\mathrm{Max}(R)$ consists of only one element (of only finitely many elements).

Remark 1.10. For any semilocal ring R, $J(R) = \mathrm{Max}(R)$ and

$$\dim J(R) = \dim \mathrm{Max}(R) = 0.$$

A prime ideal that is the intersection of finitely many maximal ideals contains one of these maximal ideals and is therefore itself maximal. For a semilocal ring R we therefore have $J(R) = \mathrm{Max}(R)$. The irreducible components of $J(R)$ are precisely the points of $J(R)$, whence $\dim J(R) = 0$.

Exercises

1. Any nonempty Hausdorff space has Krull dimension 0.

2. An affine or projective variety of dimension 0 consists of only finitely many points.

3. A local ring of dimension 0 consists of only units and nilpotent elements, a semilocal ring of dimension 0 of only units and zero divisors.

4. In a local ring $R, 0$ and 1 are the only idempotent elements (in particular, $\mathrm{Spec}(R)$ is connected).

5. Let $K[\|X_1, \ldots, X_n\|]$ denote the ring of formal power series in the indeterminates $X_1, \ldots, X_n$ over a field K.

   a) A formal power series $\sum a_{\nu_1, \ldots, \nu_n} X_1^{\nu_1} \ldots X_n^{\nu_n}$ $(a_{\nu_1 \ldots \nu_n} \in K)$ is a unit in $K[\|X_1, \ldots, X_n\|]$ if and only if the "constant term" $a_{0 \ldots 0}$ is $\neq 0$.

   b) $K[\|X_1, \ldots, X_n\|]$ is a local ring.

6. Let $\alpha_k : R_k \to P$ $(k = 1, 2)$ be two ring homomorphisms. By the *fiber product* $R_1 \times_P R_2$ of R_1 and R_2 over P (with respect to α_1, α_2) we understand a triple (S, β_1, β_2), where the $\beta_k : S \to R_k$ $(k = 1, 2)$ are ring homomorphisms with $\alpha_1 \circ \beta_1 = \alpha_2 \circ \beta_2$ and the following universal property holds: If (T, γ_1, γ_2) is any triple like (S, β_1, β_2), then there is exactly one ring homomorphism $\delta : T \to S$ with $\gamma_k = \beta_k \circ \delta$ $(k = 1, 2)$.

   a) In the direct product $R_1 \times R_2$ consider the subring S of all (r_1, r_2) with $\alpha_1(r_1) = \alpha_2(r_2)$. Let $\beta_k : S \to R_k$ be the restriction of the canonical projection $R_1 \times R_2 \to R_k$ $(k = 1, 2)$. Then (S, β_1, β_2) is the fiber product of R_1 and R_2 over P.

   b) Let I_1, I_2 be ideals of a ring, $\alpha_k : R/I_k \to R/I_1 + I_2$ and $\beta_k : R/I_1 \cap I_2 \to R/I_k$ the canonical epimorphisms $(k = 1, 2)$. Then $(R/I_1 \cap I_2, \beta_1, \beta_2)$ is the fiber product of R/I_1 and R/I_2 over $R/I_1 + I_2$. (This generalizes 1.7 in the case $n = 2$.)

2. Prime ideal chains and integral ring extensions

The propositions of this section serve as preparation for the study of the dimension of algebraic varieties, but they are also of great significance for ring theory.

Let S/R be an extension of rings, where $R \neq \{0\}$, and let I be an ideal of R. ($I = R$ is the most important special case of the following constructions.)

Definition 2.1. $x \in S$ is called *integral over I* if there is a polynomial $f \in R[X]$ of the form

$$f = X^n + a_1 X^{n-1} + \cdots + a_n \qquad (n > 0, a_i \in I, i = 1, \ldots, n) \qquad (1)$$

such that $f(x) = 0$. S/R is called an *integral ring extension* if every $x \in S$ is integral over R. (If in (1) one requires that $a_i \in I^i$ for $i = 1, \ldots, n$, then x is called *integrally dependent* on I. This concept will play no role for us.)

Proposition 2.2. For $x \in S$ the following statements are equivalent.

a) x is integral over I.

b) $R[x]$ is finitely generated as an R-module and $x \in \mathrm{Rad}(IR[x])$.

c) There is a subring S' of S with $R[x] \subset S'$ such that S' is finitely generated as an R-module and $x \in \mathrm{Rad}(IS')$.

Proof.

a)→b). Let f be given as in (1). Every $g \in R[X]$ can be divided by f with remainder: $g = q \cdot f + r$ with $q, r \in R[X]$, $\deg r < \deg f$. Since $g(x) = r(x)$ we see that $\{1, x, \ldots, x^{n-1}\}$ is a system of generators of the R-module $R[x]$. From $f(x) = 0$ it follows that $x^n \in IR[x]$, so $x \in \mathrm{Rad}(IR[x])$.

b)→c). This follows by putting $S' = R[x]$.

c)→a). If $\{w_1, \ldots, w_l\}$ is a system of generators of the R-module S' and $x^m \in IS'$, then we can write

$$x^m w_i = \sum_{k=1}^{l} \rho_{ik} w_k \text{ or } \sum_{k=1}^{l} (x^m \delta_{ik} - \rho_{ik}) w_k = 0 \qquad (i = 1, \ldots, l)$$

for some $\rho_{ik} \in I$. By Cramer's Rule, $\det(x^m \delta_{ik} - \rho_{ik}) w_k = 0$ for $k = 1, \ldots, l$. Further, $1 = \sum_{k=1}^{l} a_k w_k$ for some $a_k \in R$ and so $\det(x^m \delta_{ik} - \rho_{ik}) = 0$. The complete expansion of the determinant then leads to an equation (1), where $n = m \cdot l$.

Corollary 2.3. If S is finitely generated as an R-module, then S is integral over R. In this case $x \in S$ is integral over I if and only if $x \in \mathrm{Rad}(IS)$.

Corollary 2.4. If $x_1, \ldots, x_n \in S$ are integral over I, then $R[x_1, \ldots, x_n]$ is a finitely generated R-module and $x_i \in \mathrm{Rad}(IR[x_1, \ldots, x_n])$ for $i = 1, \ldots, n$.

This follows from 2.2 by induction.

Using 2.4 one can (according to J. David) give the following brief proof of Ch. I, Proposition 3.2. Let L/K be an extension of fields, where $L = K[x_1,\ldots,x_n]$ for some $x_i \in L$ $(i = 1,\ldots,n)$. We show by induction on n that L/K is algebraic. For $n = 1$ this is clear. Suppose that $n \geq 2$ and the assertion has been proved for $n-1$ elements, but that it is false for n elements. Say x_1 is transcendental over K. Since $L = K(x_1)[x_2,\ldots,x_n]$, L is algebraic over $K(x_1)$ by the induction hypothesis. Let $u_i \in K[x_1]$ be the leading coefficient of an algebraic equation of x_i over $K[x_1]$ $(i = 2,\ldots,n)$ and $u := \prod_{i=2}^{n} u_i$. Then by 2.4 L is integral over $K[x_1, 1/u]$. Let p be a prime polynomial in $K[x_1]$ that does not divide u. $1/p$ satisfies an equation

$$\left(\frac{1}{p}\right)^m + a_1 \left(\frac{1}{p}\right)^{m-1} + \cdots + a_m = 0 \quad (m > 0, a_i \in K[x_1, 1/u]).$$

After multiplying by p^m and a suitable power of u, we get an equation

$$u^\rho + b_1 p + \cdots + b_m p^m = 0 \quad (\rho \in \mathbb{N}, b_i \in K[x_1]).$$

But then p is a divisor of u^ρ in $K[x_1]$, a contradiction.

Corollary 2.5. If S/R and T/S are integral ring extensions, then so is T/R.

Let $x \in T$ satisfy an equation $x^n + s_1 x^{n-1} + \cdots + s_n = 0$ with $s_i \in S$ $(i = 1,\ldots,n)$. Since $s_1,\ldots,s_n$ are integral over R, $R[s_1,\ldots,s_n]$ is finitely generated as an R-module. But then $R[s_1,\ldots,s_n,x]$ is a finitely generated R-module and x is integral over R by 2.3.

Corollary 2.6. The set $\overline{R}$ of all elements of S that are integral over R is a subring of S. $\mathrm{Rad}(I\overline{R})$ is the set of all elements of S that are integral over I.

For $x, y \in \overline{R}$, $R[x,y]$ is finitely generated as an R-module. By 2.2, $x + y$, $x - y$, and $x \cdot y$ are in $\overline{R}$. If $x \in S$ is integral over I, then $x \in \mathrm{Rad}(I\overline{R})$ by 2.2.

Conversely, if $x \in \mathrm{Rad}(I\overline{R})$, then $x^m \in IR[x_1,\ldots,x_n]$ for suitable $x_1,\ldots,x_n \in \overline{R}$. Since $R[x_1,\ldots,x_n]$ is a finitely generated R-module, by 2.2 it follows that x is integral over I.

Definition 2.7. $\overline{R}$ is called the *integral closure* of R in S. R is called *integrally closed* in S if $\overline{R} = R$. An integral domain that is integrally closed in its field of fractions is called *normal*.

Example 2.8. Any factorial ring R is normal (in particular, $\mathbb{Z}$ is normal, as is $R[X_1,\ldots,X_n]$ when R itself is factorial (e.g. a field)).

Let K be the field of fractions of R and let $x \in K$ be integral over R. We consider an equation

$$x^n + r_1 x^{n-1} + \cdots + r_n = 0 \quad (n > 0, r_i \in R)$$

and a representation $x = \frac{r}{s}$, with $r, s \in R$, in lowest terms. After multiplying by s^n the equation assumes the form

$$r^n + r_1 s r^{n-1} + \cdots + r_n s^n = 0.$$

If there were a prime element of R that divides s, then it would also divide r, a contradiction. Therefore, s is a unit of R and $x \in R$.

Lemma 2.9. Let S/R be an integral ring extension, J an ideal of S, $I := J \cap R$. Then:

a) S/J is integral over R/I. (R/I is canonically a subring of S/J.)

b) If J contains a non-zerodivisor of S, then $I \neq (0)$.

Proof.

a) follows at once from Definition 2.1.

b) If $x \in J$ is not a zero divisor of S and if x satisfies the equation $x^n + r_1 x^{n-1} + \cdots + r_n = 0$ ($n > 0, r_i \in R$), then we can assume that $r_n \neq 0$. Otherwise, we could divide by x and lower the degree of the equation. It then follows that $r_n \in J \cap R = I$ and $I \neq (0)$.

For an integral ring extension S/R there is a close relation between the chains of prime ideals of R and those of S. This relation is given by the theorems of Cohen–Seidenberg, which we will now derive. Let

$$\varphi : \operatorname{Spec}(S) \to \operatorname{Spec}(R) \qquad (\mathfrak{P} \mapsto \mathfrak{P} \cap R)$$

be the continuous mapping of spectra belonging to S/R. If $\mathfrak{P} \in \operatorname{Spec}(S)$, $\mathfrak{p} := \mathfrak{P} \cap R$, then we say that "$\mathfrak{P}$ lies over $\mathfrak{p}$."

Proposition 2.10. Let S/R be an integral ring extension. Then:

a) $\varphi : \operatorname{Spec}(S) \to \operatorname{Spec}(R)$ is surjective (over a prime ideal of R there lies a prime ideal of S).

b) φ is closed (the image of any closed subset of $\operatorname{Spec}(S)$ is closed in $\operatorname{Spec}(R)$).

c) If $\mathfrak{P}_1$, $\mathfrak{P}_2 \in \operatorname{Spec}(S)$ with $\mathfrak{P}_1 \subset \mathfrak{P}_2$ are given, it follows from $\varphi(\mathfrak{P}_1) = \varphi(\mathfrak{P}_2)$ that $\mathfrak{P}_1 = \mathfrak{P}_2$.

d) φ maps $\operatorname{Max}(S)$ onto $\operatorname{Max}(R)$, and $\varphi^{-1}(\operatorname{Max}(R)) = \operatorname{Max}(S)$. ($\mathfrak{P} \in \operatorname{Spec}(S)$ is maximal if and only if $\mathfrak{P} \cap R$ is maximal.)

Proof.

a) For $\mathfrak{p} \in \operatorname{Spec}(R)$ let $N := R \setminus \mathfrak{p}$. By 2.6 any $x \in \mathfrak{p}S$ satisfies an equation

$$x^n + r_1 x^{n-1} + \cdots + r_n = 0 \qquad (n > 0, r_i \in \mathfrak{p}).$$

If x is an element of $\mathfrak{p}S \cap N$, so in particular $x \in R$, then $x^n \in \mathfrak{p}$ and so $x \in \mathfrak{p}$, contradicting $x \in N$. Since $\mathfrak{p} \cap N = \emptyset$, we can apply I.4.4: There is a $\mathfrak{P} \in \operatorname{Spec}(S)$ with $\mathfrak{p}S \subset \mathfrak{P}$, $\mathfrak{P} \cap N = \emptyset$. But then $\mathfrak{P} \cap R = \mathfrak{p}$.

b) Let $A := \mathfrak{V}(J)$ be a closed subset of $\operatorname{Spec}(S)$ with an ideal J of S and $I := J \cap R$. By 2.9, S/J is integral over R/I and hence by a) the mapping $\operatorname{Spec}(S/J) \to \operatorname{Spec}(R/I)$ is surjective. Therefore, φ maps A onto the closed subset $\mathfrak{V}(I)$ of $\operatorname{Spec}(R)$, for by I.4.12 $\mathfrak{V}(J)$ is the image of $\operatorname{Spec}(S/J) \to \operatorname{Spec}(S)$ and $\mathfrak{V}(I)$ is the image of $\operatorname{Spec}(R/I) \to \operatorname{Spec}(R)$.

c) Let $\mathfrak{p} := \mathfrak{P}_1 \cap R = \mathfrak{P}_2 \cap R$. Then $S/\mathfrak{P}_1$ is integral over $R/\mathfrak{p}$ and $\mathfrak{P}_2/\mathfrak{P}_1$ is a prime ideal of $S/\mathfrak{P}_1$ that intersects $R/\mathfrak{p}$ in (0). By 2.9b) we must have $\mathfrak{P}_1 = \mathfrak{P}_2$.

d) Let $\mathfrak{P} \in \mathrm{Spec}(S)$ be given and $\mathfrak{p} := \mathfrak{P} \cap R$. If $R/\mathfrak{p}$ is a field, then so is $S/\mathfrak{P}$, for $S/\mathfrak{P}$ arises from $R/\mathfrak{p}$ by adjoining algebraic elements. If $S/\mathfrak{P}$ is a field, then (0) is the only element of $\mathrm{Spec}(S/\mathfrak{P})$ and by a) (0) is the only element of $\mathrm{Spec}(R/\mathfrak{p})$. Therefore, $R/\mathfrak{p}$ is also a field.

When in what follows we speak of a "prime ideal chain" $\mathfrak{P}_0 \subset \mathfrak{P}_1 \subset \cdots \subset \mathfrak{P}_n$ of a ring S, the $\mathfrak{P}_i$ are to be in $\mathrm{Spec}(S)$ and the inclusions are to be proper.

Corollary 2.11. If $\mathfrak{P}_0 \subset \mathfrak{P}_1 \subset \cdots \subset \mathfrak{P}_n$ is a prime ideal chain in S and $\mathfrak{p}_i := \mathfrak{P}_i \cap R$ $(i = 0, \ldots, n)$, then $\mathfrak{p}_0 \subset \mathfrak{p}_1 \subset \cdots \subset \mathfrak{p}_n$ is a prime ideal chain in R.

Corollary 2.12. ("Going-up" Theorem of Cohen–Seidenberg). For any prime ideal chain $\mathfrak{p}_0 \subset \mathfrak{p}_1 \subset \cdots \subset \mathfrak{p}_n$ in R and for any $\mathfrak{P}_0$ lying over $\mathfrak{p}_0$, S contains a prime ideal chain $\mathfrak{P}_0 \subset \mathfrak{P}_1 \subset \cdots \subset \mathfrak{P}_n$ with $\mathfrak{P}_i \cap R = \mathfrak{p}_i$ $(i = 0, \ldots, n)$.

Proof. If a prime ideal chain $\mathfrak{p}_0 \subset \cdots \subset \mathfrak{p}_i$ lying over $\mathfrak{P}_0 \subset \cdots \mathfrak{P}_i$ has already been constructed, then in $S/\mathfrak{P}_i$ we consider a prime ideal lying over $\mathfrak{p}_{i+1}/\mathfrak{p}_i$. Then its inverse image $\mathfrak{P}_{i+1}$ in S lies over $\mathfrak{p}_{i+1}$, and $\mathfrak{P}_i \cap R = \mathfrak{p}_i$ $(i = 0, \ldots, n)$.

Corollary 2.13.

a) $\dim R = \dim S$.

b) For any $\mathfrak{P} \in \mathrm{Spec}(S)$ we have $h(\mathfrak{P}) \leq h(\mathfrak{P} \cap R), \dim(\mathfrak{P}) = \dim(\mathfrak{P} \cap R)$.

This follows from 2.11, 2.12, and the definitions of dimension and of height.

Corollary 2.14. If $\mathrm{Spec}(S)$ is Noetherian, then φ is a finite mapping, i.e. over any $\mathfrak{p} \in \mathrm{Spec}(R)$ lie only finitely many $\mathfrak{P} \in \mathrm{Spec}(S)$.

Proof. For $\mathfrak{p} \in \mathrm{Spec}(R), \mathfrak{p}S$ is contained in a prime ideal of S that lies over $\mathfrak{p}$ and so $\mathfrak{p}S \cap R = \mathfrak{p}$. The prime ideals of S lying over $\mathfrak{p}$ are minimal prime divisors of $\mathfrak{p}S$ (2.10c)). Since $\mathrm{Spec}(S)$ is Noetherian, by I.4.9a) the number of them is finite.

The assertions about prime ideal chains given above can be sharpened if we assume that R is a normal ring.

Lemma 2.15. Let R be a normal ring with field of fractions $K, L/K$ a field extension and I a prime ideal of R. If $x \in L$ is integral over I, then the minimal polynomial m of x over K has the form

$$m = X^n + a_1 X^{n-1} + \cdots + a_n \qquad \text{with} \qquad a_i \in I \qquad (i = 1, \ldots, n).$$

Proof. Let $x = x_1, x_2, \ldots, x_n$ be the zeros of m in the algebraic closure $\overline{K}$ of K. Since x can be mapped into any of the x_i $(i = 1, \ldots, n)$ by a K-automorphism of $\overline{K}$, the x_i are also zeros of a polynomial (1) of which x is a zero; that is, the x_i are also integral over I $(i = 1, \ldots, n)$. The coefficients a_k of m are the elementary symmetric functions of the x_i; by 2.6 they belong to $\mathrm{Rad}(I\overline{R})$, where $\overline{R}$ is the integral closure of R in K. Since $\overline{R} = R$ and I is a prime ideal, it follows that $a_k \in I$ for $k = 1, \ldots, n$.

Proposition 2.16. ("Going-Down" Theorem of Cohen–Seidenberg) Let S/R be an integral ring extension, where R and S are integral domains and R is normal. Let $\mathfrak{p}_0 \subset \mathfrak{p}_1$ be a prime ideal chain in $\mathrm{Spec}(R)$ and $\mathfrak{P}_1$ a prime ideal of S lying over $\mathfrak{p}_1$. Then there is a $\mathfrak{P}_0 \in \mathrm{Spec}(S)$ with $\mathfrak{P}_0 \subset \mathfrak{P}_1$ and $\mathfrak{P}_0 \cap R = \mathfrak{p}_0$.

Proof. The sets $N_0 := R \setminus \mathfrak{p}_0, N_1 := S \setminus \mathfrak{P}_1$ and $N := N_0 \cdot N_1 = \{rs \mid r \in N_0, s \in N_1\}$ are multiplicatively closed and $N_i \subset N$ ($i = 1, 2$). We shall prove that $\mathfrak{p}_0 S \cap N = \emptyset$. By I.4.4 there is then a $\mathfrak{P}_0 \in \mathrm{Spec}(S)$ with $\mathfrak{p}_0 S \subset \mathfrak{P}_0$ and $\mathfrak{P}_0 \cap N = \emptyset$. Since $\mathfrak{P}_0 \cap N_1 = \emptyset$, we have $\mathfrak{P}_0 \subset \mathfrak{P}_1$, and from $\mathfrak{P}_0 \cap N_0 = \emptyset$ it follows that $\mathfrak{P}_0 \cap R = \mathfrak{p}_0$.

Suppose there is an $x \in \mathfrak{p}_0 S \cap N$. x is then integral over $\mathfrak{p}_0$ and by 2.15 its minimal polynomial m over the field of fractions K of R has the form $m = X^n + a_1 X^{n-1} + \cdots + a_n$ with $a_i \in \mathfrak{p}_0$ ($i = 1, \ldots, n$). From $x \in N$ it further follows that $x = r \cdot s$ with $r \in N_0, s \in N_1$. The minimal polynomial over K of $s = x/r$ is

$$X^n + \frac{a_1}{r} X^{n-1} + \cdots + \frac{a_n}{r^n},$$

whose coefficients lie in R by 2.15, since s is integral over R. If we put $a_i = r^i \rho_i$ with $\rho_i \in R$ ($i = 1, \ldots, n$), it then follows from $a_i \in \mathfrak{p}_0, r \notin \mathfrak{p}_0$, that $\rho_i \in \mathfrak{p}_0$ for $i = 1, \ldots, n$. But then s is integral over $\mathfrak{p}_0$ and therefore $s \in \mathrm{Rad}(\mathfrak{p}_0 S) \subset \mathfrak{P}_1$, contradicting $s \in N_1$.

Corollary 2.17. Under the hypotheses of 2.16, $h(\mathfrak{P}) = h(\mathfrak{P} \cap R)$ for any $\mathfrak{P} \in \mathrm{Spec}(S)$.

Proof. $h(\mathfrak{P}) \leq h(\mathfrak{P} \cap R)$ was proved in 2.13. The opposite inequality now follows from 2.16, because for any prime ideal chain in $\mathrm{Spec}(R)$ ending with $\mathfrak{P} \cap R$ we can construct one of equal length in $\mathrm{Spec}(S)$ that ends with $\mathfrak{P}$.

There are examples showing that the Going-Down Theorem and 2.17 do not hold if R is not normal (see [10]).

Exercises

1. Let S/R be a ring extension, where $R \neq \{0\}$ and S is an integral domain, L/K the corresponding extension of fields of fractions. If L/K is algebraic and has a primitive element, then there is one already in S.

2. Under the hypotheses of Exercise 1 let R be normal, L/K finite separably algebraic, and S the integral closure of R in L. Let $s \in S$ be a primitive element of L/K; let $s_1, \ldots, s_n$ be the conjugates of s over K, and let D be the van der Monde determinant

$$\begin{vmatrix} 1 & s_1 & s_1^2 & \cdots & s_1^{n-1} \\ 1 & s_2 & s_2^2 & \cdots & s_2^{n-1} \\ \vdots & \vdots & \vdots & & \vdots \\ 1 & s_n & s_n^2 & \cdots & s_n^{n-1} \end{vmatrix}.$$

a) $D^2 \in R$ and $S \subset \frac{1}{D^2} \cdot (R + Rs + \cdots + Rs^{n-1})$.

b) If R is Noetherian, then S is finitely generated as an R-module (and so is a Noetherian ring). If R is an affine algebra over a field $k \subset R$, so is S.

3. Let K be a field. Any K-subalgebra $A \subset K[X]$ is finitely generated over K and $\dim A = 1$ if $A \neq K$. (Hint: If $f \in K[X], f \notin K$, then $K[X]$ is integral over $K[f]$.)

4. (J. David) Let K be a field, let $f \in K[X_1, \ldots, X_n]$ be written in the form $f = f_0 + \cdots + f_d$ (f_i homogeneous of degree i, $f_d \neq 0$), and let f_d be a product of pairwise unassociated prime polynomials. Then $K[f]$ is integrally closed in $K[X_1, \ldots, X_n]$, and $K(f)$ is algebraically closed in $K(X_1, \ldots, X_n)$.

5. Let R be a local (semilocal) ring, $P \subset R$ a subring, where R/P is an integral ring extension. Then P too is local (semilocal).

6. Let S/R be a ring extension, where $R \neq \{0\}$ and S is generated as an R-module by t elements. Then over any maximal ideal of R lie at most t maximal ideals of S.

3. The dimension of affine algebras and affine algebraic varieties

Basic to this section is the

Theorem 3.1. (Noether Normalization Theorem) Let A be an affine algebra over a field K, $I \subset A$ an ideal, $I \neq A$. There are natural numbers $\delta \leq d$ and elements $Y_1, \ldots, Y_d \in A$ such that:

a) $Y_1, \ldots, Y_d$ are algebraically independent over K.
b) A is finitely generated as a $K[Y_1, \ldots, Y_d]$-module.
c) $I \cap K[Y_1, \ldots, Y_d] = (Y_{\delta+1}, \ldots, Y_d)$.

If K is infinite and $A = K[x_1, \ldots, x_n]$, then we can also get

d) For $i = 1, \ldots, \delta$ Y_i is of the form $Y_i = \sum_{k=1}^{n} a_{ik} x_k$ $(a_{ik} \in K)$.

As a preparation for the proof we first show

Lemma 3.2. Let F be a nonconstant polynomial in $K[X_1, \ldots, X_n]$.

a) By a substitution of the form $X_i = Y_i + X_n^{r_i}$ $(i = 1, \ldots, n-1)$ with suitably chosen $r_i \in \mathbb{N}$, F is transformed into an element of the form

$$aX_n^m + \rho_1 X_n^{m-1} + \cdots + \rho_m \quad (a \in K^\times, \ \rho_i \in K[Y_1, \ldots, Y_{n-1}], \ m > 0). \quad (1)$$

b) If K has infinitely many elements, then the same result can be gotten by means of a substitution $X_i = Y_i + a_i X_n$ with suitably chosen $a_i \in K$.

Proof. Let $F = \sum a_{\nu_1 \ldots \nu_n} X_1^{\nu_1}, \ldots, X_n^{\nu_n}$.

In case a), after the substitution F assumes the form

$$F = \sum a_{\nu_1 \ldots \nu_n} (X_n^{r_1} + Y_1)^{\nu_1} \ldots (X_n^{r_{n-1}} + Y_{n-1})^{\nu_{n-1}} X_n^{\nu_n}$$
$$= \sum a_{\nu_1 \ldots \nu_n} (X_n^{\nu_n + \nu_1 r_1 + \cdots + \nu_{n-1} r_{n-1}} + \cdots),$$

where the points denote terms in which X_n occurs to a lower power. Put $r_i := k^i$ ($i = 1, \ldots, n - 1$), where $k - 1$ is the largest index for which a coefficient $a_{\nu_1 \ldots \nu_n} \neq 0$ occurs in F. The numbers $\nu_n + \nu_1 k + \cdots + \nu_{n-1} k^{n-1}$ are then distinct for different n-tuples $(\nu_1, \ldots, \nu_n)$ with $a_{\nu_1 \ldots \nu_n} \neq 0$. If m is the largest of these numbers, it follows that F does have the form (1).

In case b) let $F = F_0 + \cdots + F_m$ be the decomposition of F into homogeneous components F_i ($\deg F_i = i$, $F_m \neq 0$). After the substitution specified, F has the form $F = F_m(-a_1, \ldots, -a_{n-1}, 1) X_n^m + \cdots$. Since F_m is homogeneous and $\neq 0$, we also have $F_m(X_1, \ldots, X_{n-1}, 1) \neq 0$. Since K is infinite, we can find $a_1, \ldots, a_{n-1} \in K$ such that $F_m(-a_1, \ldots, -a_{n-1}, 1) \neq 0$ (I.1.3a)).

Proof of the Normalization Theorem.

1. Let A be a polynomial algebra $K[X_1, \ldots, X_n]$ and $I = (F)$ a principal ideal. F is not a constant.

    We then set $Y_n := F$ and choose Y_i ($i = 1, \ldots, n - 1$) as in Lemma 3.2. Then $A = K[Y_1, \ldots, Y_n][X_n]$, and because $0 = F - Y_n = aX_n^m + \rho_1(Y_1, \ldots, Y_{n-1}) X_n^{m-1} + \cdots + \rho_m(Y_1, \ldots, Y_{n-1}) - Y_n$, X_n is integral over $K[Y_1, \ldots, Y_n]$ and therefore A is finitely generated as a $K[Y_1, \ldots, Y_n]$-module.

    The elements $Y_1, \ldots, Y_n$ are algebraically independent over K, since otherwise $K(Y_1, \ldots, Y_n)$ and with it $K(X_1, \ldots, X_n)$ would have transcendence degree $< n$ over K. We now show $I \cap K[Y_1, \ldots, Y_n] = (Y_n)$.

    Any $f \in I \cap K[Y_1, \ldots, Y_n]$ can be written in the form $f = G \cdot Y_n$ with $G \in A$. Then there is an equation

$$G^s + a_1 G^{s-1} + \cdots + a_s = 0 \qquad (s > 0, a_i \in K[Y_1, \ldots, Y_n]),$$

from which we get

$$f^s + a_1 Y_n f^{s-1} + \cdots + a_s Y_n^s = 0.$$

It follows that Y_n is also a divisor of f in $K[Y_1, \ldots, Y_n]$.

2. Now let I be an arbitrary ideal in $A = K[X_1, \ldots, X_n]$. For $I = (0)$ there is nothing to prove. Hence we may assume that there is a nonconstant $F \in I$. For $n = 1$ we are already done (Case 1). Now let $n > 1$. Let $K[Y_1, \ldots, Y_n]$ with $Y_n := F$ be constructed as in Case 1. By induction we may assume that the theorem holds for $I \cap K[Y_1, \ldots, Y_{n-1}]$: There are elements $T_1, \ldots, T_{d-1} \in K[Y_1, \ldots, Y_{n-1}]$ algebraically independent over K such that $K[Y_1, \ldots, Y_{n-1}]$ is finitely generated as a $K[T_1, \ldots, T_{d-1}]$-module and $I \cap K[T_1, \ldots, T_{d-1}] = (T_{\delta+1}, \ldots, T_{d-1})$ with some $\delta < d$.

Since $K[Y_1, \ldots, Y_n]$ is finitely generated over $K[T_1, \ldots, T_{d-1}, Y_n]$, A is also finitely generated over $K[T_1, \ldots, T_{d-1}, Y_n]$ (as a module). Then $d = n$ and $T_1, \ldots, T_{n-1}, Y_n$ are algebraically independent over K. If K is infinite, we may assume that the T_i $(i = 1, \ldots, \delta)$ are linear combinations of the Y_j $(j = 1, \ldots, n-1)$ and thus are linear combinations of the X_j $(j = 1, \ldots, n)$.

Any $f \in I \cap K[T_1, \ldots, T_{n-1}, Y_n]$ can be written in the form $f = f^* + HY_n$ with $f^* \in I \cap K[T_1, \ldots, T_{n-1}] = (T_{\delta+1}, \ldots, T_{n-1})$ and $H \in K[T_1, \ldots, T_{n-1}, Y_n]$. We conclude that $I \cap K[T_1, \ldots, T_{n-1}, Y_n]$ is generated by $T_{\delta+1}, \ldots, T_{n-1}, Y_n$.

3. In the general case we write $A = K[X_1, \ldots, X_n]/J$ and as in Case 2 determine a subalgebra $K[Y_1, \ldots, Y_n]$ of $K[X_1, \ldots, X_n]$ with $J \cap K[Y_1, \ldots, Y_n] = (Y_{d+1}, \ldots, Y_n)$, where the Y_i $(i = 1, \ldots, d)$ are chosen as linear combinations of the X_k if K is infinite. The image of $K[Y_1, \ldots, Y_n]$ in A can be identified with the polynomial algebra $K[Y_1, \ldots, Y_d]$. A is then a finitely generated module over this ring. We again apply Case 2 to $I' := I \cap K[Y_1, \ldots, Y_d]$: There is a polynomial subalgebra $K[T_1, \ldots, T_d] \subset K[Y_1, \ldots, Y_d]$ over which $K[Y_1, \ldots, Y_d]$ is finitely generated as a module, so that $I' \cap K[T_1, \ldots, T_d] = (T_{\delta+1}, \ldots, T_d)$ with some $\delta \leq d$ and the T_i $(i = 1, \ldots, \delta)$ are linear combinations of the Y_j $(j = 1, \ldots, d)$, therefore also of the images x_k of the X_k in A, if K is infinite.

Since A is also finitely generated as a $K[T_1, \ldots, T_d]$-module, the elements $T_1, \ldots, T_d$ meet the requirements of the Normalization Theorem, q. e. d.

Definition 3.3. For an affine K-algebra $A \neq \{0\}$, $K[Y_1, \ldots, Y_d] \subset A$ is called a *Noetherian normalization* if $Y_1, \ldots, Y_d$ are algebraically independent over K and A is finitely generated as a $K[Y_1, \ldots, Y_d]$-module.

From the Normalization Theorem and the theorems of Cohen–Seidenberg follow important statements about the dimension of affine algebras and their prime ideal chains. We call a prime ideal chain *maximal* if there is no chain of greater length that contains all the prime ideals of the given chain.

In the following let $A \neq \{0\}$ be an affine algebra over a field K.

Proposition 3.4. If $K[Y_1, \ldots, Y_d] \subset A$ is a Noetherian normalization, then $\dim A = d$. Moreover, if A is an integral domain, then all maximal prime ideal chains in A have length d (in particular, this holds for the polynomial algebra $K[X_1, \ldots, X_d]$).

Proof. By 1.4a) and 2.13, $\dim A = \dim K[Y_1, \ldots, Y_d] \geq d$. For an arbitrary prime ideal chain

$$\mathfrak{P}_0 \subset \cdots \subset \mathfrak{P}_m \qquad (2)$$

in A it remains to be shown that $m \leq d$. We argue by induction on d.

If one puts $\mathfrak{p}_i := \mathfrak{P}_i \cap K[Y_1, \ldots, Y_d]$, then $\mathfrak{p}_0 \subset \cdots \subset \mathfrak{p}_m$ is a prime ideal chain in $K[Y_1, \ldots, Y_d]$. For $d = 0$ there is nothing to prove. Hence let $d > 0$

and suppose the assertion has been proved for polynomial algebras with fewer variables. Then there is something to be proved only for $m > 0$.

By 3.1 there is a Noetherian normalization $K[T_1, \ldots, T_d] \subset K[Y_1, \ldots, Y_d]$ with $\mathfrak{p}_1 \cap K[T_1, \ldots, T_d] = (T_{\delta+1}, \ldots, T_d)$ $(\delta \le d)$. Since $\mathfrak{p}_1 \ne (0)$, we have $\delta < d$ (2.10c)). $K[T_1, \ldots, T_\delta] \subset K[Y_1, \ldots, Y_d]/\mathfrak{p}_1$ is then a Noetherian normalization too. By the induction hypothesis, for the length of the prime ideal chain

$$(0) = \mathfrak{p}_1/\mathfrak{p}_1 \subset \mathfrak{p}_2/\mathfrak{p}_1 \subset \cdots \subset \mathfrak{p}_m/\mathfrak{p}_1 \qquad (3)$$

we have $m - 1 \le \delta < d$. It follows that $m \le d$.

If A is an integral domain and (2) a maximal chain of prime ideals in A, then $\mathfrak{P}_0 = (0)$, and $\mathfrak{P}_m$ is a maximal ideal of A. We shall prove that also $\mathfrak{p}_0 \subset \cdots \subset \mathfrak{p}_m$ is a maximal prime ideal chain in $K[Y_1, \ldots, Y_d]$. Suppose that we can "insert" another prime ideal between $\mathfrak{p}_i$ and $\mathfrak{p}_{i+1}$ $(i \in [0, m-1])$. Then choose a Noetherian normalization $K[T_1, \ldots, T_d] \subset K[Y_1, \ldots, Y_d]$ such that $\mathfrak{p}_i \cap K[T_1, \ldots, T_d] = (T_{\delta+1}, \ldots, T_d)$ for some $\delta \le d$. Then $K[T_1, \ldots, T_\delta] \subset K[Y_1, \ldots, Y_d]/\mathfrak{p}_i$ is also a Noetherian normalization. Since we can insert a prime ideal between (0) and $\mathfrak{p}_{i+1}/\mathfrak{p}_i$, this also holds for the zero ideal of $K[T_1, \ldots, T_\delta]$ and $\mathfrak{p}_{i+1}/\mathfrak{p}_i \cap K[T_1, \ldots, T_\delta]$.

But $K[T_1, \ldots, T_\delta] \subset A/\mathfrak{P}_i$ is also a Noetherian normalization. By 2.16 (Going-Down) it follows that a prime ideal can be inserted between (0) and $\mathfrak{P}_{i+1}/\mathfrak{P}_i$ and therefore also between $\mathfrak{P}_i$ and $\mathfrak{P}_{i+1}$, contradicting the maximality of the chain (2). Since $\mathfrak{p}_0 = (0)$ and $\mathfrak{p}_m$ is a maximal ideal of $K[Y_1, \ldots, Y_d]$ (2.10), this proves the maximality of the chain (3).

We now show $m = d$ by induction on d. If $d > 0$, choose a Noetherian normalization as above:

$$K[T_1, \ldots, T_d] \subset K[Y_1, \ldots, Y_d] \quad \text{with} \quad \mathfrak{p}_1 \cap K[T_1, \ldots, T_d] = (T_{\delta+1}, \ldots, T_d).$$

This ideal has height 1 (2.17). Then we must have $\delta = d - 1$ and (3) is a maximal prime ideal chain of $K[T_1, \ldots, T_{d-1}]$. By the induction hypothesis it has length $d - 1$. It follows that $m = d$.

In particular, it has been shown that affine algebras always have finite Krull dimension. For arbitrary Noetherian rings this may not be so. There are also Noetherian integral domains of finite dimension with maximal prime ideal chains of different lengths, as symbolized in Fig. 11. An example is provided by Exercise 7.

Corollary 3.5. Let $\mathfrak{P} \subset \Omega$ be prime ideals of A, $\Omega \ne A$. All maximal prime ideal chains that start with $\mathfrak{P}$ and end with Ω have the same length, namely $\dim A/\mathfrak{P} - \dim A/\Omega$.

Proof. Let $\mathfrak{P} = \mathfrak{P}_0 \subset \cdots \subset \mathfrak{P}_m = \Omega$ be such a chain and $A' := A/\mathfrak{P}$. The chain $(0) = \mathfrak{P}_0/\mathfrak{P} \subset \cdots \subset \mathfrak{P}_m/\mathfrak{P} = \Omega/\mathfrak{P}$ in A' can be lengthened to a maximal prime ideal chain of A'; by 3.4 this has length $\dim A'$. To the part of the lengthened chain that starts with $\Omega/\mathfrak{P}$ corresponds a maximal prime ideal chain in $A'' := A/\Omega$. It follows that $m = \dim A' - \dim A''$.

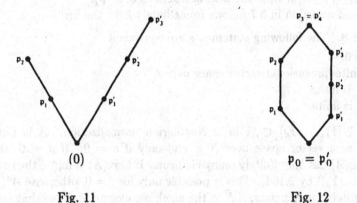

Fig. 11   Fig. 12

A ring for which the property indicated in 3.5 holds is called a chain ring. There are examples of Noetherian rings that are not chain rings (Fig. 12). For an example see [H], Vol. II, p. 327.

Corollary 3.6. Let $\mathfrak{p}_1, \ldots, \mathfrak{p}_s$ be the minimal prime ideals of A and let L_i be the field of fractions of $A/\mathfrak{p}_i$ $(i = 1, \ldots, s)$. Then:

a) $\dim A = \text{Max}_{i=1,\ldots,s}\{\text{tr deg}(L_i/K)\}$. In particular, $\dim A = \text{tr deg}(L/K)$ if A is an integral domain with field of fractions L.

b) If $\dim A/\mathfrak{p}_i$ is independent of $i \in [1, s]$, then for all $\mathfrak{p} \in \text{Spec}(A)$

$$\dim A = h(\mathfrak{p}) + \dim A/\mathfrak{p}$$

Proof. Since every maximal prime ideal chain of A starts with one of the $\mathfrak{p}_i$ $(i \in [1, s])$, it suffices to prove the assertions for integral domains. If A is· an integral domain and $K[Y_1, \ldots, Y_d] \subset A$ is a Noetherian normalization, then $d = \dim A$ is also the transcendence degree of L over K. The formula $\dim A = h(\mathfrak{p}) + \dim A/\mathfrak{p}$ follows from 3.5 with $\mathfrak{P} = (0)$, $\mathfrak{Q} = \mathfrak{p}$.

Corollary 3.7. $\dim A$ is the maximal number of elements of A that are algebraically independent over K. If $B \subset A$ is another affine K-algebra, then $\dim B \leq \dim A$.

Proof. Let $d := \dim A$. By the Normalization Theorem it suffices to show: If $Z_1, \ldots, Z_m \in A$ are algebraically independent over K, then $m \leq d$. Let $\mathfrak{p}_1, \ldots, \mathfrak{p}_s$ be the minimal prime ideals of A. By I.4.10 we have

$$(0) = (\bigcap_{i=1}^{s} \mathfrak{p}_i) \cap K[Z_1, \ldots, Z_m] = \bigcap_{i=1}^{s}(\mathfrak{p}_i \cap K[Z_1, \ldots, Z_m]),$$

since $K[Z_1, \ldots, Z_m]$ has no nilpotent elements $\neq 0$. There is then an $i \in [1, s]$ with $\mathfrak{p}_i \cap K[Z_1, \ldots, Z_m] = (0)$. From $K[Z_1, \ldots Z_m] \subset A/\mathfrak{p}_i$ it follows that $m \leq \operatorname{tr} \deg(L_i/K) \leq d$, if L_i is the field of fractions of $A/\mathfrak{p}_i$.

The second assertion in 3.7 follows immediately from the first.

Corollary 3.8. The following statements are equivalent.

a) $\dim A = 0$.

b) A is a finite-dimensional vector space over K.

c) $\operatorname{Spec}(A)$ is finite.

d) $\operatorname{Max}(A)$ is finite.

Proof. Let $K[Y_1, \ldots, Y_d] \subset A$ be a Noetherian normalization. A is finite-dimensional as a vector space over K if and only if $d = 0$. If $d = 0$, then by 1.4d) $\operatorname{Spec}(A)$ has only finitely many elements. If $\operatorname{Max}(A)$ is finite, then so is $\operatorname{Max}(K[Y_1, \ldots, Y_d])$ by 2.10d). This is possible only for $d = 0$, otherwise $\mathsf{A}^d(\overline{K})$ contains infinitely many points, if $\overline{K}$ is the algebraic closure of K, so that there are infinitely many maximal ideals $\mathfrak{m}$ in $K[Y_1, \ldots, Y_d]$, since any maximal ideal $\mathfrak{m}$ has only finitely many zeros in $\mathsf{A}^d(\overline{K})$, namely all the conjugates of a zero.

Corollary 3.9.

a) If K'/K is an extension of fields, then

$$\dim(K' \underset{K}{\otimes} A) = \dim A.$$

If A is an integral domain, then

$$\dim(K' \underset{K}{\otimes} A/\mathfrak{P}) = \dim A$$

for any minimal prime ideal $\mathfrak{P}$ of $K' \underset{K}{\otimes} A$.

b) If $A' \neq \{0\}$ is another affine K-algebra, then

$$\dim(A \underset{K}{\otimes} A') = \dim A + \dim A'.$$

If A and A' are integral domains, then

$$\dim(A \underset{K}{\otimes} A'/\mathfrak{P}) = \dim A + \dim A'$$

for any minimal prime ideal $\mathfrak{P}$ of $A \underset{K}{\otimes} A'$.

Proof. Let $K[Y_1, \ldots, Y_d] \subset A$ and $K[Z_1, \ldots, Z_\delta] \subset A'$ be Noetherian normalizations. $K' \underset{K}{\otimes} K[Y_1, \ldots, Y_d]$ is identified with $K'[Y_1, \ldots, Y_d]$ and

$$K[Y_1, \ldots, Y_d] \underset{K}{\otimes} K[Z_1, \ldots, Z_\delta]$$

with $K[Y_1, \ldots, Y_d, Z_1, \ldots, Z_\delta]$. Further,

$$K'[Y_1, \ldots, Y_d] \subset K' \underset{K}{\otimes} A \quad \text{and} \quad K[Y_1, \ldots, Y_d, Z_1, \ldots, Z_\delta] \subset A \underset{K}{\otimes} A'$$

are Noetherian normalizations. From this follows the first dimension formula in a) and b).

If A is an integral domain with field of fractions L, then we have a commutative diagram with injective ring homomorphisms

$$
\begin{array}{ccc}
K' \underset{K}{\otimes} K(Y_1, \ldots, Y_d) & \rightarrow & K' \underset{K}{\otimes} L \\
\uparrow & & \uparrow \\
K' \underset{K}{\otimes} K[Y_1, \ldots, Y_d] & \rightarrow & K' \underset{K}{\otimes} A.
\end{array}
$$

$K' \underset{K}{\otimes} L$ is a free $K' \underset{K}{\otimes} K(Y_1, \ldots, Y_d)$-module, since L has a basis over $K(Y_1, \ldots, Y_d)$. Therefore, no element $\neq 0$ in $K' \underset{K}{\otimes} K(Y_1, \ldots, Y_d)$ can be a zero divisor in $K' \underset{K}{\otimes} L$. It follows that $\mathfrak{P} \cap K' \underset{K}{\otimes} K[Y_1, \ldots, Y_d] = (0)$, for the elements of a minimal prime ideal $\mathfrak{P}$ of $K' \underset{K}{\otimes} A$ are zero divisors of this ring by I.4.10. We get the second formula in a).

The second formula in b) follows similarly with the aid of the diagram

$$
\begin{array}{ccc}
K(Y_1, \ldots, Y_d) \underset{K}{\otimes} K(Z_1, \ldots, Z_\delta) & \rightarrow & L \underset{K}{\otimes} L' \\
\uparrow & & \uparrow \\
K[Y_1, \ldots, Y_d] \underset{K}{\otimes} K[Z_1, \ldots, Z_\delta] & \rightarrow & A \underset{K}{\otimes} A',
\end{array}
$$

in which L' denotes the field of fractions of A'.

Corollary 3.10. Let A be factorial, $I \neq (0)$, $I \neq A$, an ideal with $\mathrm{Rad}(I) = I$. Then the following statements are equivalent.

a) For any minimal prime divisor $\mathfrak{p}$ of I,

$$
\dim A/\mathfrak{p} = \dim A - 1.
$$

b) I is a principal ideal.

Proof.

a)$\rightarrow$b). By 3.6 we have $h(\mathfrak{p}) = 1$ and by 1.4b) it follows that $\mathfrak{p} = (\pi)$ with a prime element π of A. If $\mathfrak{p}_1, \ldots, \mathfrak{p}_s$ ($\mathfrak{p}_i \neq \mathfrak{p}_j$ for $i \neq j$) are all minimal prime divisors of I, $\mathfrak{p}_i = (\pi_i)$ ($i = 1, \ldots, s$), then $I = \mathrm{Rad}(I) = \mathfrak{p}_1 \cap \cdots \cap \mathfrak{p}_s = (\pi_1 \cdot \ldots \cdot \pi_s)$.

b)$\rightarrow$a). If $I = (a)$ is a principal ideal and $a = \pi_1 \cdot \ldots \cdot \pi_s$ is a decomposition of a into prime elements π_i ($i = 1, \ldots, s$), then the $\mathfrak{p}_i := (\pi_i)$ are precisely the minimal prime divisors of I. By 1.4 b) we have $h(\mathfrak{p}_i) = 1$ and so $\dim A/\mathfrak{p}_i = \dim A - 1$.

If we apply the foregoing statements on prime ideal chains and the dimension of affine algebras to the coordinate rings of affine varieties, then, on the basis of the relation between the subvarieties of a variety and the ideals of its coordinate ring, we immediately get propositions on the dimension of affine varieties and on chains of irreducible subvarieties. We collect together the most important of these statements.

Proposition 3.11. For any nonempty K-variety $V \subset \mathsf{A}^n(L)$ we have:

a) $\dim V$ is independent of the choice of the field of definition K. If K'/K is a field extension with $K' \subset L$ and if V is irreducible in the K-topology, then all irreducible components of V in the K'-topology have the same dimension.

b) $\dim V \leq n$. $\dim V = n$ if and only if $V = \mathsf{A}^n(L)$.

c) If all irreducible components of V have the same dimension d, then all maximal chains

$$V_0 \subset V_1 \subset \cdots \subset V_d \qquad (V_0 \neq \emptyset, V_i \neq V_{i+1})$$

of irreducible subvarieties V_i of V have length d. Moreover, if V is irreducible,

$$\dim V = \operatorname{tr} \deg(K(V)/K),$$

where $K(V)$ is the field of fractions of $K[V]$.

d) If all the irreducible components of V have the same dimension and if $W \subset V$ is an irreducible subvariety, $W \neq \emptyset$, then

$$\dim(V) = \dim(W) + \operatorname{codim}_V(W).$$

e) If $W \subset \mathsf{A}^m(L)$ is another nonempty K-variety, then

$$\dim(V \times W) = \dim(V) + \dim(W).$$

If V and W are irreducible, then all the irreducible components of $V \times W$ have the same dimension.

f) $\dim V = 0$ if and only if V consists of only finitely many points.

g) Let the coordinate ring $K[V]$ be factorial, $W \subset V$ a subvariety, $W \neq V, W \neq \emptyset$. All the irreducible components of W have codimension 1 in V if and only if the ideal $\mathfrak{I}_V(W)$ of W in $K[V]$ is a principal ideal. (In particular, V is a hypersurface in $\mathsf{A}^n(L)$ if and only if all the irreducible components of V have codimension 1 in $\mathsf{A}^n(L)$.)

Proof.

a) follows from 3.9a): By I.3.12b) we have $K'[V] = (K' \underset{K}{\otimes} K[V])_{\mathrm{red}}$. Since the spectra of $K' \underset{K}{\otimes} K[V]$ and $(K' \underset{K}{\otimes} K[V])_{\mathrm{red}}$ are homeomorphic (I.4.12), the assertions now follow at once.

b) We write $K[V] = K[X_1, \ldots, X_n]/\mathfrak{I}(V)$. Then $\dim V = n$ if and only if $\mathfrak{I}(V) = 0$, since the prime ideal chains in $K[V]$ correspond bijectively to those in $K[X_1, \ldots, X_n]$ in which only prime ideals that contain $\mathfrak{I}(V)$ occur.

   Statements c), d), and g) are immediate translations of 3.4, 3.6, and 3.10; further, e) follows from 3.9b) with the aid of I.3.12c).

f) If $\dim V = 0$, then by 3.8 $\operatorname{Spec}(K[V])$ has only finitely many points, which are all maximal ideals. Then V has only finitely many points, which all lie in $\mathsf{A}^n(\overline{K})$, where $\overline{K}$ is the algebraic closure of K in L. If V has only finitely many points, then $\operatorname{Spec}(K[V])$ is finite, and it follows that $\dim V = 0$ by 3.8.

A K-variety $V \subset \mathsf{A}^n(L)$ is called an *affine algebraic curve (surface)* if all its irreducible components have dimension 1 (dimension 2). Special examples are the plane algebraic curves ($n = 2$).

We now consider the Krull dimension of linear varieties.

Examples 3.12. Let $\Lambda \subset \mathsf{A}^n(L)$ be a nonempty linear K-variety defined by a system of linear equations $\sum_{k=1}^n a_{ik}X_k = b_i$ ($i = 1,\ldots,m$) of rank r. Then $\dim \Lambda = n - r$.

After a coordinate transformation we may assume that Λ is given by the system $X_i = 0$ ($i = 1,\ldots,r$). Then $\mathfrak{I}(\Lambda) = (X_1,\ldots,X_r)$ and $K[\Lambda] \cong K[X_{r+1},\ldots,X_n]$, therefore $\dim \Lambda = n - r$.

From statement d) of the Normalization Theorem the following geometrical consequences can be derived.

Proposition 3.13. Let K be algebraically closed, $V \subset \mathsf{A}^n(L)$ a K-variety of dimension $d \geq 0$. Then there is a d-dimensional linear K-variety $\Lambda \subset \mathsf{A}^n(K)$ and a parallel projection $\pi : \mathsf{A}^n(K) \to \Lambda$ with the following properties (Fig. 13).

a) $\pi(V) = \Lambda$.

b) For all $x \in \Lambda, \pi^{-1}(\{x\}) \cap V$ is finite.

Proof. There is a Noetherian normalization $K[Y_1,\ldots,Y_n] \subset K[X_1,\ldots,X_n]$ with $\mathfrak{I}(V) \cap K[Y_1,\ldots,Y_n] = (Y_{\delta+1},\ldots,Y_n)$, where the Y_i ($i = 1,\ldots,\delta$) are linear homogeneous polynomials in $X_1,\ldots,X_n$. Then $K[Y_1,\ldots,Y_\delta] \subset K[V]$ is a Noetherian normalization, so $\delta = d$.

After a coordinate transformation we may assume that $Y_i = X_i$ for $i = 1,\ldots,d$. Let Λ be the d-dimensional linear variety in $\mathsf{A}^n(K)$ given by $X_{d+1} = \cdots = X_n = 0$. For any point $(a_1,\ldots,a_d,0,\ldots,0)$ there is on V at least one, at most finitely many, points whose first d coordinates coincide with $a_1,\ldots,a_d$, since over the maximal ideal $(X_1 - a_1,\ldots,X_d - a_d)$ of $K[X_1,\ldots,X_d]$ lies at least one, at most finitely many, maximal ideals of $K[V]$ (2.14 and 2.10). These are precisely the maximal ideals of the points of V named.

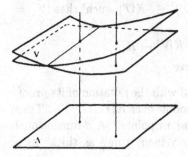

Fig. 13

Corollary 3.14. For any d-dimensional variety $V \subset A^n(K)$ $(d \geq 0)$ there is a linear variety Λ' of dimension $n - d$ that cuts V in only finitely many points, and a linear variety Λ'' of dimension $n - d - 1$ that does not meet V.

Proof. Take $\Lambda' := \pi^{-1}(\{x\})$ for some $x \in \Lambda$. Further, if H is a hyperplane that contains none of the finitely many points in $\pi^{-1}(\{x\}) \cap V$, take $\Lambda'' := \Lambda' \cap H$.

One can show that "almost all" the $(n - d)$-dimensional linear varieties in $A^n(K)$ (in a sense to be made precise) cut the variety V in only finitely many points and that the number of the intersection points is "almost always" the same.

Exercises

1. Deduce Hilbert's Nullstellensatz from the Noether Normalization Theorem.

2. With the notation of 3.1, let A be an integral domain. Then $d - \delta$ is the height of I.

3. Let K be a field.

   a) In the polynomial ring $K[X_1, X_2]$ there are infinitely many prime ideals of height 1 that are contained in (X_1, X_2).

   b) Let A be an affine K-algebra. For $\mathfrak{P} \in \mathrm{Spec}(A)$ let $h(\mathfrak{P}) \geq 2$. Then there are infinitely many $\mathfrak{p} \in \mathrm{Spec}(A)$ with $\mathfrak{p} \subset \mathfrak{P}$ and $h(\mathfrak{p}) = 1$.

4. An affine variety that is given by a polynomial parametrization with m parameters (Ch. I, §2. Exercise 3) has dimension $\leq m$. If, with the notation of that exercise, $L = K$ is algebraically closed and $K[T_1, \ldots, T_m]$ is integral over $K[f_1, \ldots, f_n]$, then V_0 is closed in the Zariski topology and has dimension m.

5. Let K be an algebraically closed field, $V \subset A^n(K)$ a K-variety that does not consist of a single point. The following statements are equivalent.

   a) There are polynomials $f_1(T), \ldots, f_n(T) \in K[T]$ such that $V = \{(f_1(t), \ldots, f_n(t)) \mid t \in K\}$.

   b) There is an injective K-homomorphism $K[V] \to K[T]$.

   In this case V is an irreducible algebraic curve.

6. Under the hypotheses of Proposition 3.13 and with the notation of its proof, let $K[V]$ be generated by m elements as a module over $K[Y_1, \ldots, Y_d]$. Then for any $x \in \Lambda, \pi^{-1}(x) \cap V$ consists of at most m points. (A d-dimensional algebraic variety is therefore at most "finitely many times as thick" as a linear variety of dimension d.)

7. Let K be a field and let $K[\|X\|][Y]$ be the polynomial ring in Y over the ring of formal power series in X over K. $\mathfrak{m}_1 := (XY - 1)$ and $\mathfrak{m}_2 := (X, Y)$ are maximal ideals of $K[\|X\|][Y]$ with $h(\mathfrak{m}_1) = 1, h(\mathfrak{m}_2) = 2$.

4. The dimension of projective varieties

Many statements on the dimension of affine varieties can immediately be transferred to projective varieties. To do this we consider the embedding given in I.§5:

$$\mathbf{A}^n(L) \to \mathbf{P}^n(L), \qquad (x_1, \ldots, x_n) \mapsto (1, x_1, \ldots, x_n).$$

Proposition 4.1. If $V \subset \mathbf{A}^n(L)$ is a K-variety and $\overline{V} \subset \mathbf{P}^n(L)$ is its projective closure, then $\dim \overline{V} = \dim V$.

Proof. By I.5.17a) it suffices to consider irreducible varieties $V \neq \emptyset$. For such a V there is a chain

$$V_0 \subset V_1 \subset \cdots \subset V_d = V \subset V_{d+1} \subset \cdots \subset V_n = \mathbf{A}^n(L) \quad (V_0 \neq \emptyset, V_i \neq V_{i+1})$$

of irreducible K-varieties, where $d = \dim V$. By I.5.17 the projective closures $\overline{V}_i$ of the V_i form a corresponding chain

$$\overline{V}_0 \subset \overline{V}_1 \subset \cdots \subset \overline{V}_d = \overline{V} \subset \overline{V}_{d+1} \subset \cdots \subset \overline{V}_n = \mathbf{P}^n(L)$$

to which corresponds in the polynomial ring $K[Y_0, \ldots, Y_n]$ a chain of homogeneous prime ideals

$$\mathfrak{P}_0 \supset \mathfrak{P}_1 \supset \cdots \supset \mathfrak{P}_d = \mathfrak{I}(\overline{V}) \supset \mathfrak{P}_{d+1} \supset \cdots \supset \mathfrak{P}_n = (0),$$

where $\mathfrak{P}_0 \neq (Y_0, \ldots, Y_n)$, since $\overline{V}_0 \neq \emptyset$. Since the length of an arbitrary prime ideal chain in $K[Y_0, \ldots, Y_n]$ is at most $n + 1$, it follows that

$$\mathfrak{P}_0 \supset \mathfrak{P}_1 \supset \cdots \supset \mathfrak{P}_d = \mathfrak{I}(\overline{V})$$

is a chain of homogeneous prime ideals of maximal length that starts with $\mathfrak{I}(\overline{V})$ and ends with a prime ideal that is properly contained in $(Y_0, \ldots, Y_n)$. Therefore, $\dim \overline{V} = d$, q. e. d.

If

$$\overline{V}_0 \subset \overline{V}_1 \subset \cdots \subset \overline{V}_m \quad (\overline{V}_0 \neq \emptyset, \overline{V}_i \neq \overline{V}_{i+1}) \tag{1}$$

is a chain of irreducible K-varieties in $\mathbf{P}^n(L)$, then the coordinate system can be chosen so that $\overline{V}_0$ does not entirely lie in the hyperplane at infinity. Then $\overline{V}_i$ is the projective closure of the affine part V_i of $\overline{V}_i$ $(i = 0, \ldots, m)$. The chain $V_0 \subset V_1 \subset \cdots \subset V_m$ can be refined in affine space to a chain of irreducible varieties of length n. Going over to the projective closures, we get a refinement of (1) to a chain of irreducible projective K-varieties of length n.

Thus we have

Proposition 4.2. Every chain (1) of irreducible projective K-varieties in $\mathbf{P}^n(L)$ is contained in a maximal such chain. All maximal chains have length n.

For prime ideals we obtain from this

Corollary 4.3. Every chain

$$\mathfrak{P}_0 \subset \mathfrak{P}_1 \subset \cdots \subset \mathfrak{P}_m \tag{2}$$

of relevant prime ideals in $K[Y_0, \ldots, Y_n]$ can be refined to a maximal such chain. All the maximal chains (2) have length n.

In speaking of chains of homogeneous prime ideals, we must include the irrelevant prime ideal $(Y_0, \ldots, Y_n)$. All maximal prime ideal chains consisting of only homogeneous prime ideals of $K[Y_0, \ldots, Y_n]$ have length $n+1$.

Proposition 4.4.

a) For any K-variety $V \subset \mathbf{P}^n(L)$ we have $\dim V \leq n$, and $\dim V = n$ if and only if $V = \mathbf{P}^n(L)$.

b) If $\tilde{V} \subset \mathbf{A}^n(L)$ is the affine cone of V, then

$$\dim \tilde{V} = \dim V + 1.$$

Moreover, $\dim V = \dim K[V] - 1 = g\text{-}\dim K[V]$.

c) $\dim V$ is independent of the choice of the field of definition K.

d) If all the irreducible components of V have the same dimension and if $W \subset V$ is an irreducible subvariety, $W \neq \emptyset$, then

$$\dim V = \dim W + \text{codim}_V W.$$

e) A projective variety V has dimension 0 if and only if it consists of only finitely many points.

f) A K-variety $V \subset \mathbf{P}^n(L)$ is a hypersurface if and only if all its irreducible components have codimension 1 in $\mathbf{P}^n(L)$.

Proof.

a) follows immediately from 4.2.

b) We have $\dim \tilde{V} = \dim V + 1$, since in the affine case we must also count the irrelevant prime ideal $(Y_0, \ldots, Y_n)$ (corresponding to the vertex of the cone). Since $K[\tilde{V}] = K[V]$, we also get the second statement in b).

c) For $\tilde{V}$ the fact that the dimension is independent of the field of definition has already been shown (3.11), hence the assertion follows from b).

d) follows from 4.2, since W is contained in an irreducible component of V.

e) Let $\dim V = 0$. For any choice of the hyperplane at infinity, by 3.11f) the affine part of V has only finitely many points. There are $n+1$ hyperplanes in $\mathbf{P}^n(L)$ with empty intersection. If we choose these in sequence to be the hyperplane at infinity, the finiteness of V follows.

Conversely, if V is finite, then the hyperplane at infinity can be placed so that it contains no point of V. It then follows that $\dim V = 0$ by 4.1.

f) A homogeneous ideal $I \subset K[Y_0, \ldots, Y_n], I \neq (0), I \neq (1)$ with $I = \text{Rad}(I)$ is a principal ideal if and only if its minimal prime divisors are principal ideals generated by homogeneous polynomials. From this statement f) follows.

It is also clear that a linear variety in $\mathbf{P}^n(L)$ has dimension $n-r$ if it is described by a linear homogeneous system of equations of rank r.

The Noether Normalization Theorem admits (like 3.13 and with a like proof) the following application in projective space.

Proposition 4.5. Let K be an algebraically closed field, $V \subset \mathbf{P}^n(K)$ a K-variety of dimension $d \geq 0$. Then in $\mathbf{P}^n(K)$ there are linear varieties Λ_d and Λ'_{n-d-1} of dimension d and $n-d-1$ respectively with

$$\Lambda_d \cap \Lambda'_{n-d-1} = \emptyset, \qquad V \cap \Lambda'_{n-d-1} = \emptyset,$$

such that under the central projection from Λ'_{n-d-1}, V is mapped *onto* Λ_d and over any point of Λ_d lie only finitely many points of V. (The projection is defined as follows. For $P \in V$ the subspace spanned by P and Λ'_{n-d-1} is a linear variety of dimension $n-d$. It cuts Λ_d in exactly one point Q, which by definition is the image of P.)

Exercises

1. Prove Proposition 4.5 in detail.
2. Let $s : \mathbf{P}^n(L) \times \mathbf{P}^m(L) \to \mathbf{P}^{(n+1)(m+1)-1}(L)$ be the mapping that assigns to $(\langle x_0, \ldots, x_n \rangle, \langle y_0, \ldots, y_m \rangle)$ the point

$$\langle x_0 y_0, \ldots, x_0 y_m, \ldots, x_i y_0, \ldots, x_i y_m, \ldots, x_n y_0, \ldots, x_n y_m \rangle.$$

   a) s is well defined and injective.
   b) $V := Im(s)$ is the projective variety described by the system of equations

$$Z_{ij} Z_{kl} - Z_{il} Z_{kj} = 0 \qquad (i, k = 0, \ldots, n; \; j, l = 0, \ldots, m).$$

   c) $\dim V = n + m$.
3. A variety that contains infinitely many points of a line contains the whole line.

References

The study of prime ideal chains in rings as the foundation of a dimension theory was begun by Krull [43] and was carried further in various works (cf. also [45]). From him come the most important theorems on the behavior of prime ideal chains under integral ring extension; they were later generalized by Cohen–Seidenberg [10]. The first proof of the Noether Normalization Theorem (for an infinite ground field) is contained in [60].

Examples of Noetherian rings of infinite Krull dimension and of rings that are not chain rings are given at the end of Nagata's book [F]. There one can find other examples of unpleasant phenomena that can occur with Noetherian rings. *Excellent rings* escape such phenomena (for the exact definition see [E]). At present many investigations are concerned with the question of the conditions under which a Noetherian ring is excellent. For example, along with affine algebras over a field, Noetherian rings R of prime characteristic p that are finitely generated as R^p-modules are excellent rings (cf. [49]); in particular, they are chain rings.

Chapter III
Regular and rational functions on algebraic varieties
Localization

As in other domains of mathematics, algebraic varieties can be investigated by studying what regular functions exist on them. Here a function is called regular if it can be written locally as the quotient of two polynomial functions. With a variety various rings of functions are associated which contain global or local information about the variety. In this chapter we first deal with the algebraic description of these function rings. In doing this we shall be led to the general investigation of modules and rings of fractions.

1. Some properties of the Zariski topology

In this section X is either an affine or a projective K-variety† or the spectrum of a ring R. Let X be endowed with the Zariski topology and $X \neq \emptyset$.

If f is an element of $K[X]$ (in case X is a variety) or of R (in case $X = \mathrm{Spec}(R)$), then $D(f)$ denotes:

a) in the case of algebraic varieties, the set of all $x \in X$ with $f(x) \neq 0$;
b) in case $X = \mathrm{Spec}(R)$, the set of the $\mathfrak{p} \in X$ with $f \notin \mathfrak{p}$.

In each case we have the easily verifiable

Rules 1.1.

a) $D(f) = X \setminus \mathfrak{V}(f)$ is open in X.
b) $D(f \cdot g) = D(f) \cap D(g)$.
c) $D(f^n) = D(f)$ for all $n \in \mathbb{N}$.
d) $D(f) = \emptyset$ if and only if f is nilpotent (I.4.5).

Proposition 1.2. The sets $D(f)$ (where, in the projective case, f varies over all homogeneous elements of $K[X]$) form a basis for the open sets of X: Every open set is even a finite union of sets of the form $D(f)$ if we assume in case $X = \mathrm{Spec}(R)$ that R is Noetherian.

Proof. If U is an open subset of X, then the closed set $A := X \setminus U$ is of the form $A = \mathfrak{V}(f_1, \ldots, f_r) = \bigcap_{i=1}^{r} \mathfrak{V}(f_i)$ with (in the projective case, homogeneous)

---

† Recall the convention that the coordinate field of a K-variety is always an algebraically closed extension field L of K.

elements $f_i \in K[X]$ or $f_i \in R$. It follows that

$$U = X \setminus \bigcap_{i=1}^{r} \mathfrak{V}(f_i) = \bigcup_{i=1}^{r} (X \setminus \mathfrak{V}(f_i)) = \bigcup_{i=1}^{r} D(f_i).$$

Proposition 1.3. In the case $X = \operatorname{Spec}(R)$ let R be Noetherian. Every subset A of X is quasicompact; that is, every open cover of A has a finite subcover.

Proof. It suffices to consider open sets, and by 1.2 it suffices to show that the sets $D(f)$ are quasicompact. We can also restrict ourselves to covers whose open sets are also of this form.

Therefore, let $D(f) = \bigcup_{\lambda \in \Lambda} D(g_\lambda)$. Let I be the ideal generated by $\{g_\lambda\}_{\lambda \in \Lambda}$. Then $\mathfrak{V}(f) = X \setminus D(f) = X \setminus \bigcup_{\lambda \in \Lambda} D(g_\lambda) = \bigcap_{\lambda \in \Lambda} (X \setminus D(g_\lambda)) = \bigcap_{\lambda \in \Lambda} \mathfrak{V}(g_\lambda)$. Since I is finitely generated, there are finitely many indices λ_i with $I = (g_{\lambda_1}, \ldots, g_{\lambda_n})$. It follows that $\mathfrak{V}(I) = \bigcap_{i=1}^{n} \mathfrak{V}(g_{\lambda_i})$ and $D(f) = \bigcup_{i=1}^{n} D(g_{\lambda_i})$.

Lemma 1.4. An open subset U of a Noetherian topological space X is dense in X if and only if U meets every irreducible component of X.

Proof. If $X = \bigcup_{i=1}^{n} X_i$ is the decomposition of X into irreducible components, then $X_i \not\subset \bigcup_{j \neq i} X_j$ and so $U_i := X \setminus \bigcup_{j \neq i} X_j \neq \emptyset$ for $i = 1, \ldots, n$. We have $U_i \subset X_i$ and U_i is open in X. If U is dense in X, then U meets each U_i and thus every irreducible component of X. If U meets each component X_i $(i = 1, \ldots, n)$ and if $U' \neq \emptyset$ is open in X, then there is an $i \in [1, n]$ with $U' \cap X_i \neq \emptyset$. Since X_i is irreducible, we also have $(U \cap X_i) \cap (U' \cap X_i) \neq \emptyset$ (I.2.10); therefore, $U \cap U' \neq \emptyset$ and U is dense in X.

Proposition 1.5. Under the same assumptions, let X be Noetherian (in case $X = \operatorname{Spec}(R)$). Every dense open subset $U \subset X$ contains a dense set of the form $D(f)$ (where, in the projective case, f is homogeneous of positive degree).

For this we need the following often useful lemma on *avoiding prime ideals*.

Lemma 1.6. Let I be an ideal; let $\mathfrak{p}_1, \ldots, \mathfrak{p}_n (n \geq 1)$ be prime ideals of a ring R. If $I \not\subset \mathfrak{p}_i$ $(i = 1, \ldots, n)$, then there is an $f \in I$ with $f \notin \mathfrak{p}_i$ $(i = 1, \ldots, n)$. If R is a positively graded ring, I a homogeneous ideal of R, and if $\mathfrak{p}_1, \ldots, \mathfrak{p}_n \in \operatorname{Proj}(R)$), then f can even be chosen to be a homogeneous element of positive degree.

Proof by induction on n. For $n = 1$ we need only consider that in the graded case we can find a homogeneous element f of positive degree with $f \in I$, $f \notin \mathfrak{p}_1$: since I is homogeneous, there is a homogeneous element $a \in I$, $a \notin \mathfrak{p}_1$. And (since $\mathfrak{p}_1 \in \operatorname{Proj}(R)$) there is a homogeneous element of positive degree $b \in R$, $b \notin \mathfrak{p}_1$. Put $f := ab$.

Now let $n > 1$ and suppose the assertion has already been proved for $n - 1$ prime ideals. We may assume that no $\mathfrak{p}_i$ is contained in a $\mathfrak{p}_j$ with $j \neq i$. Then $I \cap \mathfrak{p}_j \not\subset \mathfrak{p}_i$ for $j \neq i$, and by the induction hypothesis there is an element $x_j \in I \cap \mathfrak{p}_j, x_j \notin \bigcup_{i \neq j} \mathfrak{p}_i$, where x_j is homogeneous and $\deg x_j > 0$ in the graded case. Raising the x_j to a suitably high power if necessary, we may assume that (in the graded case) they are all of the same degree.

Put $y_j := \prod_{i \neq j} x_i$ and $f := \sum_{j=1}^n y_j$. We have $y_j \in \mathfrak{p}_i \cap I$ for $i \neq j$, but $y_j \notin \mathfrak{p}_j$ $(j = 1, \ldots, n)$; and in the graded case the y_j are of the same degree, so f is homogeneous and of positive degree. From the construction it follows that $f \in I$ but $f \notin \mathfrak{p}_i$ $(i = 1, \ldots, n)$.

Proof of Proposition 1.5. Let $X = \bigcup_{i=1}^n X_i$ be the decomposition of X into irreducible components, and $A := X \setminus U$. By 1.4 no X_i is contained in A and hence $\mathfrak{I}(A) \not\subset \mathfrak{I}(X_i)$ $(i = 1, \ldots, n)$. By 1.6 there is an $f \in \mathfrak{I}(A)$ with $f \notin \mathfrak{I}(X_i)$ $(i = 1, \ldots, n)$; in the projective case f is of positive degree. Then $D(f) \subset U$ and $D(f) \cap X_i \neq \emptyset$ for $i = 1, \ldots, n$; that is, $D(f)$ is dense in X by 1.4.

Proposition 1.7. Under the assumptions made at the start of this section, a set of the form $D(f)$ is dense in X if and only if for any non-nilpotent g in the ring from which f comes, fg is not nilpotent.

Proof. If $D(f)$ is dense in X and g is not nilpotent (so that $D(g) \neq \emptyset$), we have $D(fg) = D(f) \cap D(g) \neq \emptyset$, and fg is not nilpotent.

Conversely, if the condition in 1.7 holds and $U \neq \emptyset$ is open in X, then choose a nonempty subset of U of the form $D(g)$. Then g is not nilpotent, so neither is fg, and hence $D(f) \cap D(g) = D(fg) \neq \emptyset$; thus $D(f) \cap U \neq \emptyset$ and $D(f)$ is dense in X.

Corollary 1.8. In the case of an affine or projective K-variety, $D(f)$ is dense in X if and only if f is not a zero divisor in $K[X]$.

This corollary likewise holds for $X = \mathrm{Spec}(R)$ if R is reduced.

Exercises

1. Let R be any ring. For all $f \in R$ the sets $D(f) = \{\mathfrak{p} \in \mathrm{Spec}(R) \mid f \notin \mathfrak{p}\}$ are quasicompact.

2. Let X be a Noetherian topological space in which every nonempty irreducible closed subset has exactly one generic point (for example, the spectrum of a Noetherian ring). For $y \in X$ let X_y be the set of all $x \in X$ with $y \in \{\bar{x}\}$ (the set of "generalizations" of y), endowed with the relative topology. If $Y \subset X$ is a closed subset with $X_y \setminus \{y\} \neq \emptyset$ for all $y \in Y$, then $X \setminus Y$ is dense in X.

3. Along with the assumptions of Exercise 2 let X be connected and for every $y \in Y$ let $X_y \setminus \{y\}$ be nonempty and connected. Then $X \setminus Y$ is also connected. Hint: If $X \setminus Y = X_1 \cup X_2$ with disjoint closed (relative to $X \setminus Y$) subsets X_i $(i = 1, 2)$, consider the generic point y of an irreducible component of $\overline{X}_1 \cap \overline{X}_2$ ($\overline{X}_i$ being the closure of X_i in X); and conclude that $X_y \setminus \{y\}$ is also disconnected.

4. Let $G = \bigoplus_{i \in \mathbb{N}} G_i$ be a positively graded ring with $\mathrm{Proj}(G) \neq \emptyset$. Let $G_+ := \bigoplus_{i > 0} G_i$. For a homogeneous element $f \in G_+$ let $D(f) := \{\mathfrak{p} \in \mathrm{Proj}(G) \mid f \notin \mathfrak{p}\}$.

   a) If I is a homogeneous ideal of G with $I \cap G_+ \subset \mathfrak{p}$ for some $\mathfrak{p} \in \mathrm{Proj}(G)$, then $I \subset \mathfrak{p}$.

   b) Rules 1.1 hold for the sets $D(f), f \in G_+$ homogeneous.

   c) The sets $D(f), f \in G_+$ homogeneous, form a basis for the open sets of $\mathrm{Proj}(G)$.

   d) If $\mathrm{Proj}(G)$ is Noetherian, then every dense open subset U of $\mathrm{Proj}(G)$ contains a dense open subset $D(f)$ for some homogeneous element $f \in G_+$.

2.   The sheaf of regular functions on an algebraic variety

In this section let V be a nonempty affine or projective K-variety. Let $U \neq \emptyset$ be an open subset of V and let $r : U \to L$ be a mapping.

Definition 2.1. The function r is called *regular* at $x \in U$ if there are elements $f, g \in K[V]$, which are homogeneous and of the same degree in the projective case, such that:

1. $x \in D(g) \subset U$;
2. $r = \frac{f}{g}$ on $D(g)$; that is, for all $y \in D(g)$ we have $r(y) = \frac{f(y)}{g(y)}$.

   (In the projective case this is to be understood so that in homogeneous polynomials representing f and g, y is replaced by a system of homogeneous coordinates of y. Since f and g are homogeneous of the same degree, the result is independent of the choice of homogeneous coordinates.)

   It is clear that the concept of a function regular at a point is independent of the coordinates. If r is regular on all of U, then by 1.3 there are elements $f_1, \ldots, f_n, g_1, \ldots, g_n \in K[V]$, where in the projective case f_i and g_i are homogeneous of the same degree, and where:

1. $U = \bigcup_{i=1}^n D(g_i)$,
2. $r = \frac{f_i}{g_i}$ on $D(g_i)$ for $i = 1, \ldots, n$.

   In particular the elements of K, considered as "constant functions," are regular on U. We denote the set of regular functions on U by $\mathcal{O}(U)$. To the empty set we assign the zero ring: $\mathcal{O}(\emptyset) := \{0\}$.

Proposition 2.2. The regular functions define a sheaf of K-algebras on V, that is:

1. $\mathcal{O}(U)$ is a K-algebra for any open subset U of V.
2. If $U' \subset U$ are open in V and nonempty, then for every $r \in \mathcal{O}(U)$ the restriction $r|_{U'}$ is regular on U' and

$$\rho_{U'}^U : \mathcal{O}(U) \to \mathcal{O}(U') \quad (r \mapsto r|_{U'})$$

is a K-algebra homomorphism. If one puts $\rho_\emptyset^U := 0$, then $\rho_{U''}^{U'} \circ \rho_{U'}^U = \rho_{U''}^U$
if $U'' \subset U' \subset U$ are open subsets of V and $\rho_U^U = \mathrm{id}_{\mathcal{O}(U)}$.

3. If $U = \bigcup_{\lambda \in \Lambda} U_\lambda$ with open sets U_λ and if for every λ there is given $r_\lambda \in \mathcal{O}(U_\lambda)$ such that $\rho_{U_\lambda \cap U_{\lambda'}}^{U_\lambda}(r_\lambda) = \rho_{U_\lambda \cap U_{\lambda'}}^{U_{\lambda'}}(r_{\lambda'})$ for all $\lambda, \lambda' \in \Lambda$, then there is exactly one $r \in \mathcal{O}(U)$ with $\rho_{U_\lambda}^U(r) = r_\lambda$ for all $\lambda \in \Lambda$.

Because regularity of a function is a local property, the proof is immediate from the following remarks.

a) If $r, \bar{r}$ are regular functions on U and $r = \frac{f}{g}$ on $D(g), \bar{r} = \frac{\bar{f}}{\bar{g}}$ on $D(\bar{g})$, then $r = \frac{f\bar{g}}{g\bar{g}}$ and $\bar{r} = \frac{g\bar{f}}{g\bar{g}}$ on $D(g\bar{g}) = D(g) \cap D(\bar{g})$. We now have quotient representations of r and $\bar{r}$ with the same domain of definition, and we see that their sum, difference, and product are regular at x.

b) If $U' \subset U$ is another open set with $x \in U'$, then $r|_{U'}$ is regular at x, for there is an open set $D(g')$ with $x \in D(g') \subset U'$ and we have $r = \frac{fg'}{gg'}$ on $D(gg') \subset U'$.

(The concept of a sheaf is of fundamental significance for the study of algebraic varieties, chiefly for global questions. Since in the future we shall be more interested in local problems, the sheaf concept plays no essential role in the sequel.)

For regular functions on open sets of the form $D(g)$ the following description can be given.

Proposition 2.3. Let $g \in K[V]$ be $\neq 0$ (and, in the projective case, let g be homogeneous of positive degree). Every $r \in \mathcal{O}(D(g))$ then has a representation $r = \frac{f}{g^\nu}$ on all of $D(g)$, where $\nu \in \mathbb{N}$ and $f \in K[V]$ (homogeneous with $\deg f = \nu \cdot \deg g$ in the projective case).

Proof. r has representations $r = \frac{f_i}{g_i}$ on $D(g_i)$ $(i = 1, \ldots, n)$, where $D(g) = \bigcup_{i=1}^n D(g_i)$. On $D(g_i) \cap D(g_j) = D(g_i g_j)$ we have $f_i g_j - f_j g_i = 0$, so $g_i g_j (f_i g_j - f_j g_i) = 0$ on all of V $(i, j \in [1, n])$. Writing $r = \frac{f_i g_i}{g_i^2}$ on $D(g_i)$, we can without restriction assume that $f_i g_j - f_j g_i = 0$ on all of V, that is $f_i g_j = f_j g_i$ in $K[V]$.

From $D(g) = \bigcup_{i=1}^n D(g_i)$ it follows that $g \in \mathrm{Rad}(g_1, \ldots, g_n)$ (in the projective case this holds because g is assumed nonconstant). Hence we have an equation $g^\nu = \sum_{i=1}^n h_i g_i$ with (in the projective case, homogeneous) elements $h_1, \ldots, h_n \in K[V]$ $(\deg h_i + \deg g_i = \nu \cdot \deg g)$. Putting $f := \sum_{i=1}^n h_i f_i$, we get

$$g^\nu f_j = \sum_{i=1}^n (h_i g_i) f_j = \sum_{i=1}^n (h_i f_i) g_j = f g_j,$$

and so $r = \frac{f_j}{g_j} = \frac{f}{g^\nu}$ on $D(g_j)$ for all $j \in [1, n]$. Thus $r = \frac{f}{g^\nu}$ on all of $D(g)$.

Corollary 2.4. If V is an affine K-variety, then $\mathcal{O}(V)$ is isomorphic to $K[V]$ as a K-algebra.

Since $V = D(1)$, this follows immediately from 2.3.

Proposition 2.5. (Identity Theorem) Let U_1, U_2 be open subsets of V, $r_1 \in \mathcal{O}(U_1)$, $r_2 \in \mathcal{O}(U_2)$. Assume there exists a dense open subset U of V with $U \subset U_1 \cap U_2$ such that $r_1|_U = r_2|_U$. Then $r_1|_{U_1 \cap U_2} = r_2|_{U_1 \cap U_2}$.

Proof. We put $U' := U_1 \cap U_2$ and $r := r_1|_{U_1 \cap U_2} - r_2|_{U_1 \cap U_2}$. Then $A := \{x \in U' \mid r(x) = 0\}$ is a closed subset of U'. Indeed, if $x \in U' \setminus A$ and we write $r = \frac{f}{g}$ in a neighborhood $D(g) \subset U'$ of x, then $f(x) \neq 0$ and so $r \neq 0$ in the open neighborhood $D(f) \cap D(g)$ of x. By assumption, $U \subset A$ and U is dense in V. It follows that $A = U'$ and so $r_1 = r_2$ on U'.

Corollary 2.6. Let U be a dense open subset of V, $r \in \mathcal{O}(U)$. Then there is a uniquely determined pair (U', r'), where U' is an open set with $U \subset U'$, $r' \in \mathcal{O}(U')$ and $r = r'|_U$ such that r' cannot be extended to a regular function on an open subset of V that properly contains U'.

Definition 2.7. A regular function r given on a dense open subset U of V that cannot be extended to a regular function on an open subset of V that properly contains U is called a *rational function* on V. U is called its *domain of definition*, $V \setminus U$ its *pole set*.

Rational functions are added, subtracted, and multiplied by performing these operations on the intersection of the domains of definition of the functions and extending the results to rational functions on V. Thus the set $R(V)$ of all rational functions on V becomes a K-algebra.

If $r \in R(V)$ and $W \subset V$ is a subvariety no irreducible component of which is wholly contained in the pole set of r, then the restriction $r|_W \in R(W)$ is defined as follows. The domain of definition U of r meets every irreducible component of W, hence $W \cap U$ is dense in W. $r|_{W \cap U}$ is regular on $W \cap U$, for if r is represented in the neighborhood of $x \in W \cap U$ as the quotient of two functions f, g in $K[V]$, then in the neighborhood of x on W, it can also be written as the quotient of the homomorphic images $\bar{f}, \bar{g}$ of f, g under the epimorphism $K[V] \to K[W]$. $r|_W$ is the rational function on W that results from extending $r|_{W \cap U}$.

We now propose to determine the algebraic structure of the K-algebra $R(V)$. First, we have:

Proposition 2.8. Let $V = \bigcup_{i=1}^{n} V_i$ be the decomposition of V into irreducible components. Then the mapping

$$R(V) \to R(V_1) \times \cdots \times R(V_n) \qquad (r \mapsto (r|_{V_1}, \ldots, r|_{V_n}))$$

is an isomorphism of K-algebras.

Proof. By the remark above it is clear that the mapping is well defined. It is obviously a K-algebra homomorphism. If $(r_1, \ldots, r_n) \in \prod_{i=1}^{n} R(V_i)$ is given, for all $j \in [1, n]$ let r'_j be the restriction of r_j to the complement U_j of $V_j \cap (\bigcup_{i \neq j} V_i)$ in the domain of definition of r_j. U_j is a dense open subset of V_j, and $U_j \cap U_{j'} = \emptyset$ for $j \neq j'$. The r'_j define a regular function on the dense open subset $\bigcup_{j=1}^{n} U_j$ of V. Let r be the rational function on V they determine. By the Identity Theorem it is clear that $(r_1, \ldots, r_n) \mapsto r$ defines a mapping inverse to that given in the proposition.

By 2.8 it suffices to determine the structure of $R(V)$ for an irreducible variety V. We first consider a (homogeneous) non-zerodivisor $g \in K[V]$ and an arbitrary $f \in K[V]$ (homogeneous and of the same degree as g). Since $D(g)$ is dense in V (1.8), $\frac{f}{g}$ determines a rational function on V. Conversely,

Lemma 2.9. If $r \in R(V)$ has domain of definition U, then there is a (homogeneous) non-zerodivisor $g \in K[V]$ with $D(g) \subset U$ such that $r = \frac{f}{g^\nu}$ on $D(g)$ with some $f \in K[V], \nu \in \mathbb{N}$ (where f is homogeneous, $\deg(f) = \nu \cdot \deg(g)$).

Proof. By 1.5 U contains a dense open subset of V of the form $D(g)$, where g is not a zero divisor of $K[V]$ and, in the projective case, is homogeneous of positive degree. By 2.3 $r|_{D(g)}$ has the given form.

Proposition 2.10. Let V be irreducible.

a) In the affine case $R(V)$ is isomorphic to the field of fractions $K(V)$ of the coordinate ring $K[V]$. $R(V)$ is a finitely generated extension field of K of transcendence degree $\dim V$.

b) In the projective case $R(V)$ is K-isomorphic to the subfield $K(V)$ of the field of fractions of $K[V]$, consisting of all elements that can be written as quotients of homogeneous elements of $K[V]$ of the same degree.

Proof. Let $K(V) \to R(V)$ be the mapping that assigns to each quotient $\frac{f}{g}$ the extension to a rational function of the regular function defined on $D(g)$ by $\frac{f}{g}$. This mapping is independent of the quotient representation, is a K-algebra homomorphism, and is injective. By 2.9 it is also surjective. That $\dim V$ equals the transcendence degree of $K(V)$ over K (in the affine case) has been shown in II.3.11c).

Proposition 2.11. Let V be an irreducible projective K-variety, where K is algebraically closed. Then $\mathcal{O}(V) = K$: on an irreducible projective K-variety (K algebraically closed) the constants are the only globally regular functions.

Proof. According to 2.10b) we can identify $\mathcal{O}(V)$ with a subalgebra of $Q(K[V])$. Let $K[V] = K[Y_0, \ldots, Y_n]/\mathfrak{J}(V)$, and let y_i be the image of Y_i in $K[V]$. Then $V = \bigcup_{i=0}^{n} D(y_i)$. With a suitable numbering we may assume that $y_i \neq 0$ for $i = 0, \ldots, m$ and $y_j = 0$ for $j = m+1, \ldots, n$. Let $K[V]_\nu$ denote the homogeneous component of degree ν of the graded ring $K[V]$.

By 2.3, for $i = 0, \ldots, m$ $r \in \mathcal{O}(V)$ has a representation $r = \frac{f_i}{y_i^{\nu_i}}$ ($\nu_i \in \mathbb{N}$, $f_i \in K[V]_{\nu_i}$). Let $\nu := \sum_{i=0}^{m} \nu_i$. Then $y_0^{\alpha_0} \cdot \ldots \cdot y_m^{\alpha_m} r \in K[V]_\nu$ for each monomial $y_0^{\alpha_0} \cdot \ldots \cdot y_m^{\alpha_m}$ with $\sum_{i=0}^{m} \alpha_i = \nu$, since $\alpha_i \geq \nu_i$ for at least one $i \in [0, m]$. We get $K[V]_\nu r^t \subset K[V]_\nu$ and in particular $r^t \in \frac{1}{y_0^\nu} K[V]_\nu$ for all $t \in \mathbb{N}$. Since $K[V][r]$ is therefore a $K[V]$-submodule of the finitely generated $K[V]$-module $K[V] + \frac{1}{y_0^\nu} K[V]$, it follows from I.2.17 and II.2.2 that r is integral over $K[V]$:

$$r^\rho + a_1 r^{\rho-1} + \cdots + a_\rho = 0 \qquad (a_i \in K[V], \rho > 0). \tag{*}$$

From $f_0^\rho + a_1 y_0^{\nu_0} f_0^{\rho-1} + \cdots + a_\rho y_0^{\nu_0 \rho} = 0$, by comparing coefficients we see that we can replace the a_i in (*) by their homogeneous components of degree 0. Hence r is algebraic over K, that is $r \in K$, since K is algebraically closed, q. e. d.

Under the hypotheses of 2.11 every nonconstant rational function r has a nonempty pole set and hence also a nonempty zero set (otherwise $\frac{1}{r}$ would have an empty pole set).

We get more information on projective varieties from the following considerations. Let $V \subset A^n(L)$ be a nonempty affine K-variety and $\overline{V} \subset P^n(L)$ its projective closure in the sense of I.§5.

V is then a dense open subset of $\overline{V}$ and every (dense) open subset $U \subset V$ is also open (and dense) in $\overline{V}$.

Let $\mathcal{O}_V(U)$ and $\mathcal{O}_{\overline{V}}(U)$ denote the regular functions on U in the affine and in the projective case.

Every $r \in \mathcal{O}_V(U)$ can be considered as an element of $\mathcal{O}_{\overline{V}}(U)$. Namely, if $r = \frac{f}{g}$ on $D(g)$ with $f, g \in K[V]$, then choose polynomials $F, G \in K[X_1, \ldots, X_n]$ representing f and g. Put $F^* := Y_0^d \cdot F(\frac{Y_1}{Y_0}, \ldots, \frac{Y_n}{Y_0}), G^* := Y_0^d \cdot G(\frac{Y_1}{Y_0}, \ldots, \frac{Y_n}{Y_0})$ with $d := \text{Max}(\deg(F), \deg(G))$. Let f^*, g^* be the images of F^*, G^* in $K[\overline{V}]$. Then $r = \frac{f^*}{g^*}$ on $D(g^*)$.

Through the process of dehomogenization (I.§5) we immediately find that every $r \in \mathcal{O}_{\overline{V}}(U)$ belongs to $\mathcal{O}_V(U)$. Thus:

Lemma 2.12. Let U be an open subset of an affine variety V, $\overline{V}$ the projective closure of V. Then
$$\mathcal{O}_V(U) = \mathcal{O}_{\overline{V}}(U).$$

It follows that every rational function on V can by uniquely extended to a rational function on $\overline{V}$, and the restriction of a rational function on $\overline{V}$ to V is a rational function on V. These two mappings are inverses.

Proposition 2.13. Let $\overline{V}$ be the projective closure of a nonempty affine variety V. Then there is a K-algebra isomorphism $R(V) \cong R(\overline{V})$. In particular, for irreducible V, $R(\overline{V})$ is a finitely generated extension field of K of transcendence degree $\dim \overline{V} = \dim V$.

A finitely generated extension field F/K is also called an "algebraic function field over K." An important method of investigating such fields is to consider them as fields of rational functions of suitable projective varieties. This is always possible: if $F = K(z_1, \ldots, z_m)$, then F is the function field of the irreducible affine variety V with coordinate ring $K[z_1, \ldots, z_m]$ and by 2.13 also of $\overline{V}$, the projective closure of V. Such a $\overline{V}$ is called a *projective model* of F/K. Using such models one can assign to F/K invariants with whose aid, for example, it can often be shown that F/K is not purely transcendental or that F/K is not K-isomorphic to another algebraic function field F'/K.

Whereas up to now we have been occupied with the maximal extension of regular functions, regular functions will now be considered "locally" too. As at the start let V be an affine or projective variety, W a nonempty irreducible subset of V (which need not be closed). By $\mathfrak{A}(W)$ we denote the set of all open subsets U of V with $U \cap W \neq \emptyset$. If $U_1, U_2 \in \mathfrak{A}(W)$, then $U_1 \cap U_2 \in \mathfrak{A}(W)$ (I.2.10).

Definition 2.14. If $U_1, U_2 \in \mathfrak{A}(W)$, two functions $r_1 \in \mathcal{O}(U_1), r_2 \in \mathcal{O}(U_2)$ are called *equivalent in W* if there is a $U \in \mathfrak{A}(W)$ with $U \subset U_1 \cap U_2$ such that $r_1|_U = r_2|_U$.

Obviously this defines an equivalence relation on $\bigcup_{U \in \mathfrak{A}(W)} \mathcal{O}(U)$. An equivalence class with respect to this equivalence relation is called a *regular function germ in W*. Let $\mathcal{O}_{V,W}$ be the set of regular function germs in W. Then $K \subset \mathcal{O}_{V,W}$ if the elements of K are identified with the germs of the constant functions.

Function germs are added and multiplied by adding or multiplying representatives on the intersection of their domains and then passing over to the germs. The result is independent of the choice of representatives.

Remark 2.15. $\mathcal{O}_{V,W}$ is a local K-algebra. Its maximal ideal is the set $\mathfrak{m}_{V,W}$ of all $\rho \in \mathcal{O}_{V,W}$ of which a representative r vanishes on a nonempty open subset of W.

Proof. It is clear that $\mathcal{O}_{V,W}$ is a commutative ring with 1 and the set $\mathfrak{m}_{V,W}$ described in the remark is an ideal of $\mathcal{O}_{V,W}$. We show that $\mathcal{O}_{V,W} \setminus \mathfrak{m}_{V,W}$ consists only of units of $\mathcal{O}_{V,W}$, whence it follows that $\mathfrak{m}_{V,W}$ is the unique maximal ideal of $\mathcal{O}_{V,W}$.

If $\rho \in \mathcal{O}_{V,W} \setminus \mathfrak{m}_{V,W}$ is given, then ρ can be represented by a quotient $\frac{f}{g}$ with $D(g) \in \mathfrak{A}(W)$, where $D(f) \in \mathfrak{A}(W)$, because $\rho \notin \mathfrak{m}_{V,W}$. The function germ represented by $\frac{g}{f}$ is inverse to ρ.

In the case where W consists of a single point x we write $\mathcal{O}_{V,x}$ or simply $\mathcal{O}_x$ instead of $\mathcal{O}_{V,W}$ and call $\mathcal{O}_{V,x}$ the *local ring* of x on V or the *stalk* of the sheaf $\mathcal{O}$ at x. $\mathfrak{m}_{V,x}$ or simply $\mathfrak{m}_x$ denotes the maximal ideal of $\mathcal{O}_{V,x}$. The algebraic structure of the ring $\mathcal{O}_{V,x}$ is more complicated than that of the ring $R(V)$ of rational functions on V. It depends on whether x is a "regular or a singular point" of V (Ch. VI) and on the nature of the singularity. One is interested in the rings $\mathcal{O}_x$ because they reflect the local properties of the variety V in the neighborhood of x in ideal-theoretic properties of the ring $\mathcal{O}_x$.

If V is an affine variety with projective closure $\overline{V}$, then for any nonempty irreducible subset W of V (resp. every point $x \in V$) we have a canonical ring isomorphism

$$\mathcal{O}_{V,W} \cong \mathcal{O}_{\overline{V},W} \qquad (\text{resp. } \mathcal{O}_{V,x} \cong \mathcal{O}_{\overline{V},x}),$$

since all regular function germs in $\mathcal{O}_{\overline{V},W}$ and $\mathcal{O}_{\overline{V},x}$ can be represented by regular functions on V (through restrictions to V). The study of the rings $\mathcal{O}_{\overline{V},W}$ and $\mathcal{O}_{\overline{V},x}$ can therefore be carried out "in affine space," which is often advantageous.

If V is irreducible, then we have an injective K-algebra homomorphism $\mathcal{O}_{V,W} \to R(V)$ given by assigning to a germ with respect to W the uniquely determined rational function that represents the germ. $\mathcal{O}_{V,W}$ is thus identified with the subalgebra of $R(V)$ consisting of all rational functions whose pole set does not contain W, and $R(V)$ is the field of fractions of $O_{V,W}$.

The algebraic considerations of the next two sections can be understood as a first step in investigating the structure of the rings of regular functions defined in this section.

Exercises

Let V and W be two nonempty affine or projective K-varieties. A mapping $\varphi : V \to W$ is called K-regular (or a K-morphism) if it is continuous (in the Zariski topology) and if for every open subset $U \subset W$ with $\varphi^{-1}(U) \neq \emptyset$ we have: if $f \in O_W(U)$, then $f \circ \varphi \in O_V(\varphi^{-1}(U))$. Here O_V (resp. O_W) denotes the sheaf of K-regular functions on V (resp. W). A K-regular mapping φ is called a K-isomorphism if there is a K-regular mapping $\psi : W \to V$ with $\psi \circ \varphi = \mathrm{id}_V$, $\varphi \circ \psi = \mathrm{id}_W$.

1. The composition of regular mappings is regular. If $\varphi : V \to W$ is regular, then for any open set $U \subset W$ with $\varphi^{-1}(U) \neq \emptyset$, the mapping $O_W(U) \to O_V(\varphi^{-1}(U))$ $(f \mapsto f \circ \varphi)$ is a K-algebra homomorphism (an isomorphism if φ is an isomorphism).

2. For a regular mapping $\varphi : V \to W, x \in V$, and $f \in O_{W,\varphi(x)}$, let $\varphi_x(f) \in O_{V,x}$ be defined as follows. f is represented in some neighborhood of $\varphi(x)$ by a regular function F. Then $\varphi_x(f)$ is the germ of $F \circ \varphi$ at x. $\varphi_x : O_{W,\varphi(x)} \to O_{V,x}$ is a well-defined K-algebra homomorphism with $\varphi_x(\mathfrak{m}_{W,\varphi(x)}) \subset \mathfrak{m}_{V,x}$; it is an isomorphism if φ is an isomorphism.

In what follows let $V \subset \mathbb{A}^m(L)$, $W \subset \mathbb{A}^n(L)$ be nonempty affine K-varieties.

3. A mapping $\varphi : V \to W$ is K-regular if and only if there are polynomials $P_1, \ldots, P_n \in K[X_1, \ldots, X_m]$ such that $\varphi(x) = (P_1(x), \ldots, P_n(x))$ for all $x \in V$.

4. For a K-regular mapping $\varphi : V \to W$ let $K[\varphi] : K[W] \to K[V]$ be the K-algebra homomorphism given by $f \mapsto f \circ \varphi$. If $\psi : W \to Z$ is another K-regular mapping into a K-variety Z, then $K[\psi \circ \varphi] = K[\varphi] \circ K[\psi]$. Moreover, $K[\mathrm{id}] = \mathrm{id}_{K[V]}$. $\varphi \mapsto K[\varphi]$ defines a bijection of the set of all K-regular mappings from V to W onto the set of all K-algebra homomorphisms $K[W] \to K[V]$. Here the K-isomorphisms of V onto W correspond bijectively to the K-algebra isomorphisms $K[W] \xrightarrow{\sim} K[V]$.

5. Let K be a field of characteristic $p > 0$ and $F : \mathbb{A}^n(L) \to \mathbb{A}^n(L)$ the mapping given by $F(x_1, \ldots, x_n) = (x_1^p, \ldots, x_n^p)$ (the Frobenius morphism). F is a bijective K-regular mapping but is not an isomorphism.

6. Let $\varphi : V \to W$ be K-regular. If $Z \subset V$ is a subvariety, $\overline{\varphi(Z)}$ the closure of $\varphi(Z)$ in the K-topology on W, then $K[\varphi]^{-1}(\mathfrak{I}(Z))$ is the ideal of $\overline{\varphi(Z)}$ in $K[W]$, if $\mathfrak{I}(Z)$ is the ideal of Z in $K[V]$. Moreover:

   a) If Z is irreducible, so is $\overline{\varphi(Z)}$.

   b) $\dim \overline{\varphi(V)} \leq \dim V$.

   c) $\overline{\varphi(V)} = W$ if and only if $K[\varphi]$ is injective. (In this case φ is called a *dominant* morphism.)

7. For a regular mapping $\varphi : V \to W$ the following statements are equivalent.

   a) $K[\varphi]$ is surjective.

   b) $\varphi(V)$ is a subvariety of W and $\varphi : V \to \varphi(V)$ is an isomorphism.

(In this case φ called a *closed immersion* or an *embedding* of V into W.)

8. The following correspond bijectively:

   a) The embeddings of V into $\mathbb{A}^m(L)$.

   b) The systems of generators of length m of the K-algebra $K[V]$.

9. Give an example of a space curve that cannot be embedded in a plane.

10. Give a geometric description of the regular mapping $\varphi : \mathbb{A}^2(L) \to \mathbb{A}^2(L)$ with $\varphi(x_1, x_2) = (x_1, x_1 x_2)$ and determine $\varphi^{-1}(C)$, where C is the curve $X_1^p - X_2^q = 0$ $(p, q \in \mathbb{N})$ or $X_1^2(1 - X_1^2) - X_2^2 = 0$.

11. Let O be the sheaf of regular functions on an algebraic variety V, $U \subset V$ a nonempty open set. Show that $O(U)$ is the *projective (inverse) limit* of the rings $O(D(g))$ with $D(g) \subset U$; that is: If R is any ring and for each g with $D(g) \subset U$ there is a ring homomorphism $\alpha_g : R \to O(D(g))$ such that for $D(g') \subset D(g)$ we have

$$\alpha_{g'} = \rho_{D(g')}^{D(g)} \circ \alpha_g,$$

then there is exactly one ring homomorphism $\alpha : R \to O(U)$ with $\alpha_g = \rho_{D(g)}^{U} \circ \alpha$ for all g with $D(g) \subset U$.

12. Under the hypotheses of Exercise 11 let $x \in V$. For an open set $U \subset V$ with $x \in U$ denote by $\rho_x^U : O(U) \to O_x$ the mapping that assigns to each regular function on U its germ at x. ρ_x^U is a ring homomorphism with $\rho_x^U = \rho_x^{U'} \circ \rho_{U'}^U$ for all open subsets U' of V with $x \in U' \subset U$. O_x is the *inductive (direct) limit* of the rings $O(U)$ with $x \in U$; that is, the following universal property holds: If R is any ring and if for each U with $x \in U$ there is a ring homomorphism $\alpha_U : O(U) \to R$ such that $\alpha_U = \alpha_{U'} \cdot \rho_{U'}^U$ for every open set U' with $x \in U' \subset U$, then there is a unique ring homomorphism $\alpha : O_x \to R$ with $\alpha_U = \alpha \circ \rho_x^U$ for all open U with $x \in U$.

3. Rings and modules of fractions. Examples

This section treats, in general form, the formation of fractions such as have already appeared in special cases in the last section.

Let R be a ring, S a multiplicatively closed subset† of R, M an R-module. In the future for any $r \in R$ we shall denote by $\mu_r : M \to M$ the (linear) mapping with $\mu_r(m) = rm$ for all $m \in M$. (It is called "multiplication by r.")

Definition 3.1. An R-module M together with a linear mapping $i : M \to M_S$ is called a *module of fractions* or *quotient module* of M *with denominator set S* (or just, *by S*) if:

1. For all $s \in S, \mu_s : M \to M$ is bijective.
2. If N is any R-module for which $\mu_s : N \to N$ is bijective for all $s \in S$, and if $j : M \to N$ is any linear mapping, then there is a unique linear mapping $l : M_S \to N$ with $j = l \circ i$.

i is called the *canonical mapping* into the module of fractions.

As always when an object is defined by a universal property, it turns out that the pair (M_S, i), if it exists, is uniquely determined up to isomorphism in the following sense: If (M_S^*, i^*) is also module of fractions of M with denominator set S, then there is a unique isomorphism $M_S \to M_S^*$ for which the diagram

$$
\begin{array}{ccc}
 & \overset{i}{\nearrow} & M_S \\
M & & \downarrow \\
 & \underset{i^*}{\searrow} & M_S^*
\end{array}
\tag{1}
$$

is commutative.

The existence proof for the module of fractions is similar to that for the field of fractions (or quotient field) of an integral domain.

a) M_S *as the set of fractions* $\frac{m}{s} (m \in M, s \in S)$. On $M \times S$ we introduce the following equivalence relation (the definition of equality for fractions): For $(m, s), (m', s') \in M \times S$ we write $(m, s) \sim (m', s')$ if and only if there is some $s'' \in S$ with $s''(s'm - sm') = 0$. An easy computation shows that this is actually an equivalence relation. The equivalence class to which (m, s) belongs will be denoted by $\frac{m}{s}$. Let M_S be the set of all equivalence classes on $M \times S$ with respect to this relation; and let $i : M \to M_S$ be the mapping with $i(m) = \frac{m}{1}$ for all $m \in M$.

b) *Addition and multiplication* on M_S are defined by the following formulas (rules for calculating with fractions):

$$\frac{m}{s} + \frac{m'}{s'} := \frac{s'm + sm'}{ss'},$$

$$r \cdot \frac{m}{s} := \frac{rm}{s} \quad (r \in R).$$

One immediately checks by computation that the results are always independent of the choice of representative (m, s) of the class $\frac{m}{s}$ and that the axioms of an R-module are fulfilled. $\frac{0}{1}$ is the neutral element for addition. Obviously, $i : M \to M_S$ is a linear mapping.

---

† Recall the convention that $1 \in S$ for any multiplicatively closed subset S of a ring R. This is convenient, though avoidable, for what follows.

For each $s \in S, \mu_s : M_S \to M_S$ is bijective, since the assignment $\frac{m'}{s'} \mapsto \frac{m'}{ss'}$ provides a well-defined linear mapping $j_s : M_S \to M_S$ that inverts μ_s.

c) To prove the universal property let N and j be given as in Definition 3.1. If a linear mapping $l : M_S \to N$ with $j = l \circ i$ exists, then for all $m \in M$ the condition $l(\frac{m}{1}) = j(m)$ must hold. For all $s \in S$ we must then also have $s \cdot l(\frac{m}{s}) = l(s \cdot (\frac{m}{s}))$, that is

$$l(\frac{m}{s}) = \mu_s^{-1}(j(m)). \qquad (2)$$

This shows that l, if it exists, is uniquely determined by the requirement $j = l \circ i$. On the other hand, formula (2), as one easily computes, determines a well-defined linear mapping $l : M_S \to N$ that meets the requirement.

On the basis of (1) we can identify any module of fractions of M by S with the M_S thus constructed. This identifies $\frac{m}{s}$ with $\mu_s^{-1}(i(m))$ for all $m \in M, s \in S$.

In the special case $M = R$ the construction above provides an R-module R_S. The formula

$$\frac{r}{s} \cdot \frac{r'}{s'} := \frac{rr'}{ss'}$$

defines a well-defined multiplication on R through which R becomes a commutative ring with unit element $\frac{1}{1}$. $i : R \to R_S$ is then a ring homomorphism. That, for $s \in S$, the mapping $\mu_s : R \to R$ is bijective, is equivalent to: $i(s) = \frac{s}{1}$ is a unit in R_S.

If T is any ring and $j : R \to T$ is any ring homomorphism such that $j(s)$ is a unit in T for all $s \in S$, then T is an R-module with scalar multiplication $r \cdot t := j(r) \cdot t$ $(r \in R, t \in T)$ and $\mu_s : T \to T$ is bijective for all $s \in S$. The mapping $l : R_S \to T$ with $j = l \circ i$ is a ring homomorphism, as a computation immediately shows.

We call R_S the *ring of fractions* or *quotient ring* of R with denominator set S and $i : R \to R_S$ the canonical homomorphism into the quotient ring. The discussion above shows that R can also be defined by a universal property:

Proposition 3.2. For all $s \in S, i(s)$ is a unit in R_S. If T is any ring, $j : R \to T$ a ring homomorphism for which $j(s)$ is a unit in T for all $s \in S$, then there is a unique ring homomorphism $l : R_S \to T$ with $j = l \circ i$.

Each R-module N for which μ_s is bijective for all $s \in S$ can be made into an R_S-module by defining a (well-defined) scalar multiplication by the formula

$$\frac{r}{s} \cdot n := \mu_s^{-1}(rn) \qquad (\frac{r}{s} \in R_S, n \in N).$$

In particular, the quotient module M_S of an R-module M is an R_S-module with scalar multiplication

$$\frac{r}{s} \cdot \frac{m}{s'} := \frac{rm}{ss'}.$$

M_S will always be so considered in the future. Conversely, for any R_S-module N, $\mu_s : N \to N$ $(n \mapsto \frac{s}{1}n)$ is bijective for all $s \in S$, since $\frac{s}{1}$ is a unit in R_S.

Remark 3.3. The R_S-module M_S is generated by $i(M)$. The canonical mapping $i : M \to M_S$ (resp. $i : R \to R_S$) is an isomorphism if and only if for all $s \in S$ the mapping $\mu_s : M \to M$ is bijective (resp. if any $s \in S$ is a unit in R). (In these cases the formation of fractions is of course superfluous, and we identify M_S with M (resp. R_S with R).)

Because $\frac{m}{s} = \frac{1}{s} \cdot \frac{m}{1}$, the first statement is trivial. The second results because, under the given condition, (M, id_M) (resp. (R, id_R)) already satisfies the universal property for S in 3.1 (resp. 3.2).

Examples.

a) If $R \neq \{0\}$ is an integral domain and $S := R \setminus \{0\}$, then R_S is the field of fractions, or quotient field, of R, and $i : R \to R_S$ is the embedding of R into the field of fractions that identifies the $r \in R$ with the "improper fractions" $\frac{r}{1}$.

b) Let $R \neq \{0\}$ be any ring, S the multiplicatively closed set of all non-zerodivisors of R. In this case R_S is called the *full ring of fractions* or quotient ring of R. In the future it will always be denoted by $Q(R)$.

In particular, if $R = K[V]$ is the coordinate ring of an affine K-variety and $R(V)$ is the ring of rational functions on V, then we have an injective ring homomorphism $j : K[V] \to R(V)$ assigning to any $f \in K[V]$ the function given on V by f. If f is not a zero divisor in $K[V]$, then $j(f)$ is a unit in $R(V)$, for by 1.8 $D(f)$ is dense in V and the extension of $\frac{1}{f}$ to a rational function on V is an inverse of $j(f)$. j thus induces a ring homomorphism (3.2)

$$l : Q(K[V]) \to R(V).$$

Here $l(\frac{f}{g})$ is the extension of the function on $D(g)$ given by $\frac{f}{g}$ to a rational function on V. From 2.9 it follows that l is surjective. l is also injective, for if $\frac{f}{g} = 0$ on $D(g)$, then $gf = 0$ on all of V and so $\frac{f}{g} = 0$ as an element of $Q(K[V])$. Thus we have generalized 2.10a) to

Proposition 3.4. The ring $R(V)$ of rational functions on an affine K-variety $V \neq \emptyset$ is K-isomorphic to the full quotient ring $Q(K[V])$ of the affine coordinate ring $K[V]$.

c) Let R be any ring, g an element of R. $S := \{1, g, g^2, \ldots\}$ is a multiplicatively closed subset of R. In this case the module of fractions M_S of an R-module M is denoted by M_g, and the ring of fractions by R_g.

If $R = K[V]$ is the coordinate ring of an affine K-variety and if $g \neq 0$, then from 2.3 we get:

Proposition 3.5. The ring $\mathcal{O}(D(g))$ of regular functions on $D(g)$ is K-isomorphic to $K[V]_g$.

d) Let R be any ring, $\mathfrak{p} \in \mathrm{Spec}(R)$. $S := R \setminus \mathfrak{p}$ is multiplicatively closed. The ring of fractions R_S will also be denoted by $R_\mathfrak{p}$; it is called the *local ring of the prime ideal* $\mathfrak{p}$ of R or the *localization* of R at $\mathfrak{p}$.

$R_\mathfrak{p}$ is indeed a local ring. Its maximal ideal $\mathfrak{m}_\mathfrak{p}$ consists of all elements $\frac{p}{s}$ with $p \in \mathfrak{p}, s \in S$; these elements evidently form an ideal in R. Further, if $\frac{r}{s} \in R_\mathfrak{p} \setminus \mathfrak{m}_\mathfrak{p}$ is given, then $r \notin \mathfrak{p}$ and $\frac{r}{s}$ is a unit of R with inverse $\frac{s}{r}$. Therefore $\mathfrak{m}_\mathfrak{p}$ is a maximal ideal of R, and there is no other maximal ideal.

As in b) let $R = K[V]$ be the coordinate ring of an affine K-variety and $W \subset V$ a nonempty irreducible subset of V, $\overline{W}$ its closure in V.

$\mathfrak{I}(W) = \mathfrak{I}(\overline{W})$ is a prime ideal of $K[V]$, which we shall here denote by $\mathfrak{p}_W$. Let $j : K[V] \to \mathcal{O}_{V,W}$ be the K-algebra homomorphism assigning to each $f \in K[V]$ the germ of f in W. For $f \notin \mathfrak{p}_W$ we have $j(f) \notin \mathfrak{m}_{V,W}$, so it is a unit of $\mathcal{O}_{V,W}$. Therefore, j induces a K-algebra homomorphism $K[V]_{\mathfrak{p}_W} \to \mathcal{O}_{V,W}$; one easily checks that it is an isomorphism.

Proposition 3.6. Let W be a nonempty irreducible subset of an affine K-variety V. The ring $\mathcal{O}_{V,W}$ of regular function germs in W is K-isomorphic to the local ring $K[V]_{\mathfrak{p}_W}$ of the coordinate ring $K[V]$ at the prime ideal $\mathfrak{p}_W = \mathfrak{I}(W)$.

In particular, for $x \in V$ the local ring $\mathcal{O}_{V,x} \cong K[V]_{\mathfrak{p}_x}$ with $\mathfrak{p}_x := \{f \in K[V] \mid f(x) = 0\}$.

e) Let $G = \bigoplus_{i \in \mathbb{Z}} G_i$ be a graded ring, S a multiplicatively closed subset of G consisting of homogeneous elements alone. G_S can be endowed with a natural grading; $(G_S)_i$ consists of all quotients $\frac{g}{s} \in G_S$, where $g \in G$ is homogeneous and $\deg g - \deg s = i$. (The last equation holds for all possible representations of g/s as a quotient of homogeneous elements.) We have $G_S = \bigoplus_{i \in \mathbb{Z}} (G_S)_i$.

Of particular interest is the subring $(G_S)_0$ of homogeneous elements of degree 0. If $V \neq \emptyset$ is a projective K-variety with coordinate ring $K[V]$ and S is the set of all homogeneous non-zerodivisors of $K[V]$, then the K-algebra of rational functions on V can be identified with $(K[V]_S)_0$ (2.10). If g is a homogeneous element of positive degree in $K[V]$, then $\mathcal{O}(D(g)) = (K[V]_g)_0$. Further, if W is a nonempty irreducible subset of V and S is the set of all homogeneous elements not in the prime ideal $\mathfrak{I}(W)$, then $\mathcal{O}_{V,W} \cong (K[V]_S)_0$.

In general let $\mathfrak{p} \neq G$ be a homogeneous prime ideal of a graded ring $G = \bigoplus_{i \in \mathbb{Z}} G_i$, and let S be the set of homogeneous elements in $G \setminus \mathfrak{p}$. Then $(G_S)_0$ is a local ring, which is also denoted by $G_{(\mathfrak{p})}$ and called the *homogeneous localization* of G at $\mathfrak{p}$. Its maximal ideal is the set of all quotients $\frac{p}{s}$, where $p \in \mathfrak{p}, s \in S$ are homogeneous of the same degree. If G is an integral domain and $\mathfrak{p} = (0)$, then $(G_S)_0$ is a field.

Exercises

Let S be a multiplicatively closed subset of a ring R. The rings and modules of fractions with denominator set S can be described in a way different from that of the text, namely:

1. Let $\{X_s\}_{s \in S}$ be a family of indeterminates and $R' := R[\{X_s\}]/I$, where I is the ideal generated by all the polynomials $sX_s - 1$ with $s \in S$. Let $i : R \to R'$ be the composition of the canonical injection $R \to R[\{X_s\}]$ with the canonical epimorphism $R[\{X_s\}] \to R'$. Then (R', i) is the ring of fractions of R with denominator set S.

2. For any R-module M, $R_S \otimes_R M$ together with the canonical R-linear mapping $M \to R_S \otimes_R M$ ($m \mapsto 1 \otimes m$) is the module of fractions of M with denominator set S.

4. Properties of rings and modules of fractions

Let R be a ring, M an R-module, $S \subset R$ multiplicatively closed.

If $\rho : P \to R$ is a ring homomorphism, then we can consider M as a P-module with the following scalar multiplication: For $p \in P, m \in M$, we put $pm := \rho(p)m$. In particular, any R_S-module N is also an R-module via the canonical mapping $i : R \to R_S$. In the future we shall tacitly consider R_S-modules as also being R-modules in this way.

For $m \in M$ we call $\text{Ann}(m) := \{r \in R \mid rm = 0\}$ the *annihilator* of m. $\text{Ann}(M) := \{r \in R \mid rm = 0 \text{ for all } m \in M\}$ is called the *annihilator* of M. $\text{Ann}(m)$ and $\text{Ann}(M)$ are ideals of R.

For any ideal $I \subset \text{Ann}(M)$ we can consider M as an R/I-module: For $r + I \in R/I, m \in M$, let $(r + I)m := rm$. Because $IM = 0$, this is independent of the choice of representatives.

Rule 4.1. If $i : M \to M_S$ is the canonical mapping, then

$$\text{Ker}(i) = \{m \in M \mid \text{ there is an } s \in S \text{ with } sm = 0\}.$$

Proof. For $m \in M, i(m) = \frac{m}{1} = \frac{0}{1}$ if and only if there is an $s \in S$ with $sm = 0$ (definition of equality of fractions).

Definition 4.2. The *torsion submodule* $T(M)$ is the set of all $m \in M$ for which there is a non-zerodivisor $s \in R$ with $sm = 0$. M is called *torsion-free* if $T(M) = \langle 0 \rangle$, a *torsion module* if $T(M) = M$.

If S is the set of all non-zerodivisors of R, then $T(M) = \text{Ker}(i)$. M is torsion-free if and only if i is injective, and M is a torsion module if and only if $M_S = \langle 0 \rangle$.

For general $S, i : R \to R_S$ is injective if and only if S contains no zero divisor of R.

Rule 4.3. $M_S = \langle 0 \rangle$ if and only if for any $m \in M$ there is an $s \in S$ with $sm = 0$. $R_S = \{0\}$ if and only if $0 \in S$.

The first statement is clear from the definition of equality of fractions; for the second consider that 1 can be annihilated only by 0.

For $g \in R$ we have, for example, $R_g = \{0\}$ if and only if g is nilpotent.

Proposition 4.4. $M = \langle 0 \rangle$ if and only if $M_{\mathfrak{m}} = \langle 0 \rangle$ for all $\mathfrak{m} \in \mathrm{Max}(R)$.

Proof. If $M_{\mathfrak{m}} = \langle 0 \rangle$ for all $\mathfrak{m} \in \mathrm{Max}(R)$ and $m \in M$ is given, then by 4.3 $\mathrm{Ann}(m)$ is contained in no maximal ideal of R; therefore, $\mathrm{Ann}(m) = R$ and so $1 \in \mathrm{Ann}(m)$. But this means that $m = 0$.

We have thus gotten our first "local–global statement": A module vanishes if and only if it vanishes "locally" for all maximal ideals.

Definition 4.5. By the *support of* M we understand the set

$$\mathrm{Supp}(M) := \{ \mathfrak{p} \in \mathrm{Spec}(R) \mid M_{\mathfrak{p}} \neq \langle 0 \rangle \}.$$

Proposition 4.6. If M is finitely generated, then $\mathrm{Supp}(M) = \mathfrak{V}(\mathrm{Ann}(M)) = \{ \mathfrak{p} \in \mathrm{Spec}(R) \mid \mathfrak{p} \supset \mathrm{Ann}(M) \}$. In particular, $\mathrm{Supp}(M)$ is a closed subset of $\mathrm{Spec}(R)$.

Proof. Let $M = \langle m_1, \ldots, m_t \rangle$ and $\mathfrak{p} \notin \mathrm{Supp}(M)$, so $M_{\mathfrak{p}} = \langle 0 \rangle$. By 4.3 there are then elements $s_i \in R \setminus \mathfrak{p}$ with $s_i m_i = 0$ $(i = 1, \ldots, t)$. $s := \prod_{i=1}^{t} s_i$ is then in $\mathrm{Ann}(M)$ and not in $\mathfrak{p}$, therefore $\mathfrak{p} \notin \mathfrak{V}(\mathrm{Ann}(M))$.

Conversely, if $\mathfrak{p} \notin \mathfrak{V}(\mathrm{Ann}(M))$, then there is an $s \in \mathrm{Ann}(M), s \notin \mathfrak{p}$, and by 4.3 it follows that $M_{\mathfrak{p}} = \langle 0 \rangle$.

Now let N be another R-module and $l : M \to N$ a linear mapping, $M \to N_S$ the composition of l and the canonical mapping $i_N : N \to N_S$.

On the basis of the universal property 3.1 there is a unique R-linear mapping $l_S : M_S \to N_S$ such that the diagram

$$
\begin{array}{ccc}
M & \xrightarrow{l} & N \\
\downarrow & & \downarrow \\
M_S & \xrightarrow{l_S} & N_S
\end{array}
$$

commutes. Here we have $s l_S(\frac{m}{s}) = l_S(\frac{m}{1}) = \frac{l(m)}{1}$, so $l_S(\frac{m}{s}) = \frac{l(m)}{s}$ for all $\frac{m}{s} \in M_S$.

l_S is called the mapping of the module of fractions induced by l. It is easy to see that

$$\mathrm{Hom}_R(M, N) \to \mathrm{Hom}_{R_S}(M_S, N_S) \qquad (l \mapsto l_S)$$

is an R-linear mapping: $(r_1 l_1 + r_2 l_2)_S = r_1 (l_1)_S + r_2 (l_2)_S$ for $r_1, r_2 \in R, l_1, l_2 \in \mathrm{Hom}_R(M, N)$. Moreover, $(l' \circ l)_S = l'_S \circ l_S$ if $l' : N \to P$ is another R-linear mapping.

Rule 4.7. If l is injective (resp. surjective, bijective), then so is l_S.

If l is injective and $l_S(\frac{m}{s}) = \frac{l(m)}{s} = 0$ for some $\frac{m}{s} \in M_S$, then there is an $s' \in S$ with $s' l(m) = 0 = l(s'm)$. It follows that $\frac{m}{s} = 0$. That l_S is surjective (resp. bijective) if l is results at once.

If $U \subset M$ is a submodule, then by 4.7 we can consider U_S in a canonical way as a submodule of the R_S-module M_S, identifying U_S with the set of all fractions $\frac{u}{s} \in M_S$ with $u \in U, s \in S$. We shall always tacitly do this in the future. In particular, for an ideal I of R, I_S will be considered an ideal of R_S. In the case of a localization $R_{\mathfrak{p}}$ of R at some $\mathfrak{p} \in \mathrm{Spec}(R)$ we also write $IR_{\mathfrak{p}}$ instead of $I_{\mathfrak{p}}$.

We have the following, easily verifiable

Rules 4.8.

a) If $\{U_\lambda\}_{\lambda \in \Lambda}$ is a family of submodules of M, then

$$\left(\bigcap_{\lambda \in \Lambda} U_\lambda\right)_S = \bigcap_{\lambda \in \Lambda} (U_\lambda)_S \qquad \text{if } \Lambda \text{ is finite,}$$

and

$$\left(\sum_{\lambda \in \Lambda} U_\lambda\right)_S = \sum_{\lambda \in \Lambda} (U_\lambda)_S.$$

If M is the direct sum of the U_λ, then M_S is the direct sum of the $(U_\lambda)_S$.
b) $\mathrm{Ann}(M)_S = \mathrm{Ann}_{R_S}(M_S)$ if M is a finitely generated R-module.
c) For any ideal I of R,

$$(I^n)_S = (I_S)^n, \qquad (\mathrm{Rad}\, I)_S = \mathrm{Rad}(I_S).$$

If R is reduced, so is R_S.
d) If I is an ideal of R with $I \cap S \neq \emptyset$, then $I_S = R_S$.

To give an overview of the submodules of a module of fractions (resp. the ideals of a ring of fractions), it is convenient to introduce the following concept.

Definition 4.9. If $U \subset M$ is a submodule, the set $S(U)$ of all $m \in M$ for which there exists an $s \in S$ with $sm \in U$ is called the *S-component* of U.

$S(U)$ is a submodule of M containing U, and $S(S(U)) = S(U)$. Moreover, $S(\bigcap_{i=1}^n U_i) = \bigcap_{i=1}^n S(U_i)$ for submodules $U_i \subset M$ $(i = 1, \ldots, n)$.

Let $\mathfrak{A}(M_S)$ be the set of all submodules of the R_S-module M_S and $\mathfrak{A}_S(M)$ the set of all submodules $U \subset M$ with $S(U) = U$.

Proposition 4.10. The mapping

$$\alpha : \mathfrak{A}_S(M) \to \mathfrak{A}(M_S) \qquad (U \mapsto U_S)$$

is an inclusion-preserving bijection. Its inverse mapping assigns to any $U' \in \mathfrak{A}(M_S)$ the submodule $i^{-1}(U') \subset M$, where $i : M \to M_S$ is the canonical mapping.

Proof. For $U' \in \mathfrak{A}(M_S)$ let $U := i^{-1}(U')$. If $m \in S(U)$, so that $sm \in U$ for some $s \in S$, then $\frac{s}{1} \cdot i(m) \in U'$; therefore, $i(m) \in U'$ (since $\frac{s}{1}$ is a unit in R_S) and so $m \in U$. It follows that $S(U) = U$. Thus $U' \mapsto U$ defines a mapping $\beta : \mathfrak{A}(M_S) \to \mathfrak{A}_S(M)$.

It is clear that $U_S \subset U'$. If $\frac{m}{s} \in U'$, then $\frac{m}{1} \in U'$, so that $m \in U$ and $U' = U_S$. This shows that $\alpha \circ \beta = \mathrm{id}_{\mathfrak{A}(M_S)}$.

If $U \in \mathfrak{A}_S(M)$, then $i^{-1}(U_S) = U$, since $\frac{m}{1} = \frac{u}{s}$ ($m \in M$, $u \in U$, $s \in S$) implies $s'm \in U$ for some $s' \in S$; therefore $m \in U = S(U)$. Thus $\beta \circ \alpha = \mathrm{id}_{\mathfrak{A}_S(M)}$, q. e. d.

If $U \subset M$ is any submodule of M, then the considerations above show that $U_S = U_S^*$ for any submodule $U^* \subset M$ with $U \subset U^* \subset S(U)$. In particular, it follows from 4.10 that the ideals of R_S correspond bijectively to the ideals of R that coincide with their S-components.

Corollary 4.11. If M is a Noetherian R-module, then M_S is a Noetherian R_S-module. If R is a Noetherian ring, so is R_S.

The function rings on algebraic varieties considered in §2 are all Noetherian, since they are rings of fractions of affine algebras over fields, which are Noetherian by the Basis Theorem.

For $\mathfrak{p} \in \mathrm{Spec}(R)$ we have

$$S(\mathfrak{p}) = \{r \in R \mid \text{there is an } s \in S \text{ with } sr \in \mathfrak{p}\} = \begin{cases} \mathfrak{p}, & \text{if } \mathfrak{p} \cap S = \emptyset, \\ R, & \text{if } \mathfrak{p} \cap S \neq \emptyset. \end{cases}$$

More generally, if $I = \bigcap_{i=1}^n \mathfrak{p}_i$ with $\mathfrak{p}_i \in \mathrm{Spec}(R)$, then

$$S(I) = \bigcap_{\mathfrak{p}_i \cap S = \emptyset} \mathfrak{p}_i .$$

Proposition 4.12. Let $i : R \to R_S$ be the canonical mapping, Σ the set of all $\mathfrak{p} \in \mathrm{Spec}(R)$ with $\mathfrak{p} \cap S = \emptyset$. Then:

a) Every $\mathfrak{P} \in \mathrm{Spec}(R_S)$ is of the form $\mathfrak{P} = \mathfrak{p}_S$ with a uniquely determined $\mathfrak{p} \in \Sigma$.

b) $\mathrm{Spec}(i)$ defines a homeomorphism of $\mathrm{Spec}(R_S)$ onto Σ (endowed with the relative topology of the topology on $\mathrm{Spec}(R)$).

c) For all $\mathfrak{p} \in \Sigma$ we have $h(\mathfrak{p}_S) = h(\mathfrak{p})$, and for any ideal I of R with $I_S \neq R_S$ we have $h(I_S) \geq h(I)$.

d) $\dim R_S \leq \dim R$.

e) If R is a factorial ring and $0 \notin S$, then R_S is factorial too.

Proof.

a) follows from 4.10 and the above formula for $S(\mathfrak{p})$.

b) By a) $\mathrm{Spec}(i)$ defines a bijection of $\mathrm{Spec}(R_S)$ onto Σ: because $\mathrm{Spec}(i)$ is continuous (I.4.11) it suffices to show that $\mathrm{Spec}(i)$ is also a closed mapping of $\mathrm{Spec}(R_S)$ onto Σ. If J is an ideal of R_S, then the set of all $\mathfrak{P} \in \mathrm{Spec}(R_S)$ that contain J is mapped by $\mathrm{Spec}(i)$ onto the set of all $\mathfrak{p} \in \mathrm{Spec}(R)$ that contain $i^{-1}(J)$ with $\mathfrak{p} \cap S = \emptyset$, therefore onto a closed subset of Σ.

c) The formula $h(\mathfrak{p}_S) = h(\mathfrak{p})$ results from a); and the formula $h(I_S) \geq h(I)$ then follows, because the height of an ideal is defined as the infimum of the heights of the prime ideals containing it.

d) results, because the dimension of a ring $R \neq \{0\}$ is the supremum of the heights of the $\mathfrak{p} \in \mathrm{Spec}(R)$.

e) Since $0 \notin S$, i is injective. If π is a prime element of R and $(\pi) \cap S = \emptyset$, then $\frac{\pi}{1}$ is a prime element in R_S. If $(\pi) \cap S \neq \emptyset$, then $\frac{\pi}{1}$ is a unit in R_S. It follows that in R_S every element $\neq 0$ is either a unit or a product of prime elements.

Corollary 4.13.

a) For any $f \in R$, $\mathrm{Spec}(R_f) \to \mathrm{Spec}(R)$ defines a homeomorphism of $\mathrm{Spec}(R_f)$ onto $D(f) \subset \mathrm{Spec}(R)$.

b) For any $\mathfrak{p} \in \mathrm{Spec}(R)$, $\mathrm{Spec}(R_{\mathfrak{p}}) \to \mathrm{Spec}(R)$ defines a homeomorphism of $\mathrm{Spec}(R_{\mathfrak{p}})$ onto the set of all prime ideals contained in $\mathfrak{p}$ (the set of all "generalizations" of $\mathfrak{p}$). We have $h(\mathfrak{p}) = \dim R_{\mathfrak{p}}$.

For local rings on algebraic varieties the following statements result.

Proposition 4.14. Let W be a nonempty irreducible subset of an affine variety V. Then:

a) The elements of $\mathrm{Spec}(\mathcal{O}_{V,W})$ correspond bijectively to the irreducible subvarieties $V' \subset V$ with $W \subset V'$, the minimal prime ideals to the irreducible components V_i of V with $W \subset V_i$.

b) $\mathcal{O}_{V,W}$ is an integral domain if and only if W is contained in exactly one irreducible component of V.

c) $\dim \mathcal{O}_{V,W} = \mathrm{codim}_V(\overline{W})$, where $\overline{W}$ is the closure of W in V.

d) The ideals J of $\mathcal{O}_{V,W}$ with $J \neq \mathcal{O}_{V,W}$ and $\mathrm{Rad}\, J = J$ are in one-to-one correspondence with the subvarieties of V all of whose irreducible components contain W.

Proof. If $\mathfrak{p}_W$ is the prime ideal in $K[V]$ belonging to W, then (by 3.6) $\mathcal{O}_{V,W} \cong K[V]_{\mathfrak{p}_W}$. By 4.13b) the elements of $\mathrm{Spec}(K[V]_{\mathfrak{p}_W})$ correspond bijectively to the prime ideals of $K[V]$ contained in $\mathfrak{p}_W$, therefore to the irreducible subvarieties of V that contain W (I.3.11). Since $K[V]_{\mathfrak{p}_W}$ is reduced (4.8c)), $K[V]_{\mathfrak{p}_W}$ is an integral domain if and only if it has only one minimal prime ideal. Hence b) follows. c) follows from a) and the definitions of Krull dimension and codimension. The ideals $J \neq \mathcal{O}_{V,W}$ with $\mathrm{Rad}\, J = J$ are the intersections of their minimal prime divisors (I.4.5). They correspond bijectively to the finite intersections of prime ideals $\mathfrak{p} \in \mathrm{Spec}(K[V])$ with $\mathfrak{p} \subset \mathfrak{p}_W$. Hence d) follows.

Note that, in particular, Proposition 4.14 can be applied if $W = \{x\}$ is a point x of V. It is the first piece of evidence for our earlier stated thesis that the local ring $\mathcal{O}_{V,x}$ contains information about the behavior of V in the neighborhood of x.

A submodule $U \subset M$ is always contained in the kernel of the composite mapping

$$M \xrightarrow{i} M_S \xrightarrow{\epsilon} M_S/U_S,$$

if i and ϵ are the canonical mappings. By the universal properties of the residue module M/U and of the module of fractions $(M/U)_S$, an R_S-linear mapping is induced:

$$\rho : (M/U)_S \to M_S/U_S, \qquad \rho(\frac{m+U}{s}) = \frac{m}{s} + U_S.$$

Rule 4.15. (Permutability of forming the residue class module and the module of fractions) ρ is an isomorphism.

Proof. ρ is obviously surjective. We show $\mathrm{Ker}(\rho) = 0$. If $\rho(\frac{m+U}{s}) = 0$, then $\frac{m}{s} \in U_S$, so there are elements $u \in U, s' \in S$, with $\frac{m}{s} = \frac{u}{s'}$, and so there is an $s'' \in S$ with $s''(s'm - su) = 0$. It follows that

$$\frac{m+U}{s} = \frac{s''s'm+U}{s''ss'} = \frac{s''su+U}{s''s's} = 0.$$

Rule 4.16. Let I be an ideal in R and S' the image of S in R/I. The canonical mapping

$$\rho : (R/I)_{S'} \to R_S/I_S, \qquad \rho(\frac{r+I}{s+I}) = \frac{r}{s} + I_S$$

is a ring isomorphism.

This is proved like 4.15.

Rule 4.17. If $M \xrightarrow{\alpha} N \xrightarrow{\beta} P$ is an exact sequence of R-modules and linear mappings (i.e. $\mathrm{Im}(\alpha) = \mathrm{Ker}(\beta)$), then

$$M_S \xrightarrow{\alpha_S} N_S \xrightarrow{\beta_S} P_S$$

is an exact sequence of R_S-modules.

Proof. From $\beta \circ \alpha = 0$ it follows that $\beta_S \circ \alpha_S = (\beta \circ \alpha)_S = 0$, so $\mathrm{Im}\, \alpha_S \subset \mathrm{Ker}\, \beta_S$. If $\frac{n}{s} \in N_S$ with $\beta_S(\frac{n}{s}) = \frac{\beta(n)}{s} = 0$ is given, then there is an $s' \in S$ with $0 = s'\beta(n) = \beta(s'n)$, therefore $s'n \in \mathrm{Im}\, \alpha$. Since $\frac{n}{s} = \frac{s'n}{s's}$, it follows that $\mathrm{Ker}\, \beta_S = \mathrm{Im}\, \alpha_S$.

Examples 4.18.

a) Let I be an ideal of R, let $\mathfrak{p} \in \mathrm{Spec}(R)$ contain I, and let $\mathfrak{p}'$ be the image of $\mathfrak{p}$ in R/I. By 4.16 we have canonical isomorphism

$$(R/I)_{\mathfrak{p}'} \cong R_\mathfrak{p}/I_\mathfrak{p}.$$

In the case $I = \mathfrak{p}$ this yields an isomorphism

$$Q(R/\mathfrak{p}) \cong R_\mathfrak{p}/\mathfrak{p}R_\mathfrak{p}.$$

The residue field of the local ring $R_\mathfrak{p}$ by its maximal ideal $\mathfrak{p}R_\mathfrak{p}$ is therefore isomorphic to the field of fractions of $R/\mathfrak{p}$.

b) Let $\mathfrak{p}$ be a minimal prime ideal of a reduced ring R. By 4.8c) $R_\mathfrak{p}$ is also reduced, and $\mathfrak{p}R_\mathfrak{p}$ is the only minimal prime ideal of R (4.12). It follows that $\mathfrak{p}R_\mathfrak{p} = (0)$ by I.4.5 and so $R_\mathfrak{p} = R_\mathfrak{p}/\mathfrak{p}R_\mathfrak{p} \cong Q(R/\mathfrak{p})$. The local ring of a minimal prime ideal $\mathfrak{p}$ in a reduced ring is therefore always a field, isomorphic to the field of fractions of $R/\mathfrak{p}$.

Whereas hitherto we have, for the most part, fixed the denominator set S, we will now derive some rules in connection with changing the denominator set. Along with S let there be given another multiplicatively closed subset T of R. Let S' be the image of S in R_T and T' that of T in R_S.

By the universal property of rings of fractions we have a canonical ring homomorphism

$$i_S^T : R_T \to (R_S)_{T'} \qquad \left(\frac{r}{t} \mapsto \frac{\frac{r}{t}}{\frac{t}{1}} \right)$$

and a corresponding R_T-linear mapping

$$j_S^T : M_T \to (M_S)_{T'} \qquad \left(\frac{m}{t} \mapsto \frac{\frac{m}{1}}{\frac{t}{1}} \right),$$

where $(M_S)_{T'}$ is considered an R_T-module via i_S^T; that is, scalar multiplication is given by the formula

$$\frac{r}{t} \cdot \frac{\frac{m}{s}}{\frac{t'}{1}} = \frac{\frac{rm}{s}}{\frac{tt'}{1}}.$$

Rule 4.19. If S' consists only of units of R_T, then i_S^T and j_S^T are isomorphisms.

Proof. On the basis of the universal property we have a canonical ring homomorphism

$$R_S \to R_T \qquad \left(\frac{r}{s} \mapsto \frac{r}{1} \left(\frac{s}{1} \right)^{-1} \right)$$

and hence also a canonical ring homomorphism

$$\rho : (R_S)_{T'} \to R_T \qquad \left(\frac{\frac{r}{s}}{\frac{t}{1}} \mapsto \frac{r}{t} \left(\frac{s}{1} \right)^{-1} \right).$$

Obviously $\rho \circ i_S^T = \mathrm{id}_{R_T}$, and one immediately computes that $i_S^T \circ \rho = \mathrm{id}_{(R_S)_{T'}}$.

Corollary 4.20.

a) If $S \subset T$, we have canonical isomorphisms $R_T \cong (R_S)_{T'}$ and $M_T \cong (M_S)_{T'}$.

b) Moreover, if T' consists only of units of R_S, then we have canonical isomorphisms $R_T \cong R_S$ and $M_T \cong M_S$.

Examples 4.21.

a) For $\mathfrak{p}, \mathfrak{q} \in \mathrm{Spec}(R)$ with $\mathfrak{p} \subset \mathfrak{q}$ let $S := R \setminus \mathfrak{q}$ and $T := R \setminus \mathfrak{p}$. Then we have canonical isomorphisms

$$R_{\mathfrak{p}} \cong (R_{\mathfrak{q}})_{\mathfrak{p} R_{\mathfrak{q}}}, \qquad M_{\mathfrak{p}} \cong (M_{\mathfrak{q}})_{\mathfrak{p} R_{\mathfrak{q}}}. \qquad (1)$$

Indeed, by 4.19 we have $R_{\mathfrak{p}} \cong (R_{\mathfrak{q}})_{T'}$, where T' is the image of T in $R_{\mathfrak{q}}$. $\mathfrak{p} R_{\mathfrak{q}}$ is a prime ideal of $R_{\mathfrak{q}}$ and $T'' := R_{\mathfrak{q}} \setminus \mathfrak{p} R_{\mathfrak{q}}$ contains T'. By 4.20 the canonical mapping $(R_{\mathfrak{q}})_{T'} \to (R_{\mathfrak{q}})_{T''}$ is an isomorphism, since the images of the elements of T'' in $(R_{\mathfrak{q}})_{T'} \cong R_{\mathfrak{p}}$ are units.

Formulas (1) are often applied. For example, if $O_{V,W}$ is the local ring of an irreducible subvariety $W \neq \emptyset$ of a variety V and if $W' \subset W$ is another irreducible subvariety, $W' \neq \emptyset$, then we have a canonical ring isomorphism

$$O_{V,W} \cong (O_{V,W'})_{\mathfrak{p}},$$

where $\mathfrak{p}$ is the prime ideal in $O_{V,W'}$ corresponding to W.

b) If $R \neq \{0\}$ is an integral domain, it follows from (1) that if $\mathfrak{p} = (0)$ there is a canonical isomorphism

$$Q(R) \cong Q(R_{\mathfrak{q}}). \tag{2}$$

Here $R_{\mathfrak{q}}$ is identified with the set of all $\frac{r}{s} \in Q(R)$ with $r \in R$, $s \in R \setminus \mathfrak{q}$. Hence we can consider the local rings $R_{\mathfrak{q}}$ as subrings of $Q(R)$; naturally, they have the same field of fractions. The corresponding statement holds for all rings of fractions R_S with $0 \notin S$. (This generalizes the fact that $O_{V,W} \subset R(V)$ for nonempty irreducible varieties $W \subset V$ (§2).)

Rule 4.22. For $f, g \in R$ we have a canonical ring isomorphism

$$(R_f)_g \xrightarrow{\sim} R_{fg} \qquad \left(\frac{\frac{r}{f^{\nu}}}{\frac{g^{\mu}}{1}} \mapsto \frac{r f^{\mu} g^{\nu}}{(fg)^{\nu + \mu}} \right)$$

and a corresponding isomorphism of R_{fg}-modules given by the analogous formula. (Here $(R_f)_g$ denotes the ring of fractions of R_f by the set of powers of $\frac{g}{1}$; $(M_f)_g$ is defined correspondingly.)

Proof. Since the images of f and g are units in R_{fg}, by the universal property of rings of fractions there is a canonical homomorphism $(R_f)_g \to R_{fg}$ satisfying the given formula. Since the image of fg in $(R_f)_g$ is a unit, we also have a ring homomorphism

$$R_{fg} \to (R_f)_g \qquad \left(\frac{r}{(fg)^{\nu}} \mapsto \frac{\frac{r}{f^{\nu}}}{\frac{g^{\nu}}{1}} \right)$$

and this inverts the homomorphisms above.

The many canonical isomorphisms for the rings and modules of fractions that have hitherto been derived will often be used in the form that tacitly identifies isomorphic objects. One quickly gets accustomed to this procedure, which saves much labor of writing, by at first recalling the rules as they are applied.

For the conclusion of this section we shall derive two additional structure theorems for rings of fractions of reduced rings.

Let $R \neq \{0\}$ be a reduced ring with only finitely many minimal prime ideals $\mathfrak{p}_1, \ldots, \mathfrak{p}_t$ ($\mathfrak{p}_i \neq \mathfrak{p}_j$ for $i \neq j$), S the set of all non-zerodivisors of R. By I.4.10 $S = R \setminus \bigcup_{i=1}^{t} \mathfrak{p}_i$. By 4.12 the $(\mathfrak{p}_i)_S$ ($i = 1, \ldots, t$) are the only elements of the spectrum of $Q(R) = R_S$; they are both maximal and minimal, and $\bigcap_{i=1}^{t}(\mathfrak{p}_i)_S = (0)$. By the Chinese Remainder Theorem (II.1.7), we therefore have a canonical isomorphism

$$R_S \cong R_S/\mathfrak{p}_1 R_S \times \cdots \times R_S/\mathfrak{p}_t R_S.$$

Here $R_S/\mathfrak{p}_iR_S = (R/\mathfrak{p}_i)_{S_i}$ by 4.16, where S_i is the image of S in $R/\mathfrak{p}_i$. Since $R_S/\mathfrak{p}_iR_S$ is a field, $(R/\mathfrak{p}_i)_{S_i} \cong Q(R/\mathfrak{p}_i)$ by 4.20b). As a ring-theoretic analogue of 2.8 we get:

Proposition 4.23. If $R \neq \{0\}$ is a reduced ring with only finitely many prime ideals $\mathfrak{p}_1, \ldots, \mathfrak{p}_t$ ($\mathfrak{p}_i \neq \mathfrak{p}_j$ for $i \neq j$), then

$$Q(R) \cong Q(R/\mathfrak{p}_1) \times \cdots \times Q(R/\mathfrak{p}_t).$$

The analogue of 4.23 in the graded case requires somewhat more care. Let G be a positively graded ring, S the set of all homogeneous non-zerodivisors of G. Let S contain an element of positive degree. Further, let G be reduced and possess only finitely many minimal prime ideals $\mathfrak{p}_1, \ldots, \mathfrak{p}_t$ ($\mathfrak{p}_i \neq \mathfrak{p}_j$ for $i \neq j$). By I.5.11 they are homogeneous, and $\mathfrak{p}_i \in \text{Proj}(G)$ ($i = 1, \ldots, t$) since S contains an element of positive degree and $S \cap \mathfrak{p}_i = \emptyset$ (I.4.10). Let S_i be the set of all homogeneous elements $\neq 0$ in $G/\mathfrak{p}_i$, $\overline{S}_i$ the image of S in $G/\mathfrak{p}_i$. We endow G_S, $(G/\mathfrak{p}_i)_{S_i}$, and $(G/\mathfrak{p}_i)_{\overline{S}_i}$ with the canonical grading (given in §3, Example e)).

It will first be shown that the canonical homomorphism

$$\alpha : G_S \to G_S/(\mathfrak{p}_1)_S \times \cdots \times G_S/(\mathfrak{p}_t)_S$$

is an isomorphism of graded rings. Here a direct product of graded rings becomes a graded ring by calling an element of the product homogeneous of degree d if and only if all its components in the factors of the product are homogeneous of degree d. α maps homogeneous elements into homogeneous elements of the same degree; and it is injective since $\bigcap_{i=1}^t (\mathfrak{p}_i)_S = (0)$.

To show that α is also surjective, for each $i \in [1, t]$ choose a homogeneous element of positive degree $a_i \in \bigcap_{j \neq i} \mathfrak{p}_j$ with $a_i \notin \mathfrak{p}_i$ (1.6). We may assume that the a_i are all of the same degree. Then $s := \sum_{i=1}^t a_i \in S$, and in G_S we have an equation

$$\frac{u_1}{s} + \cdots + \frac{u_t}{s} = 1,$$

where $\deg(\frac{a_i}{s}) = 0$ and $\frac{a_i}{s} \equiv \delta_{ij} \bmod (\mathfrak{p}_j)_S (i, j = 1, \ldots, t)$. Now if we are given $(y_1, \ldots, y_t) \in G_S/(\mathfrak{p}_1)_S \times \cdots \times G_S/(\mathfrak{p}_t)_S$, then for each y_i we choose a representative $\frac{b_i}{s_i} \in G_S$ and put $y := \sum_{i=1}^t \frac{b_i}{s_i} \cdot \frac{a_i}{s}$. Then $\alpha(y) = (y_1, \ldots, y_t)$; that is, α is also surjective.

By 4.16 $G_S/(\mathfrak{p}_i)_S \cong (G/\mathfrak{p}_i)_{S_i}$. To show that the canonical homomorphism of graded rings $(G/\mathfrak{p}_i)_{\overline{S}_i} \to (G/\mathfrak{p}_i)_{S_i}$ is an isomorphism, by 4.20b) it suffices to show that the images of the elements of S_i in $(G/\mathfrak{p}_i)_{\overline{S}_i}$ are units.

For $s_i \in S_i$ we choose a homogeneous element $s' \in G$ with image s_i in $G/\mathfrak{p}_i$. If s' is contained in none of the $\mathfrak{p}_j$ ($j = 1, \ldots, t$), then $s' \in S$, so $s_i \in \overline{S}_i$ and we are done. Otherwise , let I be the intersection of the $\mathfrak{p}_j$ that do not contain s'. By 1.6 there is a homogeneous element $p \in I$ of positive degree that is contained in none of the other minimal prime ideals. If $\deg s' > 0$ we can choose p so that p and s'^ρ for suitable $\rho \in \mathbb{N}$ have the same degree. $s := s'^\rho + p$ is then contained in none of the $\mathfrak{p}_i$, that is $s \in S$. This proves $s_i^\rho \in \overline{S}_i$, therefore the image of s_i in $(G/\mathfrak{p}_i)_{\overline{S}_i}$ is a unit. If deg $s_i = 0$ we choose $s \in S$ of positive degree and multiply s_i by the image $\overline{s}$ of s in $\overline{S}_i$. Then the images of $\overline{s}$ and $s_i\overline{s}$ is $(G/\mathfrak{p}_i)_{\overline{S}_i}$ are units, therefore also the image of s_i in $(G/\mathfrak{p}_i)_{\overline{S}_i}$ is a unit, q. e. d.

As a result of this discussion we get:

Proposition 4.24. Under the hypotheses above we have an isomorphism of graded rings

$$G_S \cong (G/\mathfrak{p}_1)_{S_1} \times \cdots \times (G/\mathfrak{p}_t)_{S_t}$$

and a ring isomorphism

$$G_{(S)} \cong (G/\mathfrak{p}_1)_{(S_1)} \times \cdots \times (G/\mathfrak{p}_t)_{(S_t)},$$

where $G_{(S)}$ denotes the subring of elements of degree 0 of G_S ($(G/\mathfrak{p}_i)_{(S_i)}$ is defined likewise). The $(G/\mathfrak{p}_i)_{(S_i)}$ are fields ($i = 1, \ldots, t$).

Exercises

1. Let R be a ring. For any nonempty open set $U \subset \operatorname{Spec}(R)$ let $\widetilde{R}(U)$ be the set of elements $(r_\mathfrak{p}) \in \prod_{\mathfrak{p} \in U} R_\mathfrak{p}$ with the following property: For any $\mathfrak{p} \in U$ there exists a $g \in R$ with $\mathfrak{p} \in D(g) \subset U$ and an $f \in R$ such that $r_\mathfrak{q} = \frac{f}{g}$ (in $R_\mathfrak{q}$) for all $\mathfrak{q} \in D(g)$. Further, let $\widetilde{R}(\emptyset) := \{0\}$. $\widetilde{R}(U)$ is then a ring, and for another nonempty open set $U' \subset U$ the mapping $\rho_{U'}^U : \widetilde{R}(U) \to \widetilde{R}(U')$ induced by the canonical projection $\prod_{\mathfrak{p} \in U} R_\mathfrak{p} \to \prod_{\mathfrak{p} \in U'} R_\mathfrak{p}$ is a ring homomorphism. Put $\rho_\emptyset^U := 0$. The system $\{\widetilde{R}(U); \rho_{U'}^U\}$ is a sheaf $\widetilde{R}$ on $\operatorname{Spec}(R)$. ($\widetilde{R}$ is called the *structure sheaf* of $\operatorname{Spec}(R)$; the pair $(\operatorname{Spec}(R), \widetilde{R})$ is called the *affine scheme* of R. This is a natural generalization of the affine varieties endowed with their sheaves of regular functions.) For any open set $D(g)$ with $g \in R$ we have $\widetilde{R}(D(g)) = R_g$. For all $\mathfrak{p} \in \operatorname{Spec}(R)$, $R_\mathfrak{p}$ is the direct limit of the rings $\widetilde{R}(U)$ with $\mathfrak{p} \in U$ as in §2, Exercise 12. Many other properties of the sheaf of regular functions on an affine variety can also be generalized immediately.

2. Let $\alpha : R \to S$ be a ring homomorphism, $\varphi := \operatorname{Spec}(\alpha)$. For $\mathfrak{p} \in \operatorname{Spec}(R)$ let $S_\mathfrak{p}$ denote the ring of fractions of S with denominator set $\alpha(R \setminus \mathfrak{p})$.

    a) The elements of $\varphi^{-1}(\mathfrak{p})$ correspond bijectively with the elements of $\operatorname{Spec}(S_\mathfrak{p}/\mathfrak{p}S_\mathfrak{p})$.

    b) If S is finitely generated as an R-module, then the number of elements of $\varphi^{-1}(\mathfrak{p})$ is at most as large as the dimension of $S_\mathfrak{p}/\mathfrak{p}S_\mathfrak{p}$ as a vector space over $R_\mathfrak{p}/\mathfrak{p}R_\mathfrak{p}$. ($\operatorname{Spec}(S_\mathfrak{p}/\mathfrak{p}S_\mathfrak{p})$ is called the *fiber of φ over $\mathfrak{p}$*.)

3. Let R be a ring. For $f, g \in R$ with $D(g) \subset D(f)$ there is a canonical ring homomorphism $\rho_g^f : R_f \to R_g$, which is an isomorphism if and only if $D(g) = D(f)$.

4. Let R be a Noetherian ring, $S \subset R$ a multiplicatively closed subset. There is an $f \in S$ such that the canonical homomorphism $R_f \to R_S$ is injective.

5. Let A and B be affine algebras over a field K. For $\mathfrak{p} \in \mathrm{Spec}(A)$, $\mathfrak{q} \in \mathrm{Spec}(B)$, suppose there exists a K-algebra isomorphism $A_\mathfrak{p} \xrightarrow{\sim} B_\mathfrak{q}$. Then there are elements $f \in A \setminus \mathfrak{p}, g \in B \setminus \mathfrak{q}$, and a K-algebra isomorphism $A_f \xrightarrow{\sim} B_g$ such that the diagram

$$
\begin{array}{ccc}
A_f & \xrightarrow{\sim} & B_g \\
\downarrow & & \downarrow \\
A_\mathfrak{p} & \xrightarrow{\sim} & B_\mathfrak{q}
\end{array}
$$

commutes, where the vertical arrows denote the canonical homomorphisms. (The isomorphism of local rings comes from an isomorphism of the "function rings" in suitable neighborhoods of $\mathfrak{p}$ and $\mathfrak{q}$.)

6. Let K be any field, $S \subset K[X_1, \ldots, X_n]$ the set of all polynomials without zeros in $\mathbb{A}^n(K)$. For a K-variety $V \subset \mathbb{A}^n(K)$ let $\mathfrak{J}(V) \subset K[X_1, \ldots, X_n]_S$ be the ideal of all fractions $\frac{f}{s}$ ($f \in K[X_1, \ldots, X_n], s \in S$) with $f(x) = 0$ for all $x \in V$. Prove:

a) The assignment $V \mapsto \mathfrak{J}(V)$ provides a bijection of the set of all K-varieties in $\mathbb{A}^n(K)$ onto the set of all ideals of $K[X_1, \ldots, X_n]_S$ that can be written as the intersection of maximal ideals.

b) Under this bijection the set of nonempty irreducible K-varieties in $\mathbb{A}^n(K)$ is mapped onto the J-spectrum of $K[X_1, \ldots, X_n]_S$ (Recall Ch. I, §3, Exercise 7).

5. The fiber sum and fiber product of modules. Gluing modules

The following two constructions are special cases of the formation of inductive and projective limits of modules. Since we need only these special cases, for the sake of simplicity we restrict ourselves to them. Let R be a ring, $\alpha_i : N \to M_i$ ($i = 1, 2$) two R-module homomorphisms.

Definition 5.1. A *fiber sum* of M_1 and M_2 over N (with respect to α_1, α_2) is a triple (S, β_1, β_2), where S is an R-module, $\beta_i : M_i \to S$ is an R-linear mapping with $\beta_1 \circ \alpha_1 = \beta_2 \circ \alpha_2$, and the following universal property holds: If (T, γ_1, γ_2) is any triple like (S, β_1, β_2), then there is a unique R-linear mapping $l : S \to T$ with $\gamma_i = l \circ \beta_i$ ($i = 1, 2$).

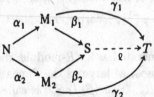

As usual the fiber sum, if it exists, is uniquely determined up to a canonical isomorphism.

If homomorphisms $\alpha_i : M_i \to N$ $(i = 1, 2)$ are given (in the opposite direction), then we define the *fiber product* (P, β_1, β_2) of M_1 and M_2 over N by the "dual" conditions: $\beta_i : P \to M_i$ $(i = 1, 2)$ are R-linear mappings with $\alpha_1 \circ \beta_1 = \alpha_2 \circ \beta_2$, and for every triple (T, γ_1, γ_2) like (P, β_1, β_2) there is a unique R-linear mapping $l : T \to P$ with $\gamma_i = \beta_i \circ l$ $(i = 1, 2)$.

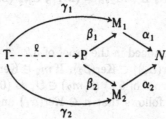

Proposition 5.2. *Fiber sums and products of modules always exist.*

Proof. For the fiber sum we consider in $M_1 \oplus M_2$ the submodule U of all elements $(\alpha_1(n), -\alpha_2(n))$ with $n \in N$, and we put $S := M_1 \oplus M_2/U$. Let β_i $(i = 1, 2)$ be the composition of the canonical injection $M_i \to M_1 \oplus M_2$ with the canonical epimorphism $M_1 \oplus M_2 \to S$. Then $\beta_1(\alpha_1(n)) = \beta_2(\alpha_2(n))$ by the construction of U. If (T, γ_1, γ_2) is given as in 5.1, then we have an R-linear mapping $h : M_1 \oplus M_2 \to T$, $(m_1, m_2) \mapsto \gamma_1(m_1) + \gamma_2(m_2)$; and we have $h(U) = 0$ since $\gamma_1 \circ \alpha_1 = \gamma_2 \circ \alpha_2$. Therefore, h induces a linear mapping $l : S \to T$ with $l \circ \beta_i = \gamma_i (i = 1, 2)$. Since $S = \beta_1(M_1) + \beta_2(M_2)$, there can be only one such mapping l.

To prove that the fiber product exists consider in $M_1 \oplus M_2$ the submodule P of all (m_1, m_2) with $\alpha_1(m_1) = \alpha_2(m_2)$. Let $\beta_i : P \to M_i$ be the restriction of the canonical projection $M_1 \oplus M_2 \to M_i$. One immediately verifies that (P, β_1, β_2) meets the requirements of the definition of the fiber product.

We write $M_1 \amalg_N M_2$ for the fiber sum and $M_1 \amalg_N M_2$ for the fiber product of M_1 and M_2 over N. We will first investigate the fiber sum:

$$
\begin{array}{ccc}
N & \xrightarrow{\alpha_1} & M_1 \\
\alpha_2 \downarrow & & \downarrow \beta_1 \\
M_2 & \xrightarrow{\beta_2} & M_1 \amalg_N M_2
\end{array}
\tag{1}
$$

Because the fiber sum is unique up to a canonical isomorphism, in proving the following rules we may assume that in (1) the module $M_1 \amalg_N M_2$ is the module S constructed in the proof of 5.2 and that $\beta_i(i=1, 2)$ is the mapping given there.

Rules 5.3.

a) $M_1 \amalg_N M_2 = \beta_1(M_1) + \beta_2(M_2)$.

b) $\mathrm{Ker}(\beta_2) = \alpha_2(\mathrm{Ker}(\alpha_1)), \mathrm{Ker}(\beta_1) = \alpha_1(\mathrm{Ker}(\alpha_2))$. In particular, if α_1 is injective, so is β_2.

c) β_1 induces an isomorphism $\mathrm{Coker}(\alpha_1) \cong \mathrm{Coker}(\beta_2)$; likewise, β_2 induces an isomorphism $\mathrm{Coker}(\alpha_2) \cong \mathrm{Coker}(\beta_1)$. In particular, α_1 is surjective if and only if β_2 is.

d) If $\alpha_1(\mathrm{resp}.\alpha_2)$ is an isomorphism, so is $\beta_2(\mathrm{resp}.\beta_1)$.

e) If $S \subset R$ is multiplicatively closed, then $\big((M_1 \amalg_N M_2)_S, (\beta_1)_S, (\beta_2)_S\big)$ is the fiber sum of $(M_1)_S$ and $(M_2)_S$ over N_S with respect to $(\alpha_1)_S, (\alpha_2)_S$:

$$(M_1 \amalg_N M_2)_S = (M_1)_S \amalg_{N_S} (M_2)_S.$$

Proof.

a) has already been mentioned in the proof of 5.2.

b) It is clear that $\alpha_2(\mathrm{Ker}(\alpha_1)) \subset \mathrm{Ker}(\beta_2)$. If $m_2 \in \mathrm{Ker}(\beta_2)$ is given, then (with the notation of the proof of 5.2) $(0, m_2) \in U$, so $(0, m_2) = (\alpha_1(n), -\alpha_2(n))$ for some $n \in N$. It follows that $n \in \mathrm{Ker}(\alpha_1)$ and that $m_2 = \alpha_2(-n) \in \alpha_2(\mathrm{Ker}(\alpha_1))$.

c) Let $C := \mathrm{Coker}(\alpha_1) = M_1/\alpha_1(N)$ and

$$C' := \mathrm{Coker}(\beta_2) = M_1 \amalg_N M_2/\beta_2(M_2).$$

Let $\overline{\beta}_1 : M_1 \to C'$ denote the composition of β_1 with the canonical epimorphism $M_1 \amalg_N M_2 \to C'$. Because of a), $\overline{\beta}_1$ is a surjection and $\overline{\beta}_1(\alpha_1(N)) = 0$. Hence β_1 induces a surjection $\beta'_1 : C \to C'$. For $m_1 \in M_1$ we have $\overline{\beta}_1(m_1) = 0$ if and only if $(m_1, 0) + U \in \beta_2(M_2)$, so $(m_1, 0) = (\alpha_1(n), -\alpha_2(n)) + (0, m_2)$ in $M_1 \oplus M_2$ with $n \in N, m_2 \in M_2$. Therefore $m_1 \in \alpha_1(N)$, and this shows that β'_1 is also injective.

d) is a consequence of b) and c).

e) holds because all the operations occurring in the constuction of the fiber sum commute with the formation of fractions.

We have analogous (dual) rules for the fiber product

$$\begin{array}{ccc} M_1 \amalg_N M_2 & \xrightarrow{\beta_1} & M_1 \\ \beta_2 \downarrow & & \downarrow \alpha_1 \\ M_2 & \xrightarrow{\alpha_2} & N \end{array} \qquad (2)$$

and they are verified just as easily, by identifying $(M_1 \amalg_N M_2, \beta_1, \beta_2)$ with the specially constructed model (P, β_1, β_2) in the proof of 5.2.

Rules 5.4.

a) $\mathrm{Ker}(\beta_1) \cap \mathrm{Ker}(\beta_2) = 0$.

b) β_1 induces an isomorphism $\mathrm{Ker}(\beta_2) \cong \mathrm{Ker}(\alpha_1)$, and β_2 induces an isomorphism $\mathrm{Ker}(\beta_1) \cong \mathrm{Ker}(\alpha_2)$. In particular, β_2 (resp. β_1) is injective if and only if α_1 (resp. α_2) is.

c) α_2 induces an injection $\mathrm{Coker}(\beta_2) \to \mathrm{Coker}(\alpha_1)$, and α_1 induces an injection $\mathrm{Coker}(\beta_1) \to \mathrm{Coker}(\alpha_2)$. In particular, if α_1 (resp. α_2) is surjective, so is β_2 (resp. β_1).

d) If α_1 (resp. α_2) is an isomorphism, so is β_2 (resp. β_1).

e) If $S \subset R$ is multiplicatively closed, then $\big((M_1 \amalg_N M_2)_S, (\beta_1)_S, (\beta_2)_S\big)$ is a fiber product of $(M_1)_S$ and $(M_2)_S$ over N_S (with respect to $(\alpha_1)_S, (\alpha_2)_S$):

$$(M_1 \amalg_N M_2)_S = (M_1)_S \amalg_{N_S} (M_2)_S.$$

Proof. Since $P = \{(m_1, m_2) \in M_1 \oplus M_2 \mid \alpha_1(m_1) = \alpha_2(m_2)\}$ and $\beta_i(m_1, m_2) = m_i (i = 1, 2)$, we get $\mathrm{Ker}(\beta_1) = \{(0, m_2) \mid \alpha_2(m_2) = 0\}$, $\mathrm{Ker}(\beta_2) = \{(m_1, 0) \mid \alpha_1(m_1) = 0\}$. From this a) and b) follow at once.

c) α_2 induces a linear mapping $\alpha_2' : M_2/\beta_2(P) \to N/\alpha_1(M_1)$ with $m_2 + \beta_2(P) \mapsto \alpha_2(m_2) + \alpha_1(M_1)$. If $m_2 + \beta_2(P) \in \mathrm{Ker}(\alpha_2')$, then $\alpha_2(m_2) = \alpha_1(m_1)$ for some $m_1 \in M_1$, and hence $m_2 = \beta_2(m_1, m_2)$ with $(m_1, m_2) \in P$; therefore, $m_2 + \beta_2(P) = \beta_2(P)$. This shows that α_2' is injective.

d) is consequence of b) and c).

e) The submodule of $(M_1)_S \oplus (M_2)_S$ consisting of all $\left(\frac{m_1}{s_1}, \frac{m_2}{s_2}\right)$ with $\frac{\alpha_1(m_1)}{s_1} = \frac{\alpha_2(m_2)}{s_2}$ is identified with the submodule of $(M_1 \oplus M_2)_S$ consisiting of all the $\frac{(m_1, m_2)}{s}$ with $\alpha_1(m_1) = \alpha_2(m_2)$, as is easily checked. Hence e) follows.

The fiber product can be used to glue modules given over open subsets of $\mathrm{Spec}(R)$. For $f, g \in R$ let M_1 be an R_f-module, M_2 an R_g-module, and suppose there is an isomorphism of R_{fg}-modules $\alpha : (M_1)_g \xrightarrow{\sim} (M_2)_f$. Here $(M_1)_g$ and $(M_2)_f$ are considered R_{fg}-modules via the canonical isomorphisms $(R_f)_g \cong R_{fg} \cong (R_g)_f$. Let $N := (M_2)_f$; let α_1 be the composition of the canonical homomorphism $M_1 \to (M_1)_g$ with α, and let $\alpha_2 : M_2 \to (M_2)_f$ be the canonical homomorphism.

Proposition 5.5. If $P := M_1 \amalg_N M_2$ is the fiber product formed with respect to α_1, α_2 then the canonical mappings $\beta_i : P \to M_i$ $(i = 1, 2)$ induce isomorphisms $P_f \xrightarrow{\sim} M_1, P_g \xrightarrow{\sim} M_2$ (of R_f- and R_g-modules respectively).

Proof. By 5.4e) $P_f \cong (M_1)_f \amalg_{N_f} (M_2)_f$. Since $(M_2)_f \to N_f$ is an isomorphism it follows that $P_f \cong (M_1)_f \cong M_1$ by 5.4d). One argues similarly for P_g.

We say that P arises through "gluing M_1 and M_2 over $D(fg)$ with respect to α." Using the universal property of the fiber product one easily verifies that one gets a module isomorphic to P if, instead of starting from α, one uses the inverse mapping α^{-1}.

This construction is important mainly in the case where $D(f) \cup D(g) = \mathrm{Spec}(R)$ (see Ch. IV, §1).

Exercises

1. Let I be an ideal of a ring R. In the situation of 5.1 let S' (resp. P') be the fiber sum (resp. the fiber product) of M_1/IM_1 and M_2/IM_2 over N/IN with respect to the canonically induced homomorphisms. Determine the relation of S' to S (resp. P' to P).

2. Let R be a ring, Λ a nonempty set, and $\Delta \subset \Lambda \times \Lambda$ a subset with the following properties; $(\lambda, \lambda) \in \Delta$ for all $\lambda \in \Lambda$, and $(\lambda, \lambda') \in \Delta, (\lambda', \lambda'') \in \Delta$ implies $(\lambda, \lambda'') \in \Delta$. For any $\lambda \in \Lambda$ let there be given an R-module M_λ and for all $(\lambda, \lambda') \in \Delta$ a linear mapping $\varphi_{\lambda'}^\lambda : M_\lambda \to M_{\lambda'}$ such that $\varphi_\lambda^\lambda = id_{M_\lambda}$ and $\varphi_{\lambda''}^{\lambda'} \circ \varphi_{\lambda'}^\lambda = \varphi_{\lambda''}^\lambda$ when $(\lambda, \lambda') \in \Delta, (\lambda', \lambda'') \in \Delta$. In analogy to the fiber sum and fiber product, define the direct and inverse limits of the "diagram" $\{M_\lambda, \varphi_{\lambda'}^\lambda\}$ and prove its existence.

References

The formation of fractions is a very old technique of algebra. In older works it was restricted to denominator sets consisting of non-zerodivisors alone. In this case a first systematic investigation of rings of fractions was made by Grell [27]. The simple but very important generalization to the case where the denominator set contains zero divisors was made relatively late by Chevalley [9] and by Uzkov [80]. The numerous rules for the formation of fractions and the related canonical mappings will often be applied tacitly. The inexperienced reader is advised to work through the somewhat terse proofs in detail in order to acquire the needed sureness in applying the rules.

The formation of fractions is the elementary basis for introducing the concept of a scheme, which generalizes the concept of a variety, and for the theory of sheaves on schemes. In the text and the exercises we have pointed out how these concepts are related to classical concepts of varieties. The reader can now begin to work his way into the modern language of algebraic geometry by either turning to the sources [M] or studying one of the briefer expositions of the theory ([N], [U]).

Chapter IV
The local–global principle in commutative algebra

Theorems on rings and modules are often proved by first establishing them in the case of local rings or modules over local rings, where this is often simpler, and then arguing from the "local" to the "global." This chapter contains some general rules for and examples of this technique. The results most important for later applications are the Forster–Swan theorem (2.14) and the solution of Serre's problem for projective modules (3.15) according to Quillen.

1. The passage from local to global

Let R be a ring, M an R-module.

Rule 1.1. If P, Q are submodules of M, then $P = Q$ if and only if $P_\mathfrak{m} = Q_\mathfrak{m}$ for all $\mathfrak{m} \in \mathrm{Max}(R)$.

We have $(P + Q/Q)_\mathfrak{m} \cong P_\mathfrak{m} + Q_\mathfrak{m}/Q_\mathfrak{m}$ and $(P + Q/P)_\mathfrak{m} \cong P_\mathfrak{m} + Q_\mathfrak{m}/P_\mathfrak{m}$ by III.4.15. If $P_\mathfrak{m} = Q_\mathfrak{m}$ for all $\mathfrak{m} \in \mathrm{Max}(R)$, then $P + Q/Q = P + Q/P = \langle O \rangle$ by III.4.4, so $P = Q$.

Corollary 1.2. A family $\{m_\lambda\}_{\lambda \in \Lambda}$ of elements of M is a system of generators of M if and only if the images of the m_λ in $M_\mathfrak{m}$ form a generating system of the $R_\mathfrak{m}$-module $M_\mathfrak{m}$ for all $\mathfrak{m} \in \mathrm{Max}(R)$.

Example 1.3. Suppose there are elements $f, g \in R$ with $D(f) \cup D(g) = \mathrm{Spec}(R)$. If M_f is a finitely generated R_f-module and M_g is a finitely generated R_g-module, then M is a finitely generated R-module. Namely, if $x_1, \ldots, x_m \in M$ are elements whose images generate the R_f-module M_f and if the images of $y_1, \ldots, y_n \in M$ generate the R_g-module M_g, let $N := \langle x_1, \ldots, x_m, y_1, \ldots, y_n \rangle$. Then $M_\mathfrak{m} = N_\mathfrak{m}$ for all $\mathfrak{m} \in \mathrm{Max}(R)$, since $\mathfrak{m} \in D(f)$ or $\mathfrak{m} \in D(g)$. In particular, by gluing two finitely generated modules (III.§5) we get another finitely generated module.

Corollary 1.4. For ideals I, J of R we have $I = J$ if and only if $I_\mathfrak{m} = J_\mathfrak{m}$ for all $\mathfrak{m} \in \mathrm{Max}(R)$ with $\mathfrak{m} \supset I \cap J$.

Indeed, for $\mathfrak{m} \notin \mathfrak{V}(I \cap J)$ we also have $I_\mathfrak{m} = J_\mathfrak{m} = R_\mathfrak{m}$ (III.4.8d)).

Corollary 1.5. A sequence of R-modules and linear mappings

$$M \xrightarrow{\alpha} N \xrightarrow{\beta} P$$

is exact if and only if it is locally exact; that is, for all $\mathfrak{m} \in \mathrm{Max}(R)$ the sequence

$$M_\mathfrak{m} \xrightarrow{\alpha_\mathfrak{m}} N_\mathfrak{m} \xrightarrow{\beta_\mathfrak{m}} P_\mathfrak{m}$$

is exact.

If $K := \mathrm{Ker}\beta, U := \mathrm{Im}\alpha$, then $K_{\mathfrak{m}} = \mathrm{Ker}\beta_{\mathfrak{m}}$ and $U_{\mathfrak{m}} = \mathrm{Im}\,\alpha_{\mathfrak{m}}$. By 1.1 $K = U$ if and only if $K_{\mathfrak{m}} = U_{\mathfrak{m}}$ for all $\mathfrak{m} \in \mathrm{Max}(R)$.

Corollary 1.6. A linear mapping $\alpha : M \to N$ of R-modules is injective (resp. surjective, bijective) if and only if for all $\mathfrak{m} \in \mathrm{Max}(R)$ the mapping $\alpha_{\mathfrak{m}}$ is injective (resp. surjective, bijective).

For injectivity apply 1.5 to the sequence $0 \to M \overset{\alpha}{\to} N$; for surjectivity consider $M \overset{\alpha}{\to} N \to 0$.

Example 1.7. Let M be an R-module. For elements $f, g \in R$ with $D(f) \cup D(g) = \mathrm{Spec}(R)$, suppose P arises from gluing M_f and M_g with respect to the canonical mappings into M_{fg} (III.§5). Then $M \cong P$: By the universal property of the fiber product P, there is an R-linear mapping $\alpha : M \to P$, and $\alpha_f : M_f \to P_f$, $\alpha_g : M_g \to P_g$ are isomorphisms. Since $D(f) \cup D(g) = \mathrm{Spec}(R)$, by III.4.20 we also find that $\alpha_{\mathfrak{m}} : M_{\mathfrak{m}} \to P_{\mathfrak{m}}$ for all $\mathfrak{m} \in \mathrm{Max}(R)$ is an isomorphism, therefore α is too.

Rule 1.1 and the corollaries resulting from it remain valid if instead of the $\mathfrak{m} \in \mathrm{Max}(R)$ we use arbitrary prime ideals $\mathfrak{p} \in \mathrm{Spec}(R)$, for by III.4.21 we get the localization at $\mathfrak{p}$ by first localizing at a maximal ideal $\mathfrak{m}$ containing $\mathfrak{p}$ and then at $\mathfrak{p}R_{\mathfrak{m}}$.

We now come to the local–global statements which hold only under restrictive hypotheses on the modules considered.

Definition 1.8. The *presentation* of M belonging to a system of generators $\{m_\lambda\}_{\lambda \in \Lambda}$ of M is the exact sequence

$$O \to K \to R^\Lambda \overset{\alpha}{\to} M \to O,$$

where α maps the canonical basis element e_λ of R^Λ to m_λ ($\lambda \in \Lambda$) and $K := \mathrm{Ker}(\alpha)$. K is called the *module of relations* of the generating system $\{m_\lambda\}_{\lambda \in \Lambda}$; an element of K is called a *relation*.

Definition 1.9. M is called *finitely presentable* if there is an $n \in \mathbb{N}$ and an exact sequence of R-modules

$$0 \to K \to R^n \to M \to 0, \tag{1}$$

where K is finitely generated.

For example, finitely generated modules over Noetherian rings are always finitely presentable, by the Basis Theorem (I.2.17).

If $\{v_1, \ldots, v_m\}$ is a system of generators of K and if we write the v_i as the rows of a matrix A, then M is uniquely determined by A (up to isomorphism): $M \cong R^n/\langle v_1, \ldots, v_m \rangle$. In other words, M is isomorphic to the cokernel of the linear mapping $R^m \overset{A}{\to} R^n$ defined by A. (It assigns the i-th row of A to the i-th canonical basis vector of R^m.)

The matrix A is called a *relation matrix* of M. If a matrix A' is gotten from A by elementary row and column operations (multiplication of a row or column by a unit of R, addition of a multiple of a row or column to another), then A' is also a relation matrix of M (possibly with respect to a different exact sequence (1)). Later (in connection with 1.16) we shall get a necessary and sufficient condition for two matrices (even of different formats) to present isomorphic R-modules.

First, we shall derive some additional general rules for the formation of fractions. For two R-modules M, N and a multiplicatively closed subset $S \subset R$ the R-linear mapping

$$\operatorname{Hom}_R(M, N) \to \operatorname{Hom}_{R_S}(M_S, N_S) \qquad (\alpha \mapsto \alpha_S)$$

induces an R_S-linear mapping

$$h : \operatorname{Hom}_R(M, N)_S \to \operatorname{Hom}_{R_S}(M_S, N_S) \qquad (\frac{\alpha}{s} \mapsto \mu_s^{-1} \circ \alpha_S).$$

Proposition 1.10.

a) If M is finitely generated, then h is injective.

b) If M is finitely presentable, then h is an isomorphism.

Proof. Let $\{m_1, \ldots, m_t\}$ be a system of generators of M and

$$0 \to K \to R^t \xrightarrow{\epsilon} M \to 0$$

the corresponding presentation.

a) For $\alpha \in \operatorname{Hom}_R(M, N), s \in S$, let $\frac{\alpha}{s} \in \operatorname{Ker} h$. Then $\frac{\alpha(m_k)}{s} = 0 \ (k = 1, \ldots, t)$ and there is an $s' \in S$ with $s'\alpha(m_k) = 0 \ (k = 1, \ldots, t)$. From $s'\alpha = 0$ follows $\frac{\alpha}{s} = 0$. Therefore h is injective.

b) Let M be finitely presentable. We may then assume that K is finitely generated. It is to be shown that h is surjective.

Let $i_M : M \to M_S$ and $i_N : N \to N_S$ be the canonical mappings. If $l \in \operatorname{Hom}_{R_S}(M_S, N_S)$ is given, then there is an $s \in S$ such that $n'_k := s \cdot l(\frac{m_k}{1}) \in i_N(N) \ (k = 1, \ldots, t)$. Let $n'_k = \frac{n_k}{1}$ with $n_k \in N$ and $\beta : R^t \to N$ be the linear mapping with $\beta(e_k) = n_k \ (k = 1, \ldots, t)$. We shall show that $(s'\beta)(K) = 0$ for suitable $s' \in S$. Then β induces a linear mapping $\alpha : M \to N$ with $\alpha(m_k) = s'n_k \ (k = 1, \ldots, t)$ and therefore $l = \mu_{s's}^{-1} \circ \alpha_S$.

According to the construction of β the diagram

$$
\begin{array}{ccc}
R^t & \xrightarrow{\beta} & N \\
\epsilon \downarrow & & \downarrow i_N \\
M & \xrightarrow{i_M} M_S \xrightarrow{sl} & N_S
\end{array}
$$

is commutative. Hence $i_N(\beta(K)) = 0$. Since K is finitely generated, there is in fact an $s' \in S$ with $s' \cdot \beta(K) = 0$, q. e. d.

Definition 1.11. An exact sequence of R-modules and linear mappings

$$0 \to M \xrightarrow{\alpha} N \xrightarrow{\beta} P \to 0 \tag{2}$$

splits if there is a linear mapping $\gamma : P \to N$ such that $\beta \circ \gamma = \mathrm{id}$.

For example, if P is a free R-module, then the sequence (2) splits. For each basis element of P choose an inverse image under β and define γ as the mapping that assigns to each basis element the inverse image chosen.

In the general case the condition of the definition is equivalent to each of the following two statements:

1. The linear mapping

$$\mathrm{Hom}_R(P,N) \to \mathrm{Hom}_R(P,P) \qquad (\gamma \mapsto \beta \circ \gamma)$$

is surjective.

Indeed, this is the case if and only if id_P lies in the image of the mapping.

2. $\alpha(M)$ is a direct summand of N.

If (2) splits, then it immediately follows that $N = \alpha(M) \oplus \gamma(P)$. Conversely, if $\alpha(M)$ is a direct summand of N, that is $N = \alpha(M) \oplus U$ with some submodule U of N, then β maps U isomorphically onto P, and we can choose γ to be the inverse of this mapping.

Rule 1.12. Let there be given an exact sequence (2), where P is finitely presentable. The sequence splits if and only if for any $\mathfrak{m} \in \mathrm{Max}(R)$ the sequence

$$0 \to M_{\mathfrak{m}} \xrightarrow{\alpha_{\mathfrak{m}}} N_{\mathfrak{m}} \xrightarrow{\beta_{\mathfrak{m}}} P_{\mathfrak{m}} \to 0 \tag{3}$$

splits.

Proof. $\mathrm{Hom}_R(P,N) \to \mathrm{Hom}_R(P,P)$ is surjective if and only if for all $\mathfrak{m} \in \mathrm{Max}(R)$ the induced mapping

$$\mathrm{Hom}_R(P,N)_{\mathfrak{m}} \to \mathrm{Hom}_R(P,P)_{\mathfrak{m}}$$

is surjective (1.6). By 1.10 this is identified with the mapping

$$\mathrm{Hom}_{R_{\mathfrak{m}}}(P_{\mathfrak{m}}, N_{\mathfrak{m}}) \to \mathrm{Hom}_{R_{\mathfrak{m}}}(P_{\mathfrak{m}}, P_{\mathfrak{m}})$$

since P is finitely presentable. The sequence (3) splits if and only if this mapping is surjective.

Corollary 1.13. Let M be a finitely presentable R-module, $U \subset M$ a finitely generated submodule. Then M/U is also finitely presentable. U is a direct summand of M if and only if $U_{\mathfrak{m}}$ is a direct summand of $M_{\mathfrak{m}}$ for all $\mathfrak{m} \in \mathrm{Max}(R)$.

The second assertion follows from 1.12, once it is shown that $P \cong M/U$ is finitely presentable. By hypothesis, there exists an exact sequence $0 \to K \xrightarrow{\alpha} R^n \xrightarrow{\beta} M \to 0$, where K is finitely generated. Let $\beta' : R^n \to P$ be the composition of β with the canonical epimorphism $M \to P$. Then $\mathrm{Ker}(\beta') = \beta^{-1}(U)$ is finitely generated, since U and K are finitely generated. Therefore P too is finitely presentable.

Rule 1.14. For $f, g \in R$ with $D(f) \cup D(g) = \mathrm{Spec}(R)$ let M_f be a finitely presentable R_f-module and let M_g be a finitely presentable R_g-module. Then M is a finitely presentable R-module.

Proof. By hypothesis there is an exact sequence of R_f-modules $0 \to \tilde{K} \to R_f^n \xrightarrow{\tilde{\alpha}} M_f \to 0$, where $\tilde{K}$ is finitely generated. We may assume that $\tilde{\alpha}$ is induced by an R-linear mapping $R^n \to M$ (1.10). In other words, there is an exact sequence of R-modules $0 \to K \to F \xrightarrow{\alpha} M$, where F is free of finite rank, such that the induced sequence

$$0 \to K_f \to F_f \xrightarrow{\alpha_f} M_f \to 0$$

is exact and K_f is finitely generated as an R_f-module. Let $0 \to K' \to F' \xrightarrow{\alpha'} M$ be the corresponding sequence constructed for M_g. We then get an exact sequence

$$0 \to U \to F \oplus F' \xrightarrow{(\alpha, -\alpha')} M \to 0,$$

where $(\alpha, -\alpha')$ is the mapping that acts on F like α and on F' like $-\alpha'$, and $U := \mathrm{Ker}(\alpha, -\alpha')$. $(\alpha, -\alpha')$ is surjective by 1.6. We shall show that U is finitely generated.

There is an R_f-linear mapping $\varphi : F_f' \to F_f$ with $\alpha_f \circ \varphi = \alpha_f'$. We construct a commutative diagram of R_f-modules with exact rows and columns

$$
\begin{array}{ccc}
0 & & 0 \\
\downarrow & & \downarrow \\
G & = & G \\
\downarrow & & \downarrow \\
0 \longrightarrow U_f \to F_f \oplus F_f' \to M_f \longrightarrow 0 \\
\quad \downarrow \beta' \qquad \downarrow \beta \qquad \| \\
0 \longrightarrow K_f \longrightarrow F_f \longrightarrow M_f \longrightarrow 0 \\
\downarrow \qquad \downarrow \\
0 \qquad 0
\end{array}
$$

where β is given by $\beta(x,y) = x - \varphi(y)$, β' is the mapping induced by β and $G := \mathrm{Ker}\,\beta = \mathrm{Ker}\,\beta'$. Since the middle column splits, because F_f is a free R_f-module, G is a homomorphic image of $F_f \oplus F_f'$, and so is finitely generated. Since K_f is also finitely generated, so is U_f. Likewise, U_g is a finitely generated R_g-module. From 1.3 it follows that U is finitely generated.

We now want to compare presentations of isomorphic modules with one another. Let there be given two exact sequences

$$0 \to K_j \xrightarrow{\beta_j} F_j \xrightarrow{\alpha_j} M_j \to 0 \qquad (j = 1, 2)$$

of R-modules, where the F_j are free.

Proposition 1.15.

a) If there is an isomorphism $i : M_1 \to M_2$, then there also exists an $\alpha \in \text{Aut}(F_1 \oplus F_2)$ such that the diagram

$$
\begin{array}{ccc}
F_1 \oplus F_2 & \xrightarrow{(\alpha_1,0)} & M_1 \\
{\scriptstyle \alpha} \downarrow & & \downarrow {\scriptstyle i} \\
F_1 \oplus F_2 & \xrightarrow[(0,\alpha_2)]{} & M_2
\end{array}
\tag{4}
$$

commutes. We have $\alpha(K_1 \oplus F_2) = F_1 \oplus K_2$ if we identify K_j with $\beta_j(K_j) \subset F_j$ $(j = 1, 2)$.

b) If there is an $\alpha \in \text{Aut}(F_1 \oplus F_2)$ with $\alpha(K_1 \oplus F_2) = F_1 \oplus K_2$, then there is also an isomorphism $i : M_1 \to M_2$ for which (4) is commutative.

Proof.

a) Since the F_j $(j = 1, 2)$ are free, for an isomorphism $i : M_1 \to M_2$ we can find linear mappings $\gamma_1 : F_1 \to F_2$ and $\gamma_2 : F_2 \to F_1$ with $i \circ \alpha_1 = \alpha_2 \circ \gamma_1$ and $\alpha_2 = i \circ \alpha_1 \circ \gamma_2$. For $(x, y) \in F_1 \oplus F_2$ let $\alpha'(x, y) := (x, y - \gamma_1(x))$ and $\alpha''(x, y) := (x - \gamma_2(y), y)$. Then obviously $\alpha', \alpha'' \in \text{Aut}(F_1 \oplus F_2)$. We shall show that $\alpha := \alpha'^{-1} \circ \alpha''$ is the mapping sought. From

$$
(i\alpha_1, \alpha_2)(\alpha''(x,y)) = i\alpha_1(x) - i\alpha_1\gamma_2(y) + \alpha_2(y) = (i \circ (\alpha_1, 0))(x, y)
$$

and

$$
(i\alpha_1, \alpha_2)(\alpha'(x,y)) = i\alpha_1(x) + \alpha_2(y) - \alpha_2\gamma_1(x) = (0, \alpha_2)(x, y),
$$

it follows that in fact $(0, \alpha_2) \circ \alpha = i \circ (\alpha_1, 0)$ since $\alpha'' = \alpha' \circ \alpha$.

Since $K_1 \oplus F_2 = \text{Ker}(\alpha_1, 0)$ and $F_1 \oplus K_2 = \text{Ker}(0, \alpha_2)$, it also immediately follows that $\alpha(K_1 \oplus F_2) = F_1 \oplus K_2$.

b) If there exists an $\alpha \in \text{Aut}(F_1 \oplus F_2)$ with $\alpha(K_1 \oplus F_2) \cong F_1 \oplus K_2$, then we get the existence of an isomorphism $i : M_1 \to M_2$ that makes (4) commute from the fact that M_1 is the cokernel of the mapping $\beta_1 \oplus \text{id}_{F_2} : K_1 \oplus F_2 \to F_1 \oplus F_2$ and M_2 is the cokernel of the mapping $\text{id}_{F_1} \oplus \beta_2 : F_1 \oplus K_2 \to F_1 \oplus F_2$.

Corollary 1.16. Let there be given exact sequences

$$
F_j' \xrightarrow{\beta_j} F_j \xrightarrow{\alpha_j} M_j \to 0 \qquad (j = 1, 2)
\tag{5}
$$

with free R-modules F_j, F_j'. Then $M_1 \cong M_2$ if and only if there is an $\alpha \in \text{Aut}(F_1 \oplus F_2)$ and $\beta \in \text{Aut}(F_1' \oplus F_2 \oplus F_1 \oplus F_2')$ such that the diagram

$$
\begin{array}{ccc}
F_1' \oplus F_2 \oplus F_1 \oplus F_2' & \xrightarrow{(\beta_1 \oplus \text{id}_{F_2}, 0)} & F_1 \oplus F_2 \\
{\scriptstyle \beta} \downarrow & & \downarrow {\scriptstyle \alpha} \\
F_1' \oplus F_2 \oplus F_1 \oplus F_2' & \xrightarrow[(0, \text{id}_{F_1} \oplus \beta_2)]{} & F_1 \oplus F_2
\end{array}
\tag{6}
$$

commutes. (Here $(\beta_1 \oplus \mathrm{id}_{F_2}, 0)$ is the mapping that coincides with β_1 on F_1' and with id_{F_2} on F_2, and with 0 on F_1 and F_2'. $(0, \mathrm{id}_{F_1} \oplus \beta_2)$ is defined similarly.)

Proof. If such a diagram is given, it follows that $M_1 \cong M_2$, because these modules are isomorphic to the cokernels of the mappings in the rows of the diagram.

Now let $M_1 \cong M_2$ and $K_j := \mathrm{Ker}(\alpha_j) = \mathrm{Im}(\beta_j)$ $(j = 1, 2)$. By 1.15a) there exists an $\alpha \in \mathrm{Aut}(F_1 \oplus F_2)$ and an isomorphism α' such that the diagram

$$\begin{array}{ccccc} F_1' \oplus F_2 & \xrightarrow{\beta_1 \oplus \mathrm{id}} & K_1 \oplus F_2 & \longrightarrow & F_1 \oplus F_2 \\ & \alpha' \downarrow & & & \downarrow \alpha \\ F_1 \oplus F_2' & \xrightarrow{\mathrm{id} \oplus \beta_2} & F_1 \oplus K_2 & \longrightarrow & F_1 \oplus F_2 \end{array}$$

commutes. Again applying 1.15a) to the isomorphism α', we get the diagram (6) sought.

In the following we denote by $M(r \times s; R)$ the R-module of all $r \times s$-matrices with coefficients in R and by $Gl(r, R)$ the group of invertible $r \times r$-matrices with coefficients in R. These are the $r \times r$-matrices whose determinant is a unit of R. Every ring homomorphism $R \to R'$ induces in a natural way a mapping $M(r \times s; R) \to M(r \times s; R')$ and a group homomorphism $Gl(r, R) \to Gl(r, R')$. We call two matrices $A_1, A_2 \in M(r \times s; R)$ equivalent if there is an $A \in Gl(r, R)$ and a $B \in Gl(s, R)$ such that

$$A_1 = A \cdot A_2 \cdot B^{-1},$$

in symbols $A_1 \sim A_2$.

Let there be given two exact sequences (5), where $F_j = R^{n_j}$, $F_j' = R^{n_j'}$ with natural numbers n_j, n_j' $(j = 1, 2)$.

With respect to the canonical basis β_1 is then given by an $n_1' \times n_1$-matrix B_1 and β_2 by an $n_2' \times n_2$-matrix B_2. The matrices that belong to the rows of (6) are $r \times s$-matrices of the form

$$\left(\begin{array}{c|c} B_1 & 0 \\ \hline 0 & E_{n_2} \\ \hline \multicolumn{2}{c}{0} \end{array} \right) \quad \text{and} \quad \left(\begin{array}{c|c} \multicolumn{2}{c}{0} \\ \hline E_{n_1} & 0 \\ \hline 0 & B_2 \end{array} \right) \quad (6')$$

with $r := n_1 + n_2 + n_1' + n_2'$, $s := n_1 + n_2$, where E_{n_i} is the n_i-rowed unit matrix. 1.16 says that the modules M_1 and M_2 are isomorphic if and only if these two matrices are equivalent. In other words, B_1 and B_2 present (up to isomorphism) the same R-module if and only if the corresponding matrices in (6') are equivalent.

We now consider matrices with coefficients in the polynomial ring $R[X]$ in one indeterminate X over R. If $A \in M(r \times s; R[X])$ is given and y is an element of an R-algebra T, let $A(y)$ denote the image of A in $M(r \times s; T)$ under the substitution homomorphism $X \mapsto y$. In particular, let $A(0)$ be the matrix in $M(r \times s; R)$ that arises from A when X is replaced by 0 in all the coefficients of A. We call $A_1, A_2 \in M(r \times s; R[X])$ *locally equivalent for some* $\mathfrak{m} \in \mathrm{Max}(R)$ if the images of A_1, A_2 in $M(r \times s; R_{\mathfrak{m}}[X])$ are equivalent.

The next theorem gives a local–global principle for the equivalence of matrices over $R[X]$. In the course of the proof we need the

Lemma 1.17. Let there be given matrices

$$A_1 \in M(r \times s; R[X]), \qquad A_2 \in M(s \times t; R[X]), \qquad A_3 \in M(r \times t; R[X]),$$

and a multiplicatively closed subset $S \subset R$. Let $\overline{A}_i$ $(i = 1, 2, 3)$ be the matrix corresponding to A_i under the canonical homomorphism $R[X] \to R_S[X]$, and suppose $\overline{A}_1 \cdot \overline{A}_2 = \overline{A}_3$ and $A_1(0) \cdot A_2(0) = A_3(0)$. Then there is an $s \in S$ such that $A_1(sX) \cdot A_2(sX) = A_3(sX)$.

Proof. In the matrix $A_1 A_2 - A_3$ all the coefficients are divisible by X since $A_1(0) \cdot A_2(0) = A_3(0)$. In addition, under the canonical homomorphism

$$M(r \times t; R[X]) \to M(r \times t; R_S[X])$$

it is mapped to the zero matrix. Hence, by III.4.1, there is an $s \in S$ that annihilates all the coefficients of the matrix. But then

$$A_1(sX) \cdot A_2(sX) - A_3(sX) = 0.$$

Theorem 1.18. (Vaserstein) $A \in M(r \times s; R[X])$ is equivalent to $A(0)$ if and only if A is locally equivalent to $A(0)$ for all $\mathfrak{m} \in \mathrm{Max}(R)$.

Proof. Let A and $A(0)$ be locally equivalent for all $\mathfrak{m} \in \mathrm{Max}(R)$. Let I denote the set of all $a \in R$ with the following property: For all $f, g \in R[X]$ with $f - g \in aR[X]$, $A(f)$ and $A(g)$ are (globally) equivalent.

I is an ideal of R, for if $a_1, a_2 \in I$ are given and if $f - g = (r_1 a_1 + r_2 a_2)\varphi$ with $f, g, \varphi \in R[X], r_1, r_2 \in R$, then $(f - r_1 a_1 \varphi) - g \in a_2 R[X]$; therefore $A(g) \sim A(f - r_1 a_1 \varphi) \sim A(f)$. We shall show that $1 \in I$. Then $A(f) \sim A(g)$ for all $f, g \in R[X]$, in particular for $f = X, g = 0$, and this proves the theorem.

For $\mathfrak{m} \in \mathrm{Max}(R)$ there are matrices $C \in Gl(r, R_\mathfrak{m}[X])$ and $D \in Gl(s, R_\mathfrak{m}[X])$ such that

$$A(X) = C \cdot A(0) \cdot D.$$

(Here as in the sequel $A(X)$ and $A(0)$ also denote the images of these matrices in $M(r \times s; R_\mathfrak{m}[X])$.)

With another indeterminate Y we get

$$A(X+Y) = C(X+Y)A(0)D(X+Y) = C(X+Y)C(X)^{-1}A(X)D(X)^{-1}D(X+Y).$$

We put $C^* := C(X + Y)C(X)^{-1}, D^* \cong D(X)^{-1} \cdot D(X + Y)$. C^* is of the form $C^* = C_0(X) + C_1(X)Y + \cdots + C_m(X)Y^m$ with $C_i(X) \in M(r \times r; R_\mathfrak{m}[X])$ $(i = 1, \ldots, m)$, where $C_0(X)$ is the unit matrix. There are similar formulas for D^*, C^{*-1}, and D^{*-1}. Then there is $a' \in R \setminus \mathfrak{m}$ such that $C(X + a'Y)C(X)^{-1}$ is the image of a matrix in $M(r \times r; R[X, Y])$ (setting aside all the denominators in the coefficients of the matrix). a' can be chosen so that after the substitution

$Y \mapsto a'Y$ also D^*, C^{*-1}, and D^{*-1} become images of matrices with coefficients in $R[X, Y]$. By 1.17 we can arrange that

$$C(X + a'Y)C(X)^{-1} \quad \text{and} \quad D(X)^{-1}D(X + a'Y)$$

are images of *invertible* matrices $\Gamma(X, Y)$ and $\Delta(X, Y)$ respectively with coefficients in $R[X, Y]$, where $\Gamma(X, 0)$ and $\Delta(X, 0)$ is the unit matrix.

Over $R_{\mathfrak{m}}[X, Y]$ we have the equation

$$A(X + a'Y) = C(X + a'Y)C(X)^{-1}A(X)D(X)^{-1}D(X + a'Y)$$

and over $R[X]$ the equation

$$A(X) = \Gamma(X, 0)A(X)\Delta(X, 0).$$

By 1.17 there exists an $a'' \in R \setminus \mathfrak{m}$ such that with $a := a'a''$

$$A(X + aY) = \Gamma(X, a''Y)A(X)\Delta(X, a''Y)$$

is a matrix equation holding over $R[X, Y]$.

Now if $f, g, \varphi \in R[X]$ are given with $f - g = a \cdot \varphi$, then we get

$$A(f) = A(g + a\varphi) = \Gamma(g, a'' \cdot \varphi)A(g)\Delta(g, a'' \cdot \varphi),$$

where $\Gamma(g, a'' \cdot \varphi)$ and $\Delta(g, a'' \cdot \varphi)$ are invertible; then $A(f) \sim A(g)$ and so $a \in I$.

This has shown that for any $\mathfrak{m} \in \mathrm{Max}(R)$ there is an $a \in I$ with $a \notin \mathfrak{m}$. Therefore $I = R$, q. e. d.

For an R-module N we call the $R[X]$-module $N[X] := R[X] \otimes_R N$ the *extension module* of N to $R[X]$. $N[X]$ can be identified with the set of all "polynomials" $n_0 + Xn_1 + \cdots + X^d n_d$ with coefficients $n_i \in N$. Here, two such polynomials are added termwise and a polynomial in $N[X]$ is multiplied by one in $R[X]$ in the usual way, which is possible since only elements of R and N need be multiplied.

Definition 1.19. An $R[X]$-module M is called *extended* (from R) if there is an R-module N such that $M \cong N[X]$ as $R[X]$-modules. M is called *locally extended* for an $\mathfrak{m} \in \mathrm{Max}(R)$ if the $R_{\mathfrak{m}}[X]$-module $M_{\mathfrak{m}}$ is extended from $R_{\mathfrak{m}}$. (Here $M_{\mathfrak{m}}$ denotes the module of fractions of M with denominator set $R \setminus \mathfrak{m}$.)

If $M = N[X]$, then necessarily $N \cong M/XM$ as R-modules. A local–global principle also holds for the concept of an extended module.

Theorem 1.20. (Quillen [66]) A finitely presentable $R[X]$-module M is extended if and only if it is locally extended for any $\mathfrak{m} \in \mathrm{Max}(R)$.

Proof. By hypothesis there is an exact sequence of $R[X]$-modules

$$R[X]^m \xrightarrow{\beta_1} R[X]^n \xrightarrow{\alpha_1} M \to 0. \tag{7}$$

Modulo X this goes over into an exact sequence of R-modules

$$R^m \xrightarrow{\bar{\beta}_1} R^n \xrightarrow{\bar{\alpha}_1} M/XM \to 0. \tag{7'}$$

If $B \in M(m \times n, R[X])$ is the matrix of β_1 with respect to the canonical basis, then $B(0)$ is the matrix of $\bar{\beta}_1$.

From $(7')$ we get by extension an exact sequence of $R[X]$-modules (with $N := M/XM$):

$$R^m[X] \xrightarrow{\bar{\beta}_1[X]} R^n[X] \xrightarrow{\bar{\alpha}_1[X]} N[X] \to 0. \tag{8'}$$

(Here $\bar{\beta}_1[X]$, for example, maps the polynomials in $R^m[X]$ by applying $\bar{\beta}_1$ to the coefficients in R^m, while X is mapped to X.) $(8')$ can therefore be identified with an exact sequence of $R[X]$-modules

$$R[X]^m \xrightarrow{\beta_2} R[X]^n \xrightarrow{\alpha_3} N[X] \to 0, \tag{8}$$

where β_2 is described by the matrix $B(0)$.

By 1.16 we have $M \cong N[X]$ if and only if the $(2n) \times 2(n+m)$-matrices

$$A := \begin{pmatrix} \begin{array}{c|c} B & 0 \\ \hline 0 & E_n \end{array} \\ \hline 0 \end{pmatrix} \quad \text{and} \quad \begin{pmatrix} & 0 \\ \hline E_n & 0 \\ \hline 0 & B(0) \end{pmatrix}$$

are equivalent. Here the second matrix is obviously equivalent to $A(0)$ (row- and column-permutations). Therefore, by 1.18 $M \cong N[X]$ if and only if A and $A(0)$ are locally equivalent for all $\mathfrak{m} \in \text{Max}(R)$. Since the exact sequences (7) and (8) are consistent with localization at $\mathfrak{m}$ (III.4.17) and $R[X]_{\mathfrak{m}} \cong R_{\mathfrak{m}}[X]$, it follows that M is extended if and only if it is locally extended for all $\mathfrak{m} \in \text{Max}(R)$, q. e. d.

Theorem 1.20 plays a key role in Quillen's solution of Serre's problem on projective modules, which will be described in §3.

Exercises

1. A module M over an integral domain R is torsion-free if and only if $M_{\mathfrak{m}}$ is a torsion-free $R_{\mathfrak{m}}$-module for all $\mathfrak{m} \in \text{Max}(R)$.

2. Let M be a torsion-free module over an integral domain $R \neq \{0\}$, and let $S := R \setminus \{0\}$. For any $\mathfrak{m} \in \text{Max}(R)$ the canonical mapping $M_{\mathfrak{m}} \to M_S$ is injective. If $M_{\mathfrak{m}}$ is considered a subset of M_S, then $M = \bigcap_{\mathfrak{m} \in \text{Max}(R)} M_{\mathfrak{m}}$; in particular, $R = \bigcap_{\mathfrak{m} \in \text{Max}(R)} R_{\mathfrak{m}}$.

3. Let R be a ring, M a finitely generated R-module. For some $\mathfrak{p} \in \text{Spec}(R)$ suppose the $R_{\mathfrak{p}}$-module $M_{\mathfrak{p}}$ is generated by the images of the elements $m_1, \ldots, m_r \in M$ in $M_{\mathfrak{p}}$. Then there is an $f \in R \setminus \mathfrak{p}$ such that the R_f-module M_f is generated by the images of these elements in M_f.

4. For a module M over a ring R let $M^* := \operatorname{Hom}_R(M, R)$ denote the dual module and let $\alpha : M \to M^{**}$ be the canonical mapping into the bidual (it assigns to any $m \in M$ the linear form $M^* \to R$ that maps $\mathfrak{l} \in M^*$ to $\mathfrak{l}(m)$). M is called *reflexive* if α is an isomorphism.

   a) A finitely generated module M over a Noetherian integral domain R is reflexive if and only if it is locally reflexive for all $\mathfrak{m} \in \operatorname{Max}(R)$.

   b) For a finitely generated module M over a Noetherian integral domain, M^* is always reflexive.

5. Let S/R be a ring extension, $N \subset R$ a multiplicatively closed subset.

   a) If S is integral over R, then S_N is integral over R_N.

   b) If R is integrally closed in S, then R_N is integrally closed in S_N.

6. Let the hypotheses be as in Exercise 5. $f_{S/R} := \{r \in R \mid rS \subset R\}$ is called the *conductor* of S along R.

   a) $f_{S/R}$ is the largest S-ideal contained in R.

   b) $S = R$ if and only if $f_{S/R} = (1)$.

   c) If S is finitely generated as an R-module, then

$$f_{S_N/R_N} = (f_{S/R})_N.$$

7. Let $R \neq \{0\}$ be an integral domain with field of fractions K, $\overline{R}$ the integral closure of R in K. Suppose $\overline{R}$ is finitely generated as an R-module.

   a) For $\mathfrak{p} \in \operatorname{Spec}(R)$, $R_\mathfrak{p}$ is integrally closed in K if and only if $\mathfrak{p} \not\supset f_{\overline{R}/R}$. (The set of these $\mathfrak{p}$ is therefore open in $\operatorname{Spec}(R)$.)

   b) R is integrally closed in K if and only if $R_\mathfrak{m}$ is integrally closed in K for all $\mathfrak{m} \in \operatorname{Max}(R)$.

8. Let M be a finitely generated module over a ring R and $0 \to K \to R^n \to M \to 0$ the presentation belonging to a system of generators $\{m_1, \ldots, m_n\}$ of M. Choose a system of generators v_λ ($\lambda \in \Lambda$) of K and for $i = 0, \ldots, n-1$ denote by $F_i(M)$ the ideal of R generated by all the $(n-i)$-rowed subdeterminants of the matrix with rows v_λ. Further, let $F_i(M) = R$ for $i \geq n$.

   a) $F_i(M)$ does not depend on the special choice of the generating system of K.

   b) $F_i(M)$ does not depend on the choice of the system of generators $\{m_1, \ldots, m_n\}$ of M. (Hint: Compare the ideals for $\{m_1, \ldots, m_n\}$ and $\{m_1, \ldots, m_n, m\}$, where m is any element of M).

   c) $F_0(M) \subset F_1(M) \subset \cdots \subset F_n(M) = R$.

(The ideals $F_i(M)$ are called the *Fitting ideals* or *Fitting invariants* of the module M.)

2. The generation of modules and ideals

Let $R \neq \{0\}$ be a ring, M a finitely generated R-module. By $\mu(M)$ we denote the number of elements in a shortest system of generators of M. Such a system will also be called a *minimal generating system*. A generating system is called unshortenable if no proper subset of it is a generating system.

Lemma 2.1. In a free R-module M the minimal generating systems are just the bases of M.

Proof. Let $M \cong R^n$ and let $\{b_1, \ldots, b_m\}$ be a minimal generating system of R^n. Then $m \leq n$ and we have a surjective linear mapping $l : R^n \to R^n$ with $l(e_i) = b_i \ (i = 1, \ldots, m), l(e_j) = 0 \ (j = m+1, \ldots, n)$, where $\{e_1, \ldots, e_n\}$ is the canonical basis of R^n. Then there is also a linear mapping $l' : R^n \to R^n$ with $l \circ l' = \mathrm{id}$. If A is the matrix assigned to l with respect to the canonical basis and A' is the matrix assigned to l', then $A' \cdot A = \mathrm{id}$; therefore, $\det(A)$ is a unit in R. Hence it follows that $m = n$ and that the elements $b_1, \ldots, b_n$ (the rows of A) are linearly independent.

For dealing with modules over local rings the following lemma (first presented by Krull in a special case) is of fundamental importance.

Lemma 2.2. (Nakayama's Lemma) Let I be an ideal of R that is contained in the intersection of all $\mathfrak{m} \in \mathrm{Max}(R)$. Let M be an arbitrary R-module, $N \subset M$ a submodule for which M/N is finitely generated. If $M = N + IM$, then $M = N$.

Proof. $\overline{M} := M/N$ has a minimal generating system $\{m_1, \ldots, m_t\}$. Suppose $t > 0$. Since $\overline{M} = I\overline{M}$, there is an equation

$$m_t = \sum_{j=1}^{t} a_j m_j \qquad (a_j \in I, \ j = 1, \ldots, t).$$

Since a_t lies in all $\mathfrak{m} \in \mathrm{Max}(R)$ and therefore $1 - a_t$ is a unit in R, from $(1 - a_t)m_t = \sum_{j=1}^{t-1} a_j m_j$ it follows that $m_t \in \langle m_1, \ldots, m_{t-1} \rangle$. This contradicts the assumed minimality of the generating system. Therefore $t = 0$, so $M = N$.

Corollary 2.3. Let $(R, \mathfrak{m})$ be a local ring†, $k := R/\mathfrak{m}$ its residue field, and M a finitely generated R-module. For elements $m_1, \ldots, m_t \in M$ the following statements are equivalent:

a) $M = \langle m_1, \ldots, m_t \rangle$.
b) The residue classes $\overline{m}_1, \ldots, \overline{m}_t \in M/\mathfrak{m}M$ of the m_i form a system of generators of the k-vector space $M/\mathfrak{m}M$.

Proof. From $M/\mathfrak{m}M = \langle \overline{m}_1, \ldots, \overline{m}_t \rangle$ it follows that $M = \langle m_1, \ldots, m_t \rangle + \mathfrak{m}M$ and hence $M = \langle m_1, \ldots, m_t \rangle$ by 2.2.

From the corollary and well-known facts on vector spaces we at once get the following statements.

---

† This notation, which will often be used in the sequel, means that $\mathfrak{m}$ is the maximal ideal of R.

Corollary 2.4. Under the hypotheses of 2.3 we have:

a) $\mu(M) = \dim_k(M/\mathfrak{m}M)$.

b) $m_1, \ldots, m_t \in M$ form a minimal generating system of M if and only if their residue classes $\overline{m}_1, \ldots, \overline{m}_t \in M/\mathfrak{m}M$ form a basis.

c) If $\{m_1, \ldots, m_t\}$ is a minimal generating system of M and if

$$\sum_{i=1}^{t} r_i m_i = 0 \qquad (r_i \in R),$$

then $r_i \in \mathfrak{m}$ for $i = 1, \ldots, t$.

d) Any generating system of M contains a minimal one. Any unshortenable generating system is minimal.

e) Elements $m_1, \ldots, m_t \in M$ can be extended to a minimal generating system of M if and only if their residue classes $\overline{m}_1, \ldots, \overline{m}_t \in M/\mathfrak{m}M$ are linearly independent over k.

On the basis of these facts we have quite good information on the generation of modules over local rings. We now want to pass from the local to the global. In the following let M be a finitely generated module over an arbitrary ring $R \neq \{0\}$. For $\mathfrak{p} \in \mathrm{Spec}(R)$ we write $\mu_{\mathfrak{p}}(M)$ for the number of elements in a shortest generating system of the $R_{\mathfrak{p}}$-module $M_{\mathfrak{p}}$.

Further, for each $r \in \mathbb{N}$ we define an ideal

$$I(M, r) := \sum_{\{m_1, \ldots, m_r\} \subset M} \mathrm{Ann}(M/\langle m_1, \ldots, m_r \rangle),$$

where the sum is taken over all subsets of M, consisting of r elements.

We have $I(M, 0) = \mathrm{Ann}(M)$, $I(M, r) \subset I(M, r+1)$ for all $r \in \mathbb{N}$, and $I(M, r) = R$ if $r \geq \mu(M)$. Further, we have $I(M_S, r) = I(M, r)_S$ for any multiplicatively closed set $S \subset R$, since all the operations occurring in the definition of $I(M, r)$ are permutable with the formation of fractions (III.§4). From this and the definition of $I(M, r)$ we immediately get

Lemma 2.5. For $\mathfrak{p} \in \mathrm{Spec}(R)$ we have $\mu_{\mathfrak{p}}(M) \geq r+1$ if and only if $\mathfrak{p} \supset I(M, r)$.

Corollary 2.6. (Semicontinuity of $\mu_{\mathfrak{p}}$) For all $r \in \mathbb{N}$ the set of the $\mathfrak{p} \in \mathrm{Spec}(R)$ with $\mu_{\mathfrak{p}}(M) < r$ is open.

The following considerations of this section have the purpose of obtaining a local–global principle for the generation of modules and ideals.

Definition 2.7. (Swan [76]) An element $m \in M$ is called *basic* at $\mathfrak{p} \in \mathrm{Spec}(R)$ if $m \notin \mathfrak{p}M_{\mathfrak{p}}$.

Equivalently, $\mathfrak{p} \in \mathrm{Supp}(M)$ and m can be taken to lie in a minimal generating system of the $R_{\mathfrak{p}}$-module $M_{\mathfrak{p}}$. (As in 2.7, here m also denotes the image of m in $M_{\mathfrak{p}}$.)

Lemma 2.8. If M is a free R-module, then $m \in M$ is basic for all $\mathfrak{m} \in \text{Max}(R)$ if and only if $\langle m \rangle$ is a direct summand $\neq \langle O \rangle$ of M.

By 2.1 m is basic for $\mathfrak{m} \in \text{Max}(R)$ if and only if $R_{\mathfrak{m}} \, m$ is a direct summand $\neq \langle 0 \rangle$ of $M_{\mathfrak{m}}$. Hence the assertion follows from 1.13.

For the following let $X := J(R)$ be the J-spectrum of R. For an $m \in M$ let $X(m)$ denote the set of the $\mathfrak{p} \in X$ at which m is basic. For an ideal I of R let $\mathfrak{V}(I)$, as before, be the set of $\mathfrak{p} \in X$ that contain I. From 2.4b) it follows that m is basic at $\mathfrak{p} \in \mathfrak{V}(I)$ if and only if the image $\overline{m}$ of m in the R-module M/IM is basic at $\mathfrak{p}$ (which is equivalent to $\overline{m}$ being basic at $\mathfrak{p}/I$).

Lemma 2.9. Let X be Noetherian and $d := \dim X < \infty$. For any ideal I of R, $X(m) \cap \mathfrak{V}(I)$ has only finitely many minimal elements.

Proof. Let $\overline{m}$ be the image of m in the R/I-module M/IM. Since $\mathfrak{V}(I)$ can be identified with $J(R/I)$, $X(m) \cap \mathfrak{V}(I)$ being mapped to $X(\overline{m})$, we need only show that $X(\overline{m})$ has only finitely many minimal elements. We can therefore assume $I = (0)$.

If $\{\mathfrak{p}_1, \ldots, \mathfrak{p}_t\}$ are the generic points of the irreducible components of X (cf. I.4.8), then (again by the preliminary remark above)

$$X(m) = \bigcup_{i=1}^{t} X(m_i),$$

where m_i is the image of m in $M/\mathfrak{p}_i M$ $(i = 1, \ldots, t)$. Hence we may assume that R is an integral domain with $(0) \in J(R)$.

We now argue by induction on d. For $d = 0$, $J(R)$ has only one point and we are done. Hence let $d > 0$. If m is basic at (0), then (0) is the only minimal element of $X(m)$.

If m is not basic at (0), then the image of m in the vector space M_S ($S := R \setminus \{0\}$) equals 0; that is, there is an $r \in R \setminus \{0\}$ with $rm = 0$. $X(m)$ can then be identified with $X(\overline{m})$, where $\overline{m}$ is the image of m in the $R/(r)$-module M/rM. Since $\dim J(R/(r)) < d$, by the induction hypothesis we have finished.

Lemma 2.10. Under the hypotheses of 2.9 let

$$u_m := \text{Max}\{\mu_{\mathfrak{p}}(M) + \dim \mathfrak{V}(\mathfrak{p}) \mid \mathfrak{p} \in X(m)\}.$$

Then there are only finitely many $\mathfrak{p} \in X(m)$ with $\mu_{\mathfrak{p}}(M) + \dim \mathfrak{V}(\mathfrak{p}) = u_m$.

Proof. We consider a $\mathfrak{p} \in X(m)$ with $\mu_{\mathfrak{p}}(M) + \dim \mathfrak{V}(\mathfrak{p}) = u_m$. Let $\mu_{\mathfrak{p}}(M) =: r$. Then $r > 0$ and $\mathfrak{p} \supset I(M, r-1)$, $\mathfrak{p} \not\supset I(M, r)$. $\mathfrak{p}$ is a minimal element of $X(m) \cap \mathfrak{V}(I(M, r-1))$, for if there were a $\mathfrak{q} \in X(m) \cap \mathfrak{V}(I(M, r-1))$ with $\mathfrak{p} \supsetneq \mathfrak{q}$, then we would have $\mu_{\mathfrak{q}}(M) = \mu_{\mathfrak{p}}(M) = r$ and $u_m = \mu_{\mathfrak{p}}(M) + \dim \mathfrak{V}(\mathfrak{p}) < \mu_{\mathfrak{q}}(M) + \dim \mathfrak{V}(\mathfrak{q}) \leq u_m$, a contradiction. By 2.9 $X(m) \cap \mathfrak{V}(I(M, r-1))$ has only finitely many minimal elements. Since there are only finitely many distinct ideals $I(M, r)$, the assertion follows.

To be able to formulate the next lemma comfortably it is useful to introduce the following way of speaking.

Definition 2.11. A submodule $U \subset M$ is called k-*times basic* (for some $k \in \mathbb{N}$) at $\mathfrak{p} \in \mathrm{Spec}(R)$ if

$$\mu_{\mathfrak{p}}(M) - \mu_{\mathfrak{p}}(M/U) \geq k.$$

By Nakayama's lemma this is equivalent with the condition that $M_{\mathfrak{p}}$ has a minimal system of generators containing at least k elements of $U_{\mathfrak{p}}$. For $m \in M, U = \langle m \rangle$ is 1-time basic at $\mathfrak{p}$ if and only if m is basic at $\mathfrak{p}$.

Lemma 2.12. Let $\{m_1, \ldots, m_t\}$ be a system of elements of M and $\{\mathfrak{p}_1, \ldots, \mathfrak{p}_r\}$ $(r > 0)$ a finite set in $\mathrm{Spec}(R)$. Let $\langle m_1, \ldots, m_t \rangle$ be k_i-times basic at $\mathfrak{p}_i$ for some $k_i \in \mathbb{N}, k_i < t$ $(i = 1, \ldots, r)$. Then there are elements $a_1, \ldots, a_{t-1} \in R$ such that $\langle m_1 + a_1 m_t, \ldots, m_{t-1} + a_{t-1} m_t \rangle$ is k_i-times basic at $\mathfrak{p}_i$ $(i = 1, \ldots, r)$.

Proof (by induction on r). For $r = 1$ we may suppose R local with maximal ideal $\mathfrak{m} = \mathfrak{p}_1$. Let $U := \langle m_1, \ldots, m_t \rangle$, and let $\overline{m}_i$ be the image of m_i in $M/\mathfrak{m}M$ $(i = 1, \ldots, t)$. From the exact sequence

$$0 \to U + \mathfrak{m}M/\mathfrak{m}M \to M/\mathfrak{m}M \to M/U + \mathfrak{m}M \to 0$$

and 2.4 it follows that $\mu(M) - \mu(M/U) = \dim_{R/\mathfrak{m}} \langle \overline{m}_1, \ldots, \overline{m}_t \rangle \geq k_1$, where $k_1 < t$. If already $\dim_{R/\mathfrak{m}} \langle \overline{m}_1, \ldots, \overline{m}_{t-1} \rangle \geq k_1$ then the claim follows with $a_1 = \cdots = a_{t-1} = 0$. If $\dim_{R/\mathfrak{m}} \langle \overline{m}_1, \ldots, \overline{m}_{t-1} \rangle = k_1 - 1$, then $\overline{m}_t$ is linearly independent of $\{\overline{m}_1, \ldots, \overline{m}_{t-1}\}$. If say $\overline{m}_1, \ldots, \overline{m}_{k_1-1}$ are linearly independent, we have the claim with $a_i = 0$ $(i = 1, \ldots, t-1; i \neq k_1)$, $a_{k_1} = 1$.

Now let $r > 1$ and suppose the lemma has already been proved for $r-1$ prime ideals. We choose the numbering so that $\mathfrak{p}_r$ is minimal in the set $\{\mathfrak{p}_1, \ldots, \mathfrak{p}_r\}$. Then there are elements $a'_1, \ldots, a'_{t-1} \in R$ such that $\langle m_1 + a'_1 m_t, \ldots, m_{t-1} + a'_{t-1} m_t \rangle$ is k_i-times basic at the $\mathfrak{p}_i$ with $i = 1, \ldots, r-1$. Since $\bigcap_{i=1}^{r-1} \mathfrak{p}_i \not\subset \mathfrak{p}_r$, there is an $a \in \bigcap_{i=1}^{r-1} \mathfrak{p}_i, a \notin \mathfrak{p}_r$. Then $\langle m_1 + a'_1 m_t, \ldots, m_{t-1} + a'_{t-1} m_t, a m_t \rangle$ is also k_r-times basic at $\mathfrak{p}_r$, since a is mapped into a unit in $R_{\mathfrak{p}_r}$. There are also elements $a''_1, \ldots, a''_{t-1} \in R$ such that $U := \langle m_1 + a'_1 m_t + a''_1 a m_t, \ldots, m_{t-1} + a'_{t-1} m_t + a''_{t-1} a m_t \rangle$ is k_r-times basic at $\mathfrak{p}_r$. Since $a \in \mathfrak{p}_i$ for $i < r, U$ is also k_i-times basic at the $\mathfrak{p}_i$ with $i < r$, and the assertion is proved with $a_i := a'_i + a''_i a$ $(i = 1, \ldots, t-1)$.

Theorem 2.13. Let $X := J(R)$ be Noetherian and of finite dimension. Let $M = \langle m_1, \ldots, m_t \rangle$ and

$$\mu_{\mathfrak{p}}(M) + \dim \mathfrak{V}(\mathfrak{p}) < t \quad \text{for all} \quad \mathfrak{p} \in X(m_t).$$

Then there are elements $a_1, \ldots, a_{t-1} \in R$ such that

$$M = \langle m_1 + a_1 m_t, \ldots, m_{t-1} + a_{t-1} m_t \rangle.$$

Proof. If $X(m_t) = \emptyset$, then $M_{\mathfrak{p}} = \langle m_1, \ldots, m_{t-1} \rangle R_{\mathfrak{p}}$ for all $\mathfrak{p} \in X$, in particular for all $\mathfrak{p} \in \mathrm{Max}(R)$. By 1.2 it follows that $M = \langle m_1, \ldots, m_{t-1} \rangle$, and this is the assertion with $a_1 = \cdots = a_{t-1} = 0$.

If $X(m_t) \neq \emptyset$, then

$$u := \text{Max}\{\mu_\mathfrak{p}(M) + \dim \mathfrak{O}(\mathfrak{p}) \mid \mathfrak{p} \in X(m_t)\} > 0$$

and $t \geq 2$. By 2.10 there are only finitely many $\mathfrak{p} \in X(m_t)$ with $\mu_\mathfrak{p}(M) + \dim \mathfrak{V}(\mathfrak{p}) = u$. By 2.12 we can find $a_1 \in R$ such that $m_1 + a_1 m_t$ is basic at all these $\mathfrak{p}$.

$M' := M/\langle m_1 + a_1 m_t \rangle$ is generated by the images $m'_2, \ldots, m'_t$ of $m_2, \ldots, m_t$, and $X(m'_t) \subset X(m_t)$. For the $\mathfrak{p} \in X(m_t)$ with $\mu_\mathfrak{p}(M) + \dim \mathfrak{V}(\mathfrak{p}) = u$ we have $\mu_\mathfrak{p}(M') < \mu_\mathfrak{p}(M)$, since $m_1 + a_1 m_t$ is basic at these $\mathfrak{p}$, and so

$$u' := \text{Max}\{\mu_\mathfrak{p}(M') + \dim \mathfrak{V}(\mathfrak{p}) \mid \mathfrak{p} \in X(m'_t)\} < u;$$

therefore $u' < t - 1$.

For $t = 2$ we have $X(m'_t) = \emptyset$ and therefore $M' = \langle 0 \rangle$, so $M = \langle m_1 + a_1 m_t \rangle$. Now if $t > 2$ and the theorem has already been proved for smaller t, then there are elements $a_2, \ldots, a_{t-1} \in R$ with $M' = \langle m'_2 + a_2 m'_t, \ldots, m'_{t-1} + a_{t-1} m'_t \rangle$, whence follows the assertion $M = \langle m_1 + a_1 m_t, \ldots, m_{t-1} + a_{t-1} m_t \rangle$.

Corollary 2.14. (Theorem of Forster [24] and Swan [76]) Let $X := J(R)$ be Noetherian of finite Krull dimension. Then

$$\mu(M) \leq u := \text{Max}\{\mu_\mathfrak{p}(M) + \dim \mathfrak{V}(\mathfrak{p}) \mid \mathfrak{p} \in X \cap \text{Supp}(M)\}.$$

(More precisely, Theorem 2.13 says that from any system of generators of M, one with u elements can be derived by "elementary transformations.")

If $\text{Spec}(R)$ is Noetherian and of finite dimension, then the same holds for $J(R)$. The Forster–Swan theorem then also holds with

$$u' := \text{Max}\{\mu_\mathfrak{p}(M) + \dim R/\mathfrak{p} \mid \mathfrak{p} \in \text{Supp}(M)\}$$

in place of u, but in this form it is sometimes strictly weaker, e.g. for a semilocal ring R. In such a ring we have $J(R) = \text{Max}(R)$ (II.1.10), $\dim J(R) = 0$; on the other hand, there can be elements $\mathfrak{p}$ in $\text{Spec}(R)$ for which $\dim R/\mathfrak{p}$ is arbitrarily large.

In the general case, if $X := \text{Max}(R)$ is Noetherian and of finite dimension (and therefore $J(R)$ is too, cf. II.1.8), then

$$u \leq \dim X + \text{Max}\{\mu_\mathfrak{m}(M) \mid \mathfrak{m} \in X \cap \text{Supp}(M)\}.$$

Therefore, if M is locally for each $\mathfrak{m} \in \text{Max}(R)$ generated by r elements, then M is globally generated by $r + \dim X$ elements.

Now let $M = I$ be an ideal of R. Since $I_\mathfrak{p} = R_\mathfrak{p}$ for all $\mathfrak{p} \in \text{Spec}(R)$ with $I \not\subset \mathfrak{p}$, from 2.14 we get the following somewhat more precise statement for ideals.

Corollary 2.15. Let $J(R)$ be Noetherian of dimension d. Let I be an ideal of R and

$$u := \text{Max}\{\mu_{\mathfrak{p}}(I) + \dim \mathfrak{V}(\mathfrak{p}) \mid \mathfrak{p} \in \mathfrak{V}(I)\}.$$

Then I is generated by $\text{Max}\{u, d+1\}$ elements.

Examples 2.16.

a) A Dedekind ring is a Noetherian integral domain of dimension 1 which is locally a principal ideal ring. By 2.15 in such a ring any ideal can be generated by 2 elements.

b) If for a semilocal ring with maximal ideals $\mathfrak{m}_1, \ldots, \mathfrak{m}_r$ the local rings $R_{\mathfrak{m}_i}$ $(i = 1, \ldots, t)$ are principal ideal rings, then by 2.15 R is also a principal ideal ring.

For semilocal rings we have

Corollary 2.17. Let R be a semilocal ring with maximal ideals $\mathfrak{m}_1, \ldots, \mathfrak{m}_r$. Let M be a finitely generated R-module, $u := \text{Max}_{i=1,\ldots,r}\{\mu_{\mathfrak{m}_i}(M)\}$. Then M is generated by u elements.

Exercises

1. Give an example of an unshortenable generating system of a module that is not minimal.

2. Let R be a ring, M, N two R-modules, $l : M \to N$ a linear mapping. For $\mathfrak{m} \in \text{Max}(R)$ let $l(\mathfrak{m}) : M/\mathfrak{m}M \to N/\mathfrak{m}N$ denote the induced linear mapping of $R/\mathfrak{m}$-vector spaces. Decide whether the following statements are true or false.

   a) l is injective if and only if $l(\mathfrak{m})$ is injective for all $\mathfrak{m} \in \text{Max}(R)$.

   b) l is surjective if and only if $l(\mathfrak{m})$ is surjective for all $\mathfrak{m} \in \text{Max}(R)$.

3. Let $(R, \mathfrak{m})$ and $(S, \mathfrak{n})$ be Noetherian local rings, $\varphi : R \to S$ a ring homomorphism with $\varphi(\mathfrak{m}) \subset \mathfrak{n}$ (such a homomorphism is called a local homomorphism). Suppose:

   a) The mapping $R/\mathfrak{m} \to S/\mathfrak{n}$ induced by φ is bijective.

   b) The mapping $\mathfrak{m}/\mathfrak{m}^2 \to \mathfrak{n}/\mathfrak{n}^2$ induced by φ is surjective.

   c) S is finitely generated as an R-module.

   Then φ is surjective.

4. Let $G = \oplus_{i \in \mathbb{N}} G_i$ be a positively graded ring, where $G_0 = K$ is a field (so $\mathfrak{M} := \oplus_{i > 0} G_i$ is a maximal ideal). Let I be a homogeneous ideal of G.

   a) Homogeneous elements $a_1, \ldots, a_n \in I$ form a generating system of I if and only if their images in $G_{\mathfrak{M}}$ generate the ideal $I_{\mathfrak{M}}$.

   b) If I is finitely generated, then any unshortenable generating system of I consisting of homogeneous elements is minimal. Any generating system contains a minimal one.

5. Let $(R, \mathfrak{m})$ be a Noetherian local ring, $I \subset \mathfrak{m}$ an ideal; suppose $\xi \in \mathfrak{m}/I$ is not a zero divisor in R/I, and let $x \in \mathfrak{m}$ be a representative of ξ. We put $\overline{R} := R/(x)$ and for $a \in R$ denote by $\overline{a}$ the residue class of a in $\overline{R}$, by $\overline{I}$ the image of I in $\overline{R}$. Elements $a_1, \ldots, a_n \in I$ generate I if and only if $\overline{a}_1, \ldots, \overline{a}_n$ generate the ideal $\overline{I}$. In particular, $\mu(I) = \mu(\overline{I})$.

6. Give a brief proof (without using the Forster–Swan theorem) of the following statement: If a maximal ideal $\mathfrak{m}$ of a ring R is finitely generated, then $\mu(\mathfrak{m}) \leq \mu_\mathfrak{m}(\mathfrak{m}) + 1$.

7. Let M be a finitely generated module over a local ring $(R, \mathfrak{m})$. Let the $F_i(M)$ be the Fitting ideals of R (§1, Exercise 8). The following statements are equivalent.

     a) $\mu(M) = r$.

     b) $F_{r-1}(M) \subset \mathfrak{m}, F_r(M) = R$.

3. Projective modules

Projective modules are direct summands of free modules. They have many properties in common with them. In geometry projective modules correspond to vector bundles, free modules to trivial vector bundles.

In the following again let $R \neq \{0\}$ be any ring, M an R-module.

Definition 3.1.

     a) M is called *projective* (or an *algebraic vector bundle* over $\mathrm{Spec}(R)$) if there is an R-module M' such that $M \oplus M'$ is free.

     b) M is called *locally free* if $M_\mathfrak{m}$ is a free $R_\mathfrak{m}$-module for all $\mathfrak{m} \in \mathrm{Max}(R)$.

From the definition we immediately infer the following facts: If M is projective, then the R_S-module M_S is also projective for any multiplicatively closed subset $S \subset R$.

Any direct summand of a projective module is projective, and the direct sum of projective modules is also projective. If M is a projective R-module and $P \subset R$ is a subring such that R is a free P-module (e.g. a polynomial ring over P), then M is also projective as a P-module. If M is a locally free R-module, then $M_\mathfrak{p}$ is a free $R_\mathfrak{p}$-module for all $\mathfrak{p} \in \mathrm{Spec}(R)$, since $M_\mathfrak{p}$ is the localization of an $M_\mathfrak{m}$ for suitable $\mathfrak{m} \in \mathrm{Max}(R)$.

Proposition 3.2. The following statements are equivalent.

     a) M is projective.

     b) For any R-module epimorphism $\alpha : A \to B$, $\mathrm{Hom}_R(M, \alpha) : \mathrm{Hom}_R(M, A) \to \mathrm{Hom}_R(M, B)$ $(l \mapsto \alpha \circ l)$ is also an epimorphism.

     b′) For any diagram of R-modules and linear mappings with an exact row

$$M$$
$$\downarrow \beta$$
$$A \xrightarrow{\alpha} B \to 0$$

there exists a linear mapping $\gamma : M \to A$ with $\beta = \alpha \circ \gamma$.

c) Any exact sequence of R-modules of the form $0 \to C \to D \to M \to 0$ splits.

Proof.

a)→b). For any free R-module F, using a basis it is easy to see that $\mathrm{Hom}_R(F, \alpha) =: \bar{\alpha}$ is surjective. If M is a direct summand of F, we have a commutative diagram:

$$\begin{array}{ccc} \mathrm{Hom}_R(F, A) & \xrightarrow{\bar{\alpha}} & \mathrm{Hom}_R(F, B) \\ \downarrow & & \downarrow \\ \mathrm{Hom}_R(M, A) & \to & \mathrm{Hom}_R(M, B), \end{array}$$

where the vertical arrows are epimorphisms. b) follows.

b′) is a reformulation of b).

b)→c). Since $\mathrm{Hom}_R(M, D) \to \mathrm{Hom}_R(M, M)$ is surjective by b), the exact sequence in c) splits.

c)→a). Choose an exact sequence $0 \to K \to F \to M \to 0$ with a free R-module F. By c), M is a direct summand of F, hence is projective.

Corollary 3.3. Let M be a finitely generated projective R-module. Then there is a finitely generated R-module M' such that $M \oplus M'$ is free. In particular, M is finitely presentable.

Proof. Choose an exact sequence $0 \to M' \to F \to M \to 0$ with a free R-module F of finite rank. By 3.2c), $F \cong M \oplus M'$ and M' is finitely generated since it is a homomorphic image of F. Since $M \cong F/M'$, M is finitely presentable.

The question of when projective modules are free plays a role in many applications. In what follows we make some assertions about this problem.

Proposition 3.4. Let $(R, \mathfrak{m})$ be a local ring, M a finitely presentable R-module. Then the following statements are equivalent.

a) M is free.

b) There is an exact sequence of R-modules $0 \to K \xrightarrow{\alpha} P \xrightarrow{\beta} M \to 0$, where P is projective and the mapping $K/\mathfrak{m}K \to P/\mathfrak{m}P$ induced by α is an injection.

Proof. We need only show that b)→a). If $\mu(M) =: r$, then there is an exact sequence $0 \to K_0 \xrightarrow{\alpha_0} F_0 \xrightarrow{\beta_0} M \to 0$ with a free R-module F_0 of rank r. By 3.2b′) there is a linear mapping $\epsilon : P \to F_0$ with $\beta = \beta_0 \circ \epsilon$. With the induced mappings the diagram

$$\begin{array}{c} P/\mathfrak{m}P \searrow \\ \downarrow \qquad M/\mathfrak{m}M \\ F_0/\mathfrak{m}F_0 \nearrow \end{array}$$

is commutative. Here $F_0/\mathfrak{m}F_0 \to M/\mathfrak{m}M$ is bijective, for the mapping is surjective and both modules are $R/\mathfrak{m}$-vector spaces of dimension r. Then $P/\mathfrak{m}P \to F_0/\mathfrak{m}F_0$ is also surjective, that is $F_0 = \epsilon(P) + \mathfrak{m}F_0$. From Nakayama's Lemma it follows that $F_0 = \epsilon(P)$, hence ϵ is surjective.

It is easily seen that the mapping $\epsilon' : K \to K_0$ induced by ϵ is surjective and that $\operatorname{Ker} \epsilon' = \operatorname{Ker} \epsilon$ (when K is identified with $\alpha(K)$).

We thus get a commutative diagram with exact rows and columns

$$
\begin{array}{ccc}
0 & & 0 \\
\downarrow & & \downarrow \\
F_1 & = & F_1 \\
\downarrow & & \downarrow \\
0 \longrightarrow K & \xrightarrow{\alpha} P & \xrightarrow{\beta} M \longrightarrow 0 \\
\downarrow \epsilon' & \downarrow \epsilon & \parallel \\
0 \longrightarrow K_0 & \xrightarrow{\alpha_0} F_0 & \xrightarrow{\beta_0} M \longrightarrow 0 \\
\downarrow & & \downarrow \\
0 & & 0
\end{array}
$$

Since M is finitely presentable, there is an exact sequence $0 \to K_1 \to F \to M \to 0$ with a free R-module F of finite rank and a finitely generated R-module K_1. If we substitute this sequence for the sequence $0 \to K \to P \to M \to 0$ in the argument above, we find that K_0 is a homomorphic image of K_1; therefore, K_0 too is finitely generated.

We now consider the above diagram modulo $\mathfrak{m}$. We get the following commutative diagram of $R/\mathfrak{m}$-vector spaces with exact rows and columns:

$$
\begin{array}{ccc}
& & 0 \\
& & \downarrow \\
F_1/\mathfrak{m}F_1 & = & F_1/\mathfrak{m}F_1 \\
\downarrow & & \downarrow \\
0 \longrightarrow K/\mathfrak{m}K & \longrightarrow P/\mathfrak{m}P & \longrightarrow M/\mathfrak{m}M \longrightarrow 0 \\
\downarrow & & \downarrow \\
K_0/\mathfrak{m}K_0 & \longrightarrow F_0/\mathfrak{m}F_0 & \longrightarrow M/\mathfrak{m}M \longrightarrow 0 \\
\downarrow & & \downarrow \\
0 & & 0
\end{array}
$$

Here the middle row is exact by the assumption made in b); and the second column is exact because $0 \to F_1 \to P \to F_0 \to 0$ splits since F_0 is free. It is easy to check that $K_0/\mathfrak{m}K_0 \to F_0/\mathfrak{m}F_0$ is injective. On the other hand, $F_0/\mathfrak{m}F_0 \to M/\mathfrak{m}M$ is an isomorphism. It follows that $K_0/\mathfrak{m}K_0 = \langle 0 \rangle$ and, since K_0 is finitely generated, Nakayama's Lemma shows that $K_0 = \langle 0 \rangle$; therefore, $M \cong F_0$ is a free module, q. e. d.

Corollary 3.5. A finitely generated module over a local ring is projective if and only if it is free.

If M is projective, then in the exact sequence in 3.4b) we choose P to be the module M itself and $K = \langle 0 \rangle$. (One can very easily verify 3.5 directly without the detour through 3.4.)

Corollary 3.6. For a finitely generated module M over an arbitrary ring R the following statements are equivalent.

a) M is projective.

b) M is finitely presentable and locally free.

Proof. a) $\rightarrow$ b) follows from 3.3 and 3.5. Under the hypothesis b) there is an exact sequence $0 \rightarrow K \rightarrow F \rightarrow M \rightarrow 0$ with a free R-module F of finite rank and a finitely generated submodule $K \subset F$. If M is locally free, then the sequence splits locally for all $\mathfrak{m} \in \text{Max}(R)$; therefore, it also splits globally by 1.12. Then M is a direct summand of F, hence is projective.

Definition 3.7. For a finitely generated projective R-module P and a $\mathfrak{p} \in \text{Spec}(R)$, $\mu_{\mathfrak{p}}(P)$ is called the *rank of P at* $\mathfrak{p}$. P is called *of rank r* if $\mu_{\mathfrak{p}}(P) = r$ for all $\mathfrak{p} \in \text{Spec}(R)$.

Corollary 3.8. If P is a finitely generated projective R-module, then the function $\mu_{\mathfrak{p}}(P)$ is constant on any connected component of $\text{Spec}(R)$.

Proof. We write $P \oplus P' \cong R^n$ with some other finitely generated projective R-module P'. Then $\mu_{\mathfrak{p}}(P) + \mu_{\mathfrak{p}}(P') = n$ for all $\mathfrak{p} \in \text{Spec}(R)$. For any $r \in \mathbb{N}$ the set U of all $\mathfrak{p} \in \text{Spec}(R)$ with $\mu_{\mathfrak{p}}(P) \leq r$ is open (2.6). Since U is also the set of all $\mathfrak{p}$ with $\mu_{\mathfrak{p}}(P') \geq n - r$, it is also closed. Likewise, the set of all $\mathfrak{p}$ with $\mu_{\mathfrak{p}}(P) \geq r$ is both open and closed and therefore so is the set of all $\mathfrak{p}$ with $\mu_{\mathfrak{p}}(P) = r$. Hence the assertion follows.

If $J(R)$ is Noetherian and of dimension d, then, by the Forster–Swan Theorem, any finitely generated projective R-module P of rank r is generated globally by $d + r$ elements.

Corollary 3.9. For a finitely generated module M over a semilocal ring R the following statements are equivalent.

a) M is projective of constant rank on $\text{Max}(R)$.

b) M is free.

If M is of rank r at all $\mathfrak{m} \in \text{Max}(R)$, then M is globally generated by r elements. They form a basis, since they are mapped onto a basis under all localizations at the $\mathfrak{m} \in \text{Max}(R)$ (2.1).

A somewhat more precise statement than 3.8 is

Proposition 3.10. (Local triviality of projective modules) Let P be a finitely generated projective R-module, r its rank at $\mathfrak{p} \in \text{Spec}(R)$. Then there is an $f \in R \setminus \mathfrak{p}$ such that P_f is a free R_f-module of rank r.

Proof. Let $\{\omega_1, \ldots, \omega_p\}$ be a basis of the $R_{\mathfrak{p}}$-module $P_{\mathfrak{p}}$. We can choose the ω_i as images of elements ω_i^* of P, for if we multiply the ω_i by a common denominator, we again get a basis of $P_{\mathfrak{p}}$. Consider now the exact sequence $0 \rightarrow K \rightarrow R^r \xrightarrow{\alpha} P \rightarrow C \rightarrow 0$, where α maps the canonical basis element e_i onto ω_i^* $(i = 1, \ldots, r)$ and $K := \ker \alpha$, $C := \text{coker} \, \alpha$. By assumption $C_{\mathfrak{p}} = \langle 0 \rangle$ and, since C is finitely generated, there is an $f \in R \setminus \mathfrak{p}$ such that $C_f = \langle 0 \rangle$ (III.4.3.). The exact sequence $0 \to K_f \to R_f^r \to P_f \to 0$ splits, since P_f is a projective R_f-module (3.2). Therefore K_f is a finitely generated R_f-module. As $K_\mathfrak{p} = \langle 0 \rangle$ there exists a $g \in R \setminus \mathfrak{p}$ with $K_{fg} = \langle 0 \rangle$. Then P_{fg} is a free R_{fg}-module of rank r.

A trivial example of a projective module that is not free is gotten as follows. Let $R := K \times K$ with a field K, $P := (0) \times K$ is a projective module, for it is locally free though not of constant rank.

The following example is more interesting. Let

$$R := \mathbf{R}[X, Y, Z]/(X^2 + Y^2 + Z^2 - 1),$$

the coordinate ring of the 2-sphere over $\mathbf{R}$ and

$$P := RdX \oplus RdY \oplus RdZ/\langle xdX + ydY + zdZ \rangle,$$

where x, y, z denote the images of X, Y, Z in R. It can be shown that P is projective but not free. This is connected with the fact that the tangent bundle of the 2-sphere is not trivial (we say "You can't comb the hair of a sphere without creating a part or a cowlick").

A general method to produce examples of projective modules works as follows: If we are given elements f, g of R with $D(f) \cup D(g) = \operatorname{Spec}(R)$ along with a free R_f-module F_1 of finite rank and a free R_g-module F_2 of finite rank and an isomorphism $\alpha : (F_1)_g \xrightarrow{\sim} (F_2)_f$ of R_{fg}-modules, then by gluing F_1 and F_2 as in III.5.5 we get a projective R-module P, since P is finitely presentable by 1.14 and locally free, hence projective by 3.6.

We now turn to the projective modules over a polynomial ring $R[X]$ in one variable X over R.

Theorem 3.11. (Horrocks [40]) Let $(R, \mathfrak{m})$ be a local ring, M a finitely generated projective $R[X]$-module. Suppose there is a monic polynomial $f \in R[X]$ such that M_f is a free $R[X]_f$-module. Then M is a free $R[X]$-module.

Proof. Choose a basis of the $R[X]_f$-module M_f consisting of elements of M and denote by F the submodule of M it spans. For $P := M/F$ we then have $P_f = 0$, so there is an $n \in \mathbf{N}$ with $f^n P = 0$ and we have $P \cong M/F/f^n(M/F) \cong M/f^n M/(F + f^n M)/f^n M$. Here $M/f^n M$ is a finitely generated projective module over $S := R[X]/(f^n)$. Since f is monic, S is a free R-module of finite rank, so $M/f^n M$ is also a free R-module of finite rank. Since $F + f^n M/f^n M$ is also finitely generated as an R-module, it follows that as an R-module P is finitely presentable.

If $\overline{f}$ is the image of f in $R/\mathfrak{m}[X]$, then $(F/\mathfrak{m}F)_{\overline{f}} = (M/\mathfrak{m}M)_{\overline{f}}$; therefore, the canonical mapping $F/\mathfrak{m}F \to M/\mathfrak{m}M$ is injective. Since M is also projective as an R-module (because $R[X]$ is a free R-module), we can apply 3.4 to the exact sequence $0 \to F \to M \to P \to 0$. We find that P is a free R-module and $M \cong P \oplus F$ as R-modules.

In what follows we shall show that with the aid of elementary operations "F can be enlarged at the expense of P until no more of P remains."

Let $p_1, \ldots, p_s \in M$ be representatives for a basis of the R-module P and let $\{p_{s+1}, \ldots, p_t\}$ be an $R[X]$-basis of F. If $s = 0$, there is nothing to prove. Hence let $s > 0$ in what follows. For $k = 1, \ldots, s$ we then have equations.

$$-Xp_k = \sum_{i=1}^{s} \alpha_{ki}p_i + \sum_{j=s+1}^{t} b_{kj}p_j \qquad (\alpha_{ki} \in R, b_{kj} \in R[X]). \qquad (1)$$

An arbitrary relation $\sum_{i=1}^{s} a_i p_i + \sum_{j=s+1}^{t} b_j p_j = 0$ $(a_i, b_j \in R[X])$ can, with the aid of equations (1), be reduced to a relation

$$\sum_{i=1}^{s} \alpha_i p_i + \sum_{j=s+1}^{t} \tilde{b}_j p_j = 0 \qquad (\alpha_i \in R, \tilde{b}_j \in R[X]).$$

Since $M \cong P \oplus F$ it follows that $\alpha_i = \tilde{b}_j = 0$ $(i = 1, \ldots, s, j = s+1, \ldots, t)$.

Therefore, with respect to the generating system $\{p_1, \ldots, p_t\}$ the $R[X]$-module M has a relation matrix of the form

$$(A + XE \mid B) \qquad A = (\alpha_{ki}), B = (b_{kj}), \alpha_{ki} \in R, b_{kj} \in R[X] \qquad (2)$$

where E is the s-rowed unit matrix.

Lemma 3.12. There exist matrices

$$B_0 \in M(s \times (t-s); R) \quad \text{and} \quad \tilde{B} \in M(s \times (t-s); R[X])$$

such that $B = B_0 + (A + XE)\tilde{B}$.

This follows by "dividing" B (with remainder) by the "linear polynomial" $A + XE$ as one usually does in polynomial rings.

By the lemma, relations (1) can be written in the form

$$(A + XE) \cdot \left[\begin{pmatrix} p_1 \\ \vdots \\ p_s \end{pmatrix} + \tilde{B} \begin{pmatrix} p_{s+1} \\ \vdots \\ p_t \end{pmatrix} \right] + B_0 \begin{pmatrix} p_{s+1} \\ \vdots \\ p_t \end{pmatrix} = 0.$$

Therefore, by changing the p_i $(i = 1, \ldots, s)$ by suitable linear combinations of the p_j $(j = s+1, \ldots, t)$, we can always arrange that in (2) the matrix B also has coefficients in R only. This will be assumed in what follows.

Lemma 3.13. The ideal in $R[X]$ generated by the $s \times s$-minors of $(A + XE \mid B)$ coincides with $R[X]$ (cf. §2, Exercise 7).

It suffices to show this locally for all $\mathfrak{M} \in \text{Max}(R[X])$. By 3.6 $M_{\mathfrak{M}}$ is a free $R[X]_{\mathfrak{M}}$-module, and from $(M_{\mathfrak{M}})_f = (F_{\mathfrak{M}})_f$ it follows that $M_{\mathfrak{M}}$ has the same rank as $F_{\mathfrak{M}}$, namely $t - s$. The exact sequence

$$0 \to K \to R[X]_{\mathfrak{M}}^t \to M_{\mathfrak{M}} \to 0,$$

where K is the $R[X]_{\mathfrak{M}}$-submodule of $R[X]_{\mathfrak{M}}^t$ spanned by the rows of the matrix $(A + XE \mid B)$, splits. Therefore, K is a free $R[X]_{\mathfrak{M}}$-module of rank s. Since the rows of the matrix can be extended to a basis of $R[X]_{\mathfrak{M}}^t$, at least one $s \times s$ minor must be a unit in $R[X]_{\mathfrak{M}}$.

From 3.13 it follows that $R[X] = R[X] \cdot g + R[X] \cdot I$, where $g := \det(A + XE)$ and I is the ideal in R generated by the coefficients of B. The ring $T := R[X]/(g)$ is free as an R-module, since g is a monic polynomial. From $T = T \cdot I$ it follows that $I = R$. Since R is local, it follows that at least one coefficient of B is a unit in R.

By first making elementary column and then row operations, we bring (2) into a matrix of the form

$$
\left(
\begin{array}{c|c}
A' + XE' \mid B' & \begin{matrix} 0 \\ \vdots \\ 0 \end{matrix} \\
\hline
0 \ldots\ldots\ldots 0 & 1
\end{array}
\right)
$$

where A' and B' have coefficients in R and E' is the $(s-1)$-rowed unit matrix.

Since only elementary operations have been made, the statement of 3.13 also holds for the new matrix (cf. in this connection §1, Exercise 8). We can apply to $(A' + XE' | B')$ the same arguments as to (2). Finally, we bring the matrix into the form

$$(O \mid E),$$

where E is the s-rowed unit matrix. But this means that M is a free $R[X]$-module (of rank $t - s$), q. e. d.

Theorem 3.14. (Quillen-Suslin) The statement of 3.11 holds for any ground ring R: If M is a finitely generated projective $R[X]$-module, $f \in R[X]$ a monic polynomial such that M_f is a free $R[X]_f$-module, then M is a free $R[X]$-module.

Proof.

a) M is extended.

For $m \in \mathrm{Max}(R)$ let M_m be the module of fractions of the $R[X]$-module M with respect to $R \setminus m$. M_m is a finitely generated projective $R_m[X]$-module, for which $(M_m)_f$ is free. By 3.11 M_m is a free $R_m[X]$-module, so it is surely an extended module. From 1.20 it now follows that M is also globally extended: $M \cong N[X]$ with some R-module N. Here $N \cong M/XM$ and also $N[X]/(X - 1)N[X] \cong N$ as R-modules. Hence it suffices to show that $M/(X - 1)M$ is a free R-module. This is the goal of the following exposition.

b) Extension of M "into projective space."

We consider another polynomial ring $R[X^{-1}]$ in an indeterminate X^{-1}, and we identify the rings of fractions $R[X]_X$ and $R[X^{-1}]_{X^{-1}}$ using the R-isomorphism that maps X^{-1} to $\frac{1}{X}$. We then write $R[X, X^{-1}]$ for this ring. (The open set $D(X) \subset \mathrm{Spec}(R[X])$ is identified with the open set $D(X^{-1}) \subset \mathrm{Spec}(R[X^{-1}])$. The space that arises by identifying $D(X)$ in $\mathrm{Spec}(R[X])$ with $D(X^{-1})$ in $\mathrm{Spec}(R[X^{-1}])$ is called the "projective line" $\mathbf{P}_R^1$ over R.)

Let $f = X^n + a_1 X^{n-1} + \cdots + a_n$ and $g := 1 + a_1 X^{-1} + \cdots + a_n X^{-n}$ $(a_i \in R)$. Since $g = X^{-n} f$ and since X^{-n} is a unit of $R[X, X^{-1}]$, we have $R[X, X^{-1}]_f \cong R[X, X^{-1}]_g$ and $(M_X)_f \cong (M_X)_g$ is a free $R[X, X^{-1}]_g$-module, since by hypothesis M_f is a free $R[X]_f$-module.

Since the elements X^{-1} and g generate the unit ideal in $R[X^{-1}]$, we have $\mathrm{Spec}(R[X^{-1}]) = D(X^{-1}) \cup D(g)$. By III.5.5 there is an $R[X^{-1}]$-module M' for which M'_g is a free $R[X^{-1}]_g$-module of the same rank as $(M_X)_g$ and for which $M'_{X^{-1}} \cong M_X$ as $R[X, X^{-1}]$-modules (gluing M_X to a free $R[X^{-1}]_g$-module over $D(gX^{-1})$). M' is finitely presentable by 1.14 (we say that M' extends M into projective space).

c) M is free.

For $\mathfrak{m} \in \mathrm{Max}(R)$, $(M'_\mathfrak{m})_{X^{-1}} \cong (M_\mathfrak{m})_X$ is a free module over the ring $R_\mathfrak{m}[X, X^{-1}]$. Since X^{-1} is a monic polynomial in $R[X^{-1}]$, by 3.11 $M'_\mathfrak{m}$ is a free $R_\mathfrak{m}[X^{-1}]$-module. M' is finitely presentable and locally extended, so by 1.20 M' is extended: $M' = N'[X^{-1}]$ with some R-module N', where $N' \cong M'/X^{-1}M' \cong M'/(X^{-1} - 1)M'$.

Since M'_g is a free $R[X^{-1}]_g$-module and $g \equiv 1 \bmod (X^{-1})$, $M'_g/X^{-1}M'_g \cong M'/X^{-1}M'$ is a free R-module, therefore $M'/(X^{-1} - 1)M'$ is also a free R-module. But $M/(X-1)M \cong M_X/(X-1)M_X \cong M'_{X^{-1}}/(X^{-1} - 1)M'_{X^{-1}} = M'/(X^{-1} - 1)M'$, and hence also $M/(X-1)M$ is a free R-module, q. e. d.

We can now prove

Theorem 3.15. (Serre's Conjecture) If K is a principal ideal domain, then all finitely generated projective $K[X_1, \ldots, X_n]$-modules are free.

Proof. For $n = 0$ the assertion is correct, since (more generally) submodules of a free module of finite rank over a principal ideal domain are free (I.2.18).

Hence let $n > 0$ and suppose the statement has already been proved for polynomial rings in $n-1$ variables. If M is a finitely generated projective module over $K[X_1, \ldots, X_n]$ and S is the multiplicatively closed system of all monic polynomials in $K[X_1]$, then M_S is a projective module over $K[X_1, \ldots, X_n]_S = K[X_1]_S[X_2, \ldots, X_n]$.

It suffices to show that $K[X_1]_S$ is a principal ideal ring, for then M_S is a free $K[X_1, \ldots, X_n]_S$-module by the induction hypothesis. Just as in the proof of 3.10 we get an $f \in S$ such that M_f is a free $K[X_1, \ldots, X_n]_f$-module. By 3.14 M is then a free $K[X_1, \ldots, X_n]$- module.

Lemma 3.16. $K[X_1]_S$ is a principal ideal ring.

As a ring of fractions of a factorial ring $R := K[X_1]_S$ is itself factorial (III.4.12e)). For $\mathfrak{p} \in \mathrm{Spec}(R)$ with $\mathfrak{p} \cap K = (0)$, $R_\mathfrak{p}$ is a ring of fractions of $Q(K)[X_1]$. Therefore $h(\mathfrak{p}) \leq 1$ and $\mathfrak{p}$ is a principal ideal. On the other hand, if $\mathfrak{p} \cap K = (p)$ with some prime element p of K, then $R/pR := K/(p)(X_1)$ is a field, so $\mathfrak{p} = pR$. Any $\mathfrak{p} \in \mathrm{Spec}(R)$, $\mathfrak{p} \neq (0)$, is therefore generated by a prime element π of R.

From this it follows that R is a principal ring: For $a_1, a_2 \in R \setminus \{0\}$ let c be the greatest common divisor of a_1, a_2. If $\mathfrak{p} = (\pi)$ is a prime ideal of R and if π

occurs in the factorization of a_i to the power ν_i $(i = 1, 2)$, then it occurs in c to the power $\text{Min}\{\nu_1, \nu_2\}$. It follows that $(a_1, a_2)R_\mathfrak{p} = cR_\mathfrak{p}$ for all $\mathfrak{p} \in \text{Spec}(R)$ and hence $(a_1, a_2) = (c)$ by 1.1. But then R is a principal ideal ring, q. e. d.

If a projective module is not free, then we can always try to split off a free module of greatest possible rank from it as a direct summand. This problem is dealt with by Serre's Splitting-off Theorem, to which we now turn.

In what follows let $R \neq \{0\}$ again be an arbitrary ring; let P and P_0 be finitely generated projective R-modules. And let P_0 be of rank 1 (a "line bundle").

Lemma 3.17.

a) $P^* := \text{Hom}_R(P, P_0)$ is a finitely generated projective R-module with $\mu_\mathfrak{p}(P^*) = \mu_\mathfrak{p}(P)$ for all $\mathfrak{p} \in \text{Spec}(R)$.

b) The canonical R-linear mapping

$$\alpha : P \to \text{Hom}_R(\text{Hom}_R(P, P_0), P_0)$$

(which maps $m \in P$ to the linear mapping that takes any $l \in \text{Hom}_R(P, P_0)$ to $l(m)$) is an isomorphism.

Proof.

a) There are R-modules P_0' and P' such that $P_0 \oplus P_0' \cong R^r$, $P \oplus P' \cong R^s$ for some $r, s \in \mathbb{N}$. Hence $\text{Hom}_R(P, P_0)$ is a direct summand of $\text{Hom}_R(R^s, R^r) \cong R^{r \cdot s}$ and therefore a finitely presentable projective R- module. For all $\mathfrak{p} \in \text{Spec}(R)$ we have $P_\mathfrak{p}^* \cong \text{Hom}_{R_\mathfrak{p}}(P_\mathfrak{p}, (P_0)_\mathfrak{p})$ by 1.10. $P_\mathfrak{p}$ is a free $R_\mathfrak{p}$-module and $(P_0)_\mathfrak{p} \cong R_\mathfrak{p}$, since P_0 is of rank 1. It follows that $P_\mathfrak{p}^*$ has the same rank as $P_\mathfrak{p}$.

b) By 1.10 the canonical mapping α commutes with localization, because all the modules involved are finitely presentable. For all $\mathfrak{m} \in \text{Max}(R)$ the corresponding local mapping $\alpha_\mathfrak{m}$ is bijective, since $P_\mathfrak{m}$ is free and $(P_0)_\mathfrak{m} \cong R_\mathfrak{m}$, for the canonical mapping of a free module of finite rank into its bidual module is bijective. By 1.6 it follows that α is bijective.

The following is another fundamental theorem on projective modules.

Theorem 3.18. (Serre's Splitting-off Theorem) Let X be the J-spectrum of a ring R. Let X be Noetherian and of finite dimension. Let P_0 be a finitely generated projective R-module of rank 1 and P a finitely generated projective R-module with

$$\mu_\mathfrak{p}(P) > \dim X$$

for all $\mathfrak{p} \in X$. Then $P \cong P_0 \oplus P'$ with some (projective) R-module P' (P splits off P_0 as a direct summand).

We reduce the proof to an existence theorem for globally basic elements in a not necessarily projective module.

Theorem 3.19. (Eisenbud–Evans [19]) Let X be the spectrum or the J-spectrum of a ring R. Let X be Noetherian, M a finitely generated R-module. Let there be given elements $m_1, \ldots, m_t \in M$ such that for all $\mathfrak{p} \in X$:

$$\mu_\mathfrak{p}(M) - \mu_\mathfrak{p}(M/\langle m_1, \ldots, m_t \rangle) \geq \mathrm{Min}\{t, \dim \mathfrak{V}(\mathfrak{p}) + 1\}.$$

Then there are elements $a_2, \ldots, a_t \in R$ such that $m_1 + a_2 m_2 + \cdots + a_t m_t$ is basic for all $\mathfrak{p} \in X$.

We first prove the Splitting–off Theorem with the help of the Eisenbud–Evans Theorem. In the situation of 3.18 let $d := \dim X$.

a) If $\{m_1, \ldots, m_t\}$ is a generating system of P, then $t \geq d + 1$. There is a finitely generated R-module P' such that $F := P \oplus P'$ is a free R-module. From $\mu_\mathfrak{p}(F) - \mu_\mathfrak{p}(F/\langle m_1, \ldots, m_t \rangle) = \mu_\mathfrak{p}(F) - \mu_\mathfrak{p}(P') = \mu_\mathfrak{p}(P) \geq d+1$, by 3.19 it follows that there exists an element $m := m_1 + a_2 m_2 + \cdots + a_t m_t$ that is basic at all $\mathfrak{p} \in X$, in particular at the $\mathfrak{m} \in \mathrm{Max}(R)$. By 1.13 it generates a free direct summand $\neq \langle 0 \rangle$ of F. Then $F/\langle m \rangle \cong P/\langle m \rangle \oplus P'$ is projective, therefore $P/\langle m \rangle$ is also projective and $\langle m \rangle$ is a direct summand of P. Thus we have proved the Splitting-off Theorem for $P_0 = R$.

b) In the general case it follows, according to 3.17 and part a), that we have $P^* := \mathrm{Hom}_R(P, P_0) := R \oplus Q$ with an R-module Q. Then $P \cong P^{**} \cong \mathrm{Hom}_R(R, P_0) \oplus \mathrm{Hom}_R(Q, P_0)$. Since $\mathrm{Hom}_R(R, P_0) \cong P_0$, P splits off P_0 as a direct summand, q. e. d.

In the proof of 3.19 the following lemma will be used:

Lemma 3.20. Let X be the spectrum or the J-spectrum of R and let X be Noetherian. For elements $m_1, \ldots, m_t \in M$ and all $\mathfrak{p} \in X$ suppose that

$$\mu_\mathfrak{p}(M) - \mu_\mathfrak{p}(M/\langle m_1, \ldots, m_t \rangle) \geq \mathrm{Min}\{t, \dim \mathfrak{V}(\mathfrak{p}) + 1\}.$$

Then there are only finitely many $\mathfrak{p} \in X$ for which both $\dim \mathfrak{V}(\mathfrak{p}) + 1 < t$ and $\mu_\mathfrak{p}(M) - \mu_\mathfrak{p}(M/\langle m_1, \ldots, m_t \rangle) = \dim \mathfrak{V}(\mathfrak{p}) + 1$.

Proof. We put $U := \langle m_1, \ldots, m_t \rangle$, $\overline{M} := M/U$. Let there be given a $\mathfrak{p} \in X$ with $\dim \mathfrak{V}(\mathfrak{p}) + 1 < t$ and $\mu_\mathfrak{p}(M) - \mu_\mathfrak{p}(M/U) = \dim \mathfrak{V}(\mathfrak{p}) + 1$. If $r := \mu_\mathfrak{p}(\overline{M})$, then it suffices to show that $\mathfrak{p}$ is minimal in the set $A_r(\overline{M})$ of all $\mathfrak{q} \in X$ with $\mu_\mathfrak{q}(\overline{M}) \geq r$, for by 2.6 this set is closed and so has only finitely many minimal elements, since X is Noetherian. Since there are only finitely many distinct $A_r(\overline{M})$, the assertion follows.

Suppose there were some $\mathfrak{q} \in A_r(\overline{M})$ with $\mathfrak{q} \subsetneq \mathfrak{p}$. Then

$$\dim \mathfrak{V}(\mathfrak{q}) > \dim \mathfrak{V}(\mathfrak{p}), \quad \mu_\mathfrak{q}(\overline{M}) = r = \mu_\mathfrak{p}(\overline{M}), \quad \text{and} \quad \mu_\mathfrak{q}(M) \leq \mu_\mathfrak{p}(M).$$

From the relations $t > \dim \mathfrak{V}(\mathfrak{p}) + 1 = \mu_\mathfrak{p}(M) - \mu_\mathfrak{p}(\overline{M}) \geq \mu_\mathfrak{q}(M) - \mu_\mathfrak{q}(\overline{M}) \geq \mathrm{Min}\{t, \dim \mathfrak{V}(\mathfrak{q}) + 1\} > \dim \mathfrak{V}(\mathfrak{p}) + 1$, we would get a contradiction. Therefore $\mathfrak{p}$ is minimal in $A_r(\overline{M})$, q. e. d.

Proof of 3.19.

Under the hypotheses of the theorem, for each $s \in \mathbb{N}$ with $1 \leq s \leq t$ we shall construct elements $m_1^{(s)}, \ldots, m_s^{(s)} \in M$ of the form

$$m_i^{(s)} = m_i + \sum_{j=s+1}^{t} a_{ij} m_j \qquad (i = 1, \ldots, s; \ a_{ij} \in R)$$

such that for all $\mathfrak{p} \in X$

$$\mu_{\mathfrak{p}}(M) - \mu_{\mathfrak{p}}(M/\langle m_1^{(s)}, \ldots, m_s^{(s)} \rangle) \geq \mathrm{Min}\{s, \dim \mathfrak{V}(\mathfrak{p}) + 1\}.$$

For $s = 1$ we then get an element of the form $m = m_1 + \sum_{j=2}^{t} a_j m_j$ $(a_j \in R)$ with $\mu_{\mathfrak{p}}(M) - \mu_{\mathfrak{p}}(M/\langle m \rangle) \geq 1$ for all $\mathfrak{p} \in X$, i.e. an element that is basic at all $\mathfrak{p} \in X$.

For $s = t$ we take the given elements $m_1, \ldots, m_t$. If for some s with $1 < s \leq t$ the elements $m_1^{(s)}, \ldots, m_s^{(s)}$ have already been constructed in the way desired, then by 3.20 the set of $\mathfrak{p} \in X$ with $\dim \mathfrak{V}(\mathfrak{p}) + 1 < s$ and $\mu_{\mathfrak{p}}(M) - \mu_{\mathfrak{p}}(M/\langle m_1^{(s)}, \ldots, m_s^{(s)} \rangle) = \dim \mathfrak{V}(\mathfrak{p}) + 1$ is finite. Let $\mathfrak{p}_1, \ldots, \mathfrak{p}_r$ be the elements of this set.

If we put $m_i^{(s-1)} := m_i^{(s)} + a_i m_s^{(s)}$ $(i = 1, \ldots, s-1)$, then by 2.12 we can choose the a_i in R so that

$$\mu_{\mathfrak{p}_i}(M) - \mu_{\mathfrak{p}_i}(M/\langle m_1^{(s-1)}, \ldots, m_{s-1}^{(s-1)} \rangle) = \dim \mathfrak{V}(\mathfrak{p}_i) + 1 \qquad (i = 1, \ldots, r).$$

But for the $\mathfrak{p} \in X \setminus \{\mathfrak{p}_1, \ldots, \mathfrak{p}_r\}$ we have

$$\mu_{\mathfrak{p}}(M) - \mu_{\mathfrak{p}}(M/\langle m_1^{(s-1)}, \ldots, m_{s-1}^{(s-1)} \rangle) \geq \mu_{\mathfrak{p}}(M) - \mu_{\mathfrak{p}}(M/\langle m_1^{(s)}, \ldots, m_s^{(s)} \rangle) - 1$$
$$\geq \mathrm{Min}\{s - 1, \dim \mathfrak{V}(\mathfrak{p}) + 1\}.$$

The elements $m_1^{(s-1)}, \ldots, m_{s-1}^{(s-1)}$ have the desired property for all $\mathfrak{p}$, q. e. d.

The main results of this chapter, the theorem of Forster–Swan, the theorem of Quillen–Suslin, and Serre's Splitting-off Theorem will later be applied to answer questions about the number of generators of ideals of algebraic varieties.

Exercises

1. Let R be a ring. $(r_1, \ldots, r_n) \in R^n$ is called a *unimodular row* if the ideal in R generated by $r_1, \ldots, r_n$ equals R.

   a) For any unimodular row $(r_1, \ldots, r_n) \in R^n$ we have $P := \{(f_1, \ldots, f_n) \in R^n \mid \sum_{i=1}^{n} f_i r_i = 0\}$ is projective R-module of rank $n - 1$.

   b) Let $R := K[X_1, \ldots, X_m]$ with K a principal ideal domain $(m \geq 0)$. $(r_1, \ldots, r_n) \in R^n$ is a unimodular row if and only if there is a matrix $A \in M(n \times n; R)$ with first row $(r_1, \ldots, r_n)$ such that $\det(A)$ is a unit of R.

2. For a finitely generated module M over a ring R the following statements are equivalent.

   a) M is projective of rank r.

   b) For the Fitting ideals $F_i(M)$ (cf. §1, Exercise 8) we have:

$$F_0(M) = \cdots = F_{r-1}(M) = 0, \qquad F_r(M) = R.$$

3. Let M be a finitely presentable module over a reduced ring R. If $\mu_{\mathfrak{p}}(M)$ is constant on $\mathrm{Spec}(R)$, then M is projective.

4. An ideal $I \neq (0)$ of a Noetherian ring R with connected spectrum is projective as an R-module if and only if for all $\mathfrak{m} \in \mathrm{Max}(R)$ the ideal $I_{\mathfrak{m}} \subset R_{\mathfrak{m}}$ is generated by a non-zerodivisor of R. Two such ideals I_1, I_2 are isomorphic as R-modules if and only if there are non-zerodivisors $r_1, r_2 \in R$ such that $r_1 I_1 = r_2 I_2$.

5. Any finitely generated projective module P of rank 1 over a Noetherian ring R is isomorphic to an ideal $I \subset R$ that is locally a principal ideal, generated by some non-zerodivisor. Hint: If S is the set of all non-zerodivisors of R, then $P_S \cong R_S$ as R_S-modules. If $J \subset R_S$ is the image of P under this isomorphism, then there is an $s \in S$ with $sJ \subset R$.

6. Let R be a ring, K its full ring of fractions. An R-module $J \subset K$ for which there is a non-zerodivisor s of R with $sJ \subset R$ is called a *fractional R-ideal*. A fractional R-ideal J is called *invertible* if there is another fractional R-ideal J' such that $J \cdot J' = R$. Here $J \cdot J'$ is the set of all finite sums $\sum x_i y_i$ ($x_i \in J, y_i \in J'$) (ideal multiplication).

   a) The fractional principal ideals $J = xR$, where $x \in K$ is a unit, are invertible.

   b) The invertible R-ideals form (under ideal multiplication) a group $I(R)$; the fractional principal ideals considered in a) form a subgroup $\mathcal{H}(R)$. (The factor group $I(R)/\mathcal{H}(R)$ is called the *Picard group* of R; it is an important invariant of R.)

7. With the notation of Exercise 6, let R be a local ring. Then $I(R) = \mathcal{H}(R)$. Hint: To show that any $J \in I(R)$ is a fractional principal ideal, start from an equation $\sum_{i=1}^{n} a_i b_i = 1$ with $a_i \in J$ and $b_i \in J'$ ($i = 1, \ldots, n$) if $J \cdot J' = R$.

8. For a Noetherian ring R, $I(R)$ is the set of fractional R-ideals that are projective as R-modules.

9. A finitely generated module over a Dedekind ring (cf. 2.16) is projective if and only if it is torsion-free. For a Dedekind ring R the fractional R-ideals $\neq (0)$ form a group under ideal multiplication.

10. Let P be a finitely generated module of rank r over a Noetherian ring R of dimension 1. Then $P \cong I \oplus R^{r-1}$ with some ideal $I \neq (0)$ of R that is uniquely determined up to R-isomorphism by P.

References

The local–global principle is one of the most important methods of proof in commutative algebra. It first appeared in a similar form in number theory after Hensel's discovery of the p-adic numbers. Krull [46], who first investigated local rings systematically, often applied the principle in the form in which it appears in the text.

The procedure in §1 for proving Quillen's local–global statement for extended modules follows an (unpublished) suggestion of Hochster's. For the Forster–Swan theorem the following sharpening was conjectured by Eisenbud-Evans [20]: If $R = P[X]$ is a polynomial ring over a Noetherian ring P and M is a finitely generated R-module, then

$$\mu(M) \leq \mathrm{Max}\{\mu_{\mathfrak{p}}(M) + \dim R/\mathfrak{p} \mid \mathfrak{p} \in \mathrm{Spec}(R), \dim R/\mathfrak{p} < \dim R\}.$$

This was first shown by Sathaye [71] under some restrictive hypothesis and then confirmed in general by Mohan Kumar [56]. The proof uses the results of Quillen [66] and Suslin [75] in connection with the solution of Serre's problem, which was presented in §3. For another proof and related topics see: B. R. Plumstead, The conjectures of Eisenbud and Evans, *Amer. J. Math.* **105**, 1417–1433 (1983). S. Mandal, Number of generators of modules over Laurent polynomial rings, *J. Alg.* **80**, 306–313 (1983); Basic elements and cancellation over Laurent polynomial rings, *J. Alg.* **79**, 251–257 (1982). S. M. Bhatwadekar and A. Roy, Some theorems about projective modules over polynomial rings, *J. Alg.* **86**, 150–158 (1984); Stability theorems for overrings of polynomials rings, *Inv. Math.* **68**, 117–127 (1982).

The twenty-year history of the solution of Serre's problem is reported by Bass's survey article [8] (published shortly before the solution) and Lam's comprehensive exposition [50] (see also Ferrand [23] and Swan [77]). The proof of Horrock's theorem presented in the text goes back to Lindel [52]. A similar proof was also given by Swan (see [77]). At present one tries to prove the freeness of the finitely generated projective $R[X_1, \ldots, X_n]$-modules for more general rings R (see Ferrand [23], Lam [50], Lindel [53]). For more recent results on this subject see: H. Lindel, On a question of Bass, Quillen and Suslin concerning projective modules over polynomial rings, *Inv. Math.* **65**, 319–323 (1981). S. M. Bhatwadekar and R. A. Rao, On a question of Quillen, *Trans. AMS* **279**, 801–810 (1983).

The existence theorem 3.19 for basic elements is strongly related to the theorems of Bertini in classical algebraic geometry. From a generalization of 3.19 Flenner (Die Sätze von Bertini für lokale Ringe, *Math. Ann.* **229**, 253–294 (1977)) first derived Bertini theorems for local rings and from them also some of the global Bertini theorems.

Many theorems in §3 are analogues of facts from the theory of vector bundles over topological spaces. In the algebraic case the proofs are generally quite different from the topological, yet topology here proves itself a rich source of possibly correct algebraic results.

Chapter V

On the number of equations needed to describe an algebraic variety

We know that a nonempty linear variety in n-dimensional space that is described by a system of linear equations of rank r has dimension $n - r$, and that any linear variety of dimension d can always be described by $n - d$ equations. Two intersecting linear varieties L_1, L_2 satisfy the dimension formula $\dim(L_1 \cap L_2) = \dim L_1 + \dim L_2 - \dim(L_1 + L_2)$, where $L_1 + L_2$ is the join-space.

For algebraic varieties—the solution sets of systems of algebraic equations—the corresponding facts are essentially more difficult to prove, and instead of equations we generally get only estimates. This chapter is devoted to generalizations of the above results of linear algebra. As usual the problems are very closely related to ideal-theoretic inquiries, e.g. to the problem of making more precise statements about the number of generators of ideals.

We begin with an upper estimate for the number of equations needed to describe a variety.

1. Any variety in n-dimensional space is the intersection of n hypersurfaces.

In spite of the simplicity of its derivation this fact was proved just a few years ago. In 1882 Kronecker [42] noted that one can get by with $n + 1$ hypersurfaces (Ch. I, §5, Exercise 1, cf. also Exercises 1 and 2 below). The proof of the theorem in the heading uses only a few facts from Chapters I–III and some ring-theoretic statements that we shall now derive.

Lemma 1.1. Let $R = R_1 \times \cdots \times R_n$ be a direct product of rings. R is a principal ideal ring if and only if each R_j $(j = 1, \ldots, n)$ is a principal ideal ring.

Since R_j is a homomorphic image of R, if R is a principal ideal ring, so is R_j. Any ideal I of R is of the form $I = I_1 \times \cdots \times I_n$, where I_j is the image of I in R_j. If I_j is generated by r_j $(j = 1, \ldots, n)$, then I is generated by $(r_1, \ldots, r_n)$.

Lemma 1.2. Let $R \neq \{0\}$ be a reduced ring with only finitely many minimal prime ideals, K its full ring of fractions. Then the polynomial ring $K[X]$ is a principal ideal ring.

By III.4.23 we have $K = K_1 \times \cdots \times K_n$ with fields K_j, and hence $K[X] \cong K_1[X] \times \cdots \times K_n[X]$. The assertion now follows by 1.1.

There is an analogous lemma in the graded case. If R is a positively graded ring, we consider the polynomial ring $R[X]$ as a graded ring in the following way.

Let $\alpha \in \mathbb{N}_+$ be given. A polynomial $\sum r_i X^i$ is homogeneous of degree d if $r_i \in R$ is homogeneous of degree $d - i\alpha$ for all $i \in \mathbb{N}$. In particular, the variable X is of degree α.

Lemma 1.3. Suppose R is reduced and has only finitely many minimal prime ideals. Suppose the set S of all homogeneous non-zerodivisors of R contains an element of positive degree. Then the ring $R[X]_{(S)}$ of all fractions f/s, where $f \in R[X]$ and $s \in S$ are homogeneous of the same degree, is a principal ideal ring.

Proof. Let $\mathfrak{p}_1, \ldots, \mathfrak{p}_n$ be the minimal prime ideals of R. By III.4.24 we have an isomorphism of graded rings $R_S \cong (R_1)_{S_1} \times \cdots \times (R_n)_{S_n}$, where $R_j := R/\mathfrak{p}_j$ and S_j is the set of homogeneous elements $\neq 0$ of R_j $(j = 1, \ldots, n)$. Hence we have an isomorphism of graded rings

$$R[X]_S = R_S[X] \cong (R_1)_{S_1}[X] \times \cdots \times (R_n)_{S_n}[X]$$

and an isomorphism

$$R[X]_{(S)} = R_1[X]_{(S_1)} \times \cdots \times R_n[X]_{(S_n)}$$

of the subrings consisting of the elements of degree 0. By 1.1 it suffices to show that the $R_j[X]_{(S_j)}$ are principal ideal rings; that is, we may assume that R is an integral domain.

If in this case $I \subset R[X]_{(S)}$ is an ideal $\neq (0)$, then we consider an element $f/s \in I \setminus \{0\}$ for which $f = r_0 X^n + r_1 X^{n-1} + \cdots + r_n$ is of least possible degree n (here we are using the usual degree of a polynomial). If $n = 0$, then f/s is a unit and there is nothing to show. Hence let $n > 0$. If g/s' is another element of I, then $r_0^m g$ can for suitable $m \in \mathbb{N}$ be divided by f with remainder:

$$r_0^m g = q \cdot f + r \qquad (q, r \in R[X], \deg_X r < n).$$

Here q and r are homogeneous elements of $R[X]$ with respect to the given grading of $R[X]$, since r_0, f, g are homogeneous. From the equation

$$\frac{r}{r_0^m s'} = \frac{g}{s'} - \frac{sq}{r_0^m s'} \cdot \frac{f}{s},$$

it follows that $r/r_0^m s' \in I$ and so $r = 0$, because f was of minimal degree in X. Thus any element of I is a multiple of f/s.

Theorem 1.4. (Storch [79], Eisenbud–Evans [17].)

a) If R is a d-dimensional Noetherian ring $(d < \infty)$ and $I \subset R[X]$ is an ideal, then there are elements $f_1, \ldots, f_{d+1} \in I$ with

$$\mathrm{Rad}(I) = \mathrm{Rad}(f_1, \ldots, f_{d+1}).$$

b) Let $R = \bigoplus_{i \in \mathbb{N}} R_i$ be a positively graded Noetherian ring, where $R_0 =: K$ is a field. Let $g\text{-dim}\, R =: d < \infty$. $R[X]$ will be considered as a graded ring with X homogeneous of degree $\alpha \in \mathbb{N}_+$. Let $I \subset R[X]$ be a homogeneous ideal with $I \subset M \cdot R[X]$, where $M := \bigoplus_{i>0} R_i$. Then there are homogeneous elements $f_1, \ldots, f_{d+1} \in I$ with

$$\text{Rad}(I) = \text{Rad}(f_1, \ldots, f_{d+1}).$$

Proof. We may assume that R is reduced. Indeed, if $\mathfrak{n}$ is the ideal of all nilpotent elements of R (it is a homogeneous ideal in the graded case) and if the theorem has already been proved for $R_{\text{red}} := R/\mathfrak{n}$, then in the image $\overline{I}$ of I in $R_{\text{red}}[X]$ there are (homogeneous) elements $\overline{f}_1, \ldots, \overline{f}_{d+1}$ with

$$\text{Rad}(\overline{I}) = \text{Rad}(\overline{f}_1, \ldots, \overline{f}_{d+1}).$$

Note that R_{red} has the same dimension (g-dimension) as R. If for each $\overline{f}_i$ we choose a (homogeneous) representative $f_i \in I$, then

$$\text{Rad}(I) = \text{Rad}(f_1, \ldots, f_{d+1}).$$

Now let R be reduced and $d := \dim R$ ($d := g\text{-dim}\, R$ in the graded case). If $d = -1$ in case a), then the statement of the theorem is trivial. In the graded case if $d = -1$, then M is a minimal prime ideal of R and so $M = (0)$, therefore also $I = (0)$. The statement of the theorem is thus correct in this case.

Now let $d \geq 0$ and let S be the set of all (homogeneous) non-zerodivisors of R. In the graded case S contains elements of positive degree, since $d \geq 0$. By 1.2 (resp. 1.3) I_S (resp. the ideal $I_{(S)}$ of all fractions f/s with $f \in I$, $s \in S$, $\deg f = \deg s$) is a principal ideal generated by an element $f_1 \in I$ (resp. by an f_1/s, $f_1 \in I$, $s \in S$ homogeneous of the same degree). If $g_1, \ldots, g_t$ is any generating system of I (consisting of homogeneous elements in the graded case), then we have equations

$$rg_j^{\rho_j} = h_j f_1$$

with a (homogeneous) non-zerodivisor $r \in R$, (homogeneous) elements $h_j \in R[X]$ and numbers $\rho_j \in \mathbb{N}$ ($j = 1, \ldots, t$). Then

$$rI^\sigma \subset f_1 R[X] \subset I \qquad \text{for sufficiently large } \sigma \in \mathbb{N}_+$$

and

$$\mathfrak{V}(I) \subset \mathfrak{V}(f_1) \subset \mathfrak{V}(rI) = \mathfrak{V}(r) \cup \mathfrak{V}(I), \qquad (1)$$

where $\mathfrak{V}$ is the zero-set in $\text{Spec}(R[X])$ or $\text{Proj}(R[X])$.† If r is a unit of R, then we have finished. Otherwise, $\overline{R} := R/(r)$ is a (positively graded) ring of smaller

---

† In the projective case $\mathfrak{V}(I)$ is the set of all relevant prime ideals of $R[X]$ that contain I.

dimension (g-dimension) than R, since r is contained in none of the minimal prime ideals of R. If $\overline{I}$ is the image of I in $\overline{R}[X]$, then by the induction hypothesis and because we can immediately pass over to $\overline{R}_{\mathrm{red}}$, there are (homogeneous) elements $\overline{f}_2, \ldots, \overline{f}_{d+1}$ with

$$\mathrm{Rad}(\overline{f}_2, \ldots, \overline{f}_{d+1}) = \mathrm{Rad}(\overline{I}).$$

For each $\overline{f}_i$ we choose a (homogeneous) representative $f_i \in I$ $(i = 2, \ldots, d+1)$. Then

$$\mathrm{Rad}(r, f_2, \ldots, f_{d+1}) = \mathrm{Rad}(R[X]r + I)$$

and so

$$\mathfrak{V}(r) \cap \mathfrak{V}(I) = \mathfrak{V}(R[X]r + I) = \mathfrak{V}(r, f_2, \ldots, f_{d+1}). \qquad (2)$$

From (1) and (2) we now get

$$\mathfrak{V}(f_1, \ldots, f_{d+1}) = \mathfrak{V}(f_1) \cap \mathfrak{V}(f_2, \ldots, f_{d+1}) \subset (\mathfrak{V}(r) \cup \mathfrak{V}(I)) \cap \mathfrak{V}(f_2, \ldots, f_{d+1})$$
$$= \mathfrak{V}(r, f_2, \ldots, f_{d+1}) \cup \mathfrak{V}(I) = \mathfrak{V}(I).$$

And since $\mathfrak{V}(I) \subset \mathfrak{V}(f_1, \ldots, f_{d+1})$, equality follows and so we obtain $\mathrm{Rad}(I) = \mathrm{Rad}(f_1, \ldots, f_{d+1})$, q. e. d.

Corollary 1.5. Let I be an ideal of the polynomial ring $K[X_1, \ldots, X_n]$ over a field K. Then there are polynomials $f_1, \ldots, f_n \in I$ with

$$\mathrm{Rad}(I) = \mathrm{Rad}(f_1, \ldots, f_n).$$

If $V \subset \mathbf{A}^n(L)$ is a nonempty K-variety, then V is the intersection of n K-hypersurfaces.

Proof. Put $R := K[X_1, \ldots, X_{n-1}]$, $X := X_n$ and apply 1.4a.

Corollary 1.6. Let $\alpha_0, \ldots, \alpha_n$ be positive integers. Let the polynomial ring $R = K[X_0, \ldots, X_n]$ over a field K have the grading in which $\deg(X_i) = \alpha_i$ $(i = 0, \ldots, n)$. Let I be a homogeneous ideal with the property: There are homogeneous elements $Y_0, \ldots, Y_n$ that generate R as a K-algebra such that $I \subset (Y_0, \ldots, Y_{n-1})R$. Then there are homogeneous polynomials $f_1, \ldots, f_n \in I$ with

$$\mathrm{Rad}(I) = \mathrm{Rad}(f_1, \ldots, f_n).$$

This follows from 1.4b) if we consider that $g\text{-dim}(k[Y_0, \ldots, Y_{n-1}]) = n - 1$. In fact $(0) \subset (Y_0) \subset \cdots \subset (Y_0, \ldots, Y_{n-2})$ is a chain of relevant prime ideals of length $n - 1$, and there can be no longer chain.

Corollary 1.7. If $V \subset \mathbf{P}^n(L)$ is a projective K-variety that has a K-rational point (i.e. a point $\langle x_0, x_1, \ldots, x_n \rangle$ with $x_i \in K$ $(i = 0, \ldots, n)$), then V is the intersection of n projective K-hypersurfaces.

Proof. After a projective change of coordinates with coefficients in K we may assume that the K-rational point has coordinates $\langle 0, \ldots, 0, 1 \rangle$. The ideal of V in $K[X_0, \ldots, X_n]$ is then contained in $(X_0, \ldots, X_{n-1})$ and 1.6 can be applied.

Of course the existence of a K-rational point on V is guaranteed if K is algebraically closed (Hilbert's Nullstellensatz).

Remark. Theorem 1.4b) is generally not correct under the weaker hypothesis that $\mathrm{Rad}(I)$ is distinct from the irrelevant maximal ideal of $R[X]$ (Exercises 3 and 4). It seems to be unknown whether one can get by without the existence of a K-rational point in 1.7. For a special case in which this has been proved, cf. [67], Anhang.

Exercises

1. Let G be a positively graded ring, where $G_0 = K$ is a field with infinitely many elements and G is finitely generated as a K-algebra. If $\dim G =: n$, then for any homogeneous ideal $I \subset G$ there are homogeneous elements $F_1, \ldots, F_n \in I$ with $\mathrm{Rad}(I) = \mathrm{Rad}(F_1, \ldots, F_n)$. (Hint: I.§5, Exercise 1 and II.3.7.)

2. Under the hypotheses of 1.4b) let $I \subset R[X]$ be any homogeneous ideal (not necessarily $I \subset MR[X]$). Then there are homogeneous elements $f_1, \ldots, f_{d+2} \in I$ with $\mathrm{Rad}(I) = \mathrm{Rad}(f_1, \ldots, f_{d+2})$. (Hint: If $I \not\subset MR[X]$, then $I = (f) + I'$ with $f \in I$ a monic polynomial in X and a homogeneous ideal $I' \subset MR[X]$).

3. In the polynomial ring $K[T]$ over a field of K of characteristic 0, consider the graded K-subalgebra $R := K[T^2, T^3]$, and let $I \subset R[X]$ be the ideal generated by $X^2 - T^2$ and $X^3 - T^3$. Let $\deg(X) = 1$.

   a) I is a homogeneous prime ideal of height 1.

   b) There is no homogeneous element $F \in I$ with $I = \mathrm{Rad}(F)$.

   (On the other hand if K is a field of characteristic $p > 0$, it follows that $I = \mathrm{Rad}(X^p - T^p)$.)

4. Give an example analogous to that in Exercise 3, where R is generated as a K-algebra by homogeneous elements of degree 1.

2. Rings and modules of finite length

This section serves mainly to provide a fact that is used in proving Krull's Principal Ideal Theorem in §3. The length of a module M is (along with the number $\mu(M)$ discussed in IV.§2) another generalization of the concept of the dimension of a vector space which is often used.

A *normal series* in M is a chain

$$M = M_0 \supset M_1 \supset \cdots \supset M_l = \langle 0 \rangle \tag{1}$$

of submodules M_i of M with $M_i \neq M_{i+1}$ $(i = 0, \ldots, l-1)$. l is called the *length* of the normal series. A normal series (1) is called a *composition series* if M_i/M_{i+1} $(i = 0, \ldots, l-1)$ is a simple module, i.e. a module whose only proper submodule is the zero module. Therefore, a composition series cannot be "refined" by inserting more submodules.

Lemma 2.1. A module $M \neq \langle 0 \rangle$ over a ring R is simple if and only if there is an $\mathfrak{m} \in \mathrm{Max}(R)$ such that $M \cong R/\mathfrak{m}$.

Proof. It is clear that $R/\mathfrak{m}$ is a simple R-module for all $\mathfrak{m} \in \mathrm{Max}(R)$. If $M \neq \langle 0 \rangle$ is a simple R-module, choose $m \in M, m \neq 0$. Then $M = Rm$ and we have an epimorphism $R \to M$ ($r \mapsto rm$). If $\mathfrak{m}$ is its kernel, then $M \cong R/\mathfrak{m}$. $\mathfrak{m}$ is a maximal ideal; otherwise, M would contain a proper submodule $\neq \langle 0 \rangle$.

Definition 2.2. M is called an *Artinian module* if any decreasing chain of submodules

$$M = M_0 \supset M_1 \supset \cdots$$

becomes stationary. M is called *of finite length* if there is a bound on the lengths of all normal series (1) of M. The maximum of the lengths of the normal series is then called the *length* $l(M)$ of M. A ring R is called *Artinian* (resp. *of finite length*) if as an R-module it is Artinian (resp. of finite length).

In particular, any module of finite length is Artinian and Noetherian. Any normal series of such a module can be refined to a composition series. We have:

Theorem 2.3. (Jordan-Hölder) A module that has a composition series has finite length, and all its composition series have the same length.

Proof. Let $M = M_0 \supset M_1 \supset \cdots \supset M_l = \langle 0 \rangle$ be an arbitrary composition series of the module M. By induction on l we show that any normal series of M has length $\leq l$. Then this also holds for composition series and, because we started out from an arbitrary composition series, it follows that they all have length l.

For $l = 0$ or $l = 1$ the theorem is trivial. Therefore, let $l > 1$ and suppose the assertion has already been proved for modules that have a composition series of smaller length. Let $M = N_0 \supset N_1 \supset \cdots \supset N_\lambda = \langle 0 \rangle$ be a normal series of M. If $N_1 \subset M_1$, then it follows by the induction hypothesis applied to M_1 that $\lambda - 1 \leq l - 1$. If $N_1 \not\subset M_1$, then $N_1 + M_1 = M$, since M/M_1 is simple. From $M/M_1 = N_1 + M_1/M_1 \cong N_1/M_1 \cap N_1$ it follows that $N_1/M_1 \cap N_1$ is also simple.

Since M_1 has a composition series of length $l - 1$, it follows by the induction hypothesis that in the proper submodule $M_1 \cap N_1$ of M_1 all the normal series have length $\leq l - 2$. Since $N_1/M_1 \cap N_1$ is simple it follows that N_1 has a composition series of length $\leq l - 1$. Then $\lambda - 1 \leq l - 1$, q. e. d.

Corollary 2.4. (Additivity of length.) Let there be given a normal series (1) in M. M is of finite length if and only if M_i/M_{i+1} is of finite length for $i = 0, \ldots, l - 1$. Then

$$l(M) = \sum_{i=0}^{l-1} l(M_i/M_{i+1}).$$

Proof. It suffices to prove the assertion for $l = 2$; the general case then easily follows by induction. Let M be of finite length and $M \supset M_1 \supset M_2 = \langle 0 \rangle$ be a normal series. We refine it to a composition series. The modules of the composition series lying between M and M_1 then provide a composition series of

M/M_1; those lying between M_1 and M_2 a composition series of M_1. It follows that $l(M) = l(M/M_1) + l(M_1)$.

If M/M_1 and M_1 are of finite length, then we get a composition series of M by lengthening a composition series of M_1 with the inverse images in M of the modules from a composition series of M/M_1.

By 2.4 any submodule and any homomorphic image of a module of finite length are of finite length. A direct sum of finitely many modules of finite length is likewise of finite length, and the length of the sum equals the sum of the lengths of the summands.

Proposition 2.5. A ring $R \neq \{0\}$ is of finite length if and only if it is Noetherian and $\dim R = 0$.

Proof. If R is of finite length, then R is Noetherian and Artinian. For all $\mathfrak{p} \in \operatorname{Spec}(R)$. $R/\mathfrak{p}$ is also of finite length. For $a \in R/\mathfrak{p}$, $a \neq 0$, the ideal chain $(a) \supset (a^2) \supset \cdots \supset (a^n) \supset \cdots$ is stationary: $(a^n) = (a^{n+1})$ for all sufficiently large $n \in \mathbb{N}$.

We have $a^n = ba^{n+1}$ with some $b \in R/\mathfrak{p}$ and so $a^n(1 - ab) = 0$; and from $a \neq 0$ follows $ab = 1$, since $R/\mathfrak{p}$ is an integral domain. We see that $R/\mathfrak{p}$ is a field; therefore any $\mathfrak{p} \in \operatorname{Spec}(R)$ is a maximal ideal, and so $\dim R = 0$.

Conversely, if R is a Noetherian ring of dimension 0, then $\operatorname{Spec}(R)$ consists only of finitely many maximal ideals $\mathfrak{m}_1, \ldots, \mathfrak{m}_s$ and $I := \mathfrak{m}_1 \cap \cdots \cap \mathfrak{m}_s$ is a nilpotent ideal: $I^\rho = (0)$ for some $\rho \in \mathbb{N}$. It suffices to show that R/I is of finite length, since the R-modules $I^\alpha/I^{\alpha+1}$ are also of finite length, for they are homomorphic images of a finite direct sum of copies of R/I. From 2.4 it follows that R is also of finite length.

R/I is a 0-dimensional reduced Noetherian ring and hence by II.1.5 a finite direct product of fields. Its length then equals the number of fields involved.

Corollary 2.6. For an ideal $I \neq R$ of a Noetherian ring R, R/I is of finite length as an R-module (or as a ring, which is the same) if and only if $\mathfrak{V}(I) \subset \operatorname{Spec}(R)$ contains only maximal ideals.

Now we can also recognize the modules of finite length over Noetherian rings:

Proposition 2.7. Let M be a finitely generated module over a Noetherian ring R. The following statements are equivalent.

  a) M is of finite length.

  b) $\operatorname{Supp}(M) \subset \operatorname{Max}(R)$.

  c) $R/\operatorname{Ann}(M)$ is of finite length.

Proof.

a)→b). Let $M = M_0 \supset \cdots \supset M_l = (0)$ be a composition series of M. By 2.1, $M_i/M_{i+1} \cong R/\mathfrak{m}_i$ with $\mathfrak{m}_i \in \operatorname{Max}(R)$ $(i = 0, \ldots, l - 1)$. Now if $\mathfrak{p} \in \operatorname{Spec}(R) \setminus \operatorname{Max}(R)$, then $(M_i/M_{i+1})_\mathfrak{p} \cong (R/\mathfrak{m}_i)_\mathfrak{p} \cong R_\mathfrak{p}/\mathfrak{m}_i R_\mathfrak{p} = (0)$ and it follows that $M_\mathfrak{p} = (0)$. Therefore $\operatorname{Supp}(M) \subset \operatorname{Max}(R)$.

b)$\to$c). Since $\text{Supp}(M) = \mathfrak{V}(\text{Ann}(M))$ by III.4.6, it follows from 2.5 that $R/\text{Ann}(M)$ is of finite length.

c)$\to$a). If $R/\text{Ann}(M) =: R'$ is of finite length, then so is M, since M is a homomorphic image of a finite direct sum of copies of the R-module R'.

Exercises

1. Let M be a module over a ring R, $N \subset M$ a submodule for which M/N is of finite length. For an element $x \in R$ let $\mu_x : M \to M$ be injective and $M/(x)M$ of finite length. Then

$$l(M/(x)M) = l(N/(x)N).$$

2. A system of elements $a_1, \ldots, a_m (m \geq 0)$ in a ring R is called free if $I :=$ $(a_1, \ldots, a_m) \neq R$ and if $a_1 + I^2, \ldots, a_m + I^2$ is a basis of the R/I-module I/I^2.

   a) If $I \neq R$ and $m > 0$, then $a_1, \ldots, a_m$ is free if and only if $\sum r_i a_i = 0$ $(r_i \in R)$ implies $r_i \in I$ $(i = 1, \ldots, m)$.

   b) If $a_1, \ldots, a_m$ is free and $a_m = b \cdot c$ with $(a_1, \ldots a_{m-1}, b) \neq R$, then $\{a_1, \ldots, a_{m-1}, b\}$ is free.

   c) If, under the hypotheses of b), R/I is of finite length, then

$$l(R/I) = l(R/(a_1, \ldots, a_{m-1}, b)) + l(R/(a_1, \ldots, a_{m-1}, c)).$$

3. For an ideal $I = (a_1, \ldots, a_m)$ of a ring R let R/I be of finite length. For $J := (a_1^{v_1}, \ldots, a_m^{v_m})$ with $v_i \in \mathbb{N}_+$ $(i = 1, \ldots, m)$, R/J is also of finite length and

$$l(R/J) \leq l(R/I) \cdot \prod_{i=1}^{m} v_i.$$

If $\{a_1^{v_i}, \ldots, a_m^{v_m}\}$ is free, then equality holds in this relation.

4. Let I_k $(k = 1, \ldots, m)$ be ideals of a ring R, where $I_a + I_b = R$ for $a, b = 1, \ldots, m, a \neq b$. If $I := \bigcap_{k=1}^{m} I_k$ and R/I is of finite length, then $l(R/I) = \sum_{k=1}^{m} l(R/I_k)$.

3. Krull's Principal Ideal Theorem. Dimension of the intersection of two varieties

The Principal Ideal Theorem will provide a lower estimate of the number of generators of an ideal in a Noetherian ring and the number of equations needed to describe an algebraic variety.

Theorem 3.1. Let R be a Noetherian ring and $(a) \neq R$ a principal ideal of R. Then $h(\mathfrak{p}) \leq 1$ for any minimal prime divisor $\mathfrak{p}$ of (a), and $h(\mathfrak{p}) = 1$ if a is not a zero divisor of R.

Proof. The second statement follows from the first, since by I.4.10 the minimal prime ideals of R consist of zero divisors alone.

To prove the first statement we consider a minimal prime divisor $\mathfrak{p}$ of (a). We have $h(\mathfrak{p}) = \dim R_{\mathfrak{p}}$ by III.4.13 and $\mathfrak{p}R_{\mathfrak{p}}$ is a minimal prime divisor of $aR_{\mathfrak{p}}$. Therefore, we can assume that R is a local ring whose maximal ideal $\mathfrak{m}$ is a minimal prime divisor of (a). For any $\mathfrak{q} \in \mathrm{Spec}(R)$ with $\mathfrak{q} \neq \mathfrak{m}$, we must show $h(\mathfrak{q}) = 0$.

We denote by $\mathfrak{q}^{(i)}$ the inverse image of $\mathfrak{q}^i R_{\mathfrak{q}}$ in R (the i-th "symbolic power" of $\mathfrak{q}$) and form the ideal chain

$$(a) + \mathfrak{q}^{(1)} \supset (a) + \mathfrak{q}^{(2)} \supset \cdots$$

Since $\mathrm{Spec}(R/(a))$ has only one element, namely $\mathfrak{m}/(a)$, by 2.5 $R/(a)$ is of finite length. Hence there is an $n \in \mathbb{N}$ with

$$(a) + \mathfrak{q}^{(n)} = (a) + \mathfrak{q}^{(n+1)}.$$

Write $q \in \mathfrak{q}^{(n)}$ in the form $q = ra + q'$ with $r \in R, q' \in \mathfrak{q}^{(n+1)}$. By the definition of $\mathfrak{q}^{(n)}$ it then follows from $ra \in \mathfrak{q}^{(n)}, a \notin \mathfrak{q}$, that $r \in \mathfrak{q}^{(n)}$. We get

$$\mathfrak{q}^{(n)} = a\mathfrak{q}^{(n)} + \mathfrak{q}^{(n+1)}$$

and so $\mathfrak{q}^{(n)} = \mathfrak{q}^{(n+1)}$ by the Lemma of Nakayama, since $a \in \mathfrak{m}$. Then $\mathfrak{q}^n R_{\mathfrak{q}} = \mathfrak{q}^{n+1} R_{\mathfrak{q}}$ in $R_{\mathfrak{q}}$ and thus $\mathfrak{q}^n R_{\mathfrak{q}} = (0)$, again after Nakayama. Since the maximal ideal of $R_{\mathfrak{q}}$ is nilpotent, it follows that $h(\mathfrak{q}) = \dim R_{\mathfrak{q}} = 0$, q. e. d.

Corollary 3.2. In an n-dimensional affine or projection space let V be an irreducible algebraic variety of dimension d, and let H be a hypersurface. If $V \cap H \neq \emptyset$ and $V \not\subset H$, then all irreducible components of $V \cap H$ have dimension $d - 1$.

Proof. Since one can cover a projective space by affine spaces, it suffices to prove the statement for affine spaces. In the affine coordinate ring $K[V]$ of V a principal ideal (a) belongs to the hypersurface H, where $(a) \neq K[V]$, because $V \cap H \neq \emptyset$ and $a \neq 0$, since $V \not\subset H$. Further, $K[V \cap H] \cong (K[V]/(a))_{\mathrm{red}}$. By the Principal Ideal Theorem, all minimal prime divisors of (a) have height 1 and thus by II.3.6b) they have dimension $d - 1$. Therefore all minimal prime ideals of $K[V \cap H]$ have dimension $d - 1$, and the claim follows.

The corollary shows in particular that if two hypersurfaces cut each other in n-dimensional space without having any irreducible components in common, then all components of the intersection have dimension $n - 2$. This is not true in general, if the coordinate field is not algebraically closed, as is shown by two surfaces in $\mathbb{R}^3$ that intersect at only one point.

Corollary 3.3. Let R be a Noetherian ring. For $\mathfrak{p}, \mathfrak{p}', \mathfrak{q}_1, \ldots, \mathfrak{q}_s \in \operatorname{Spec}(R)$ suppose $\mathfrak{p} \not\subset \mathfrak{q}_i$ $(i = 1, \ldots, s)$ and $\mathfrak{p}' \subset \mathfrak{p}$. Further, suppose there exists $\mathfrak{q}' \in \operatorname{Spec}(R)$ with $\mathfrak{p}' \subsetneq \mathfrak{q}' \subsetneq \mathfrak{p}$. Then there is some $\mathfrak{q} \in \operatorname{Spec}(R)$ with $\mathfrak{p}' \subsetneq \mathfrak{q} \subsetneq \mathfrak{p}$ and $\mathfrak{q} \not\subset \mathfrak{q}_i$ $(i = 1, \ldots, s)$.

Proof. By III.1.6 there is an $x \in \mathfrak{p}$ with $x \notin \mathfrak{q}_i$ $(i = 1, \ldots, s), x \notin \mathfrak{p}'$. A minimal prime divisor of $x R_\mathfrak{p} + \mathfrak{p}' R_\mathfrak{p}$ in $R_\mathfrak{p}$ is of the form $\mathfrak{q} R_\mathfrak{p}$ with some $\mathfrak{q} \in \operatorname{Spec}(R), \mathfrak{p}' \subset \mathfrak{q} \subset \mathfrak{p}$. Since $h(\mathfrak{q}/\mathfrak{p}') = 1$ by 3.1 and $\dim R_\mathfrak{p}/\mathfrak{p}' R_\mathfrak{p} \geq 2$ by hypothesis, we have $\mathfrak{q} \neq \mathfrak{p}$. Further, $\mathfrak{q} \not\subset \mathfrak{q}_i$ $(i = 1, \ldots, s)$ and $\mathfrak{p}' \neq \mathfrak{q}$, since $x \in \mathfrak{q}, x \notin \mathfrak{q}_i$ $(i = 1, \ldots, s), x \notin \mathfrak{p}'$.

The corollary will be applied in the proof of

Theorem 3.4. (Generalized Krull Principal Ideal Theorem) Let R be a Noetherian ring, $I \neq R$ an ideal generated by m elements. For any minimal prime divisor $\mathfrak{p}$ of I, $h(\mathfrak{p}) \leq m$.

Proof (by induction on m). By 3.1 we may assume that $m > 1$ and the theorem has already been proved for ideals that can be generated by fewer than m elements.

Let $I = (a_1, \ldots, a_m)$ and let $\mathfrak{q}_1, \ldots, \mathfrak{q}_s$ be the minimal prime divisors of $(a_1, \ldots, a_{m-1})$. Then $h(\mathfrak{q}_i) \leq m - 1$ $(i = 1, \ldots, s)$. Let $\mathfrak{p}$ be a minimal prime divisor of I and let

$$\mathfrak{p} = \mathfrak{p}_0 \supset \mathfrak{p}_1 \supset \cdots \supset \mathfrak{p}_l \qquad (1)$$

be a prime ideal chain of length $l \geq 2$ (if there is no such chain, we are done). We may assume that $\mathfrak{p} \not\subset \bigcup_{i=1}^s \mathfrak{q}_i$; otherwise, $\mathfrak{p} \subset \mathfrak{q}_i$ for some $i \in [1, s]$, so $h(\mathfrak{p}) \leq m - 1$ and we are again done.

Applying 3.3 repeatedly shows that there is also a prime ideal chain (1) in which $\mathfrak{p}_{l-1} \not\subset \bigcup_{i=1}^s \mathfrak{q}_i$. We now set $\overline{R} := R/(a_1, \ldots, a_{m-1})$ and also denote the images in $\overline{R}$ of elements and ideals of R with an overbar. $\overline{\mathfrak{p}}$ is a minimal prime divisor of $(\overline{a}_m)$ and so $h(\overline{\mathfrak{p}}) \leq 1$ by 3.1.

We have $\overline{\mathfrak{p}}_{l-1} \not\subset \overline{\mathfrak{q}}_i$, since $\mathfrak{p}_{l-1} \not\subset \mathfrak{q}_i$ and $(a_1, \ldots, a_{m-1}) \subset \mathfrak{q}_i (i = 1, \ldots, s)$. Hence $\overline{\mathfrak{p}}$ is a minimal prime divisor of $\overline{\mathfrak{p}}_{l-1}$ and $\mathfrak{p}$ is one of $\mathfrak{p}_{l-1} + (a_1, \ldots, a_{m-1})$. Then in $R/\mathfrak{p}_{l-1}$ the ideal $\mathfrak{p}/\mathfrak{p}_{l-1}$ is a minimal prime divisor of an ideal generated by $m - 1$ elements, hence $l - 1 \leq h(\mathfrak{p}/\mathfrak{p}_{l-1}) \leq m - 1$. It follows that $h(\mathfrak{p}) \leq m$. q.e.d.

We will now give some applications of the generalized Principal Ideal Theorem to algebraic geometry.

Corollary 3.5. Let V be an irreducible affine or projective variety. Let $f_1, \ldots, f_m \in K[V]$ be (homogeneous) elements of the (homogeneous) coordinate ring of V whose zero set $W := \mathfrak{V}_V(f_1, \ldots, f_m)$ on V is not empty. Then for any irreducible component Z of W

$$\dim Z \geq \dim V - m.$$

Proof. We consider only the affine case; the projective is similar. If $\mathfrak{p}$ is a minimal prime divisor of $(f_1, \ldots, f_m)$, then by 3.4 and II.3.6

$$\dim K[V]/\mathfrak{p} = \dim K[V] - h(\mathfrak{p}) \geq \dim V - m.$$

Since the $K[V]/\mathfrak{p}$ are the coordinate rings of the irreducible components of W, the assertion follows. 3.5 can also be derived from 3.2 by induction.

Corollary 3.6.

a) The solution set in $\mathbf{A}^n(L)$ of a system of equations

$$F_i(X_1,\ldots,X_n) = 0 \qquad (F_i \in K[X_1,\ldots,X_n], i = 1,\ldots,m)$$

is either empty or is a K-variety each of whose irreducible components has dimension $\geq n - m$.

b) The same statement also holds for the solution set in $\mathbf{P}^n(L)$ of a system

$$F_i(Y_0,\ldots,Y_n) = 0 \qquad (i = 1,\ldots,m)$$

with only homogeneous polynomials $F_i \in K[Y_0,\ldots,Y_n]$.

Corollary 3.7. To describe an algebraic variety V in n-dimensional affine or projective space one needs at least $n - \delta(V)$ equations, where $\delta(V)$ is the minimum of the dimensions of the irreducible components of $V (\delta(V) \leq \dim V)$. In particular, the ideal of such a variety in the polynomial ring cannot be generated by fewer than $n - \delta(V)$ polynomials.

We now generalize 3.2.

Proposition 3.8. Let V and W be irreducible K-varieties in $\mathbf{A}^n(L)$ with $V \cap W \neq \emptyset$. Then for each irreducible component Z of $V \cap W$

$$\dim Z \geq \dim V + \dim W - n.$$

Proof. By II.3.11e) each irreducible component of $V \times W$ has dimension $\dim V + \dim W$. This is equivalent to

$$\dim(K[V] \underset{K}{\otimes} K[W]/\mathfrak{P}_0) = \dim V + \dim W$$

for any minimal prime ideal $\mathfrak{P}_0$ of $K[V] \underset{K}{\otimes} K[W]$. We have

$$K[V] \underset{K}{\otimes} K[W] \cong K[X_1,\ldots,X_n]/\mathfrak{I}(V) \underset{K}{\otimes} K[Y_1,\ldots,Y_n]/\mathfrak{I}(W)$$

$$\cong K[X_1,\ldots,X_n,Y_1,\ldots,Y_n]/(\mathfrak{I}(V),\mathfrak{I}(W)).$$

We denote by Δ the ideal generated by the elements $X_i - Y_i \ (i = 1,\ldots,n)$ in $K[X_1,\ldots,X_n,Y_1,\ldots,Y_n]$ and by $\mathfrak{d}$ its image in $K[V] \underset{K}{\otimes} K[W]$. Then

$$K[V] \underset{K}{\otimes} K[W]/\mathfrak{d} \cong K[X_1,\ldots,X_n,Y_1,\ldots,Y_n]/(\mathfrak{I}(V),\mathfrak{I}(W)) + \Delta$$

$$\cong K[X_1,\ldots,X_n]/\mathfrak{I}(V) + \mathfrak{I}(W),$$

if we now denote by $\mathfrak{J}(W)$ also the ideal of W in $K[X_1,\ldots,X_n]$ (replace Y_i by X_i). It follows that

$$(K[V] \underset{K}{\otimes} K[W]/\mathfrak{d})_{\mathrm{red}} \cong (K[X_1,\ldots,X_n]/\mathfrak{J}(V) + \mathfrak{J}(W))_{\mathrm{red}} \cong K[V \cap W].$$

The minimal prime ideals of $K[V \cap W]$ correspond bijectively to the minimal prime divisors $\mathfrak{P}$ of $\mathfrak{d}$ in $K[V] \underset{K}{\otimes} K[W]$, and corresponding prime ideals have the same dimension. Since $\mathfrak{d}$ is generated by n elements, $h(\mathfrak{P}) \leq n$ by 3.4 and so by II.3.6b) $\dim(K[V] \underset{K}{\otimes} K[W]/\mathfrak{P}) = \dim V + \dim W - h(\mathfrak{P}) \geq \dim V + \dim W - n$, q. e. d.

In the projective case the hypothesis $V \cap W \neq 0$ is superfluous.

Proposition 3.9. Let V and W be two irreducible K-varieties in $\mathbf{P}^n(L)$. Then

$$\dim Z \geq \dim V + \dim W - n$$

for any irreducible component Z of $V \cap W$ (note that the empty variety is assigned dimension -1).

Proof. Let $\widetilde{V}$ and $\widetilde{W}$ be the cones belonging to V and W in $\mathbf{A}^{n+1}(L)$. $\widetilde{V} \cap \widetilde{W}$ is then the cone of $V \cap W$, and $\widetilde{V} \cap \widetilde{W}$ contains the origin of $\mathbf{A}^{n+1}(L)$. (The cone of the empty variety is the origin.) If Z is an irreducible component of $V \cap W$, then $\widetilde{Z}$ is a component of $\widetilde{V} \cap \widetilde{W}$. By II.4.4b) and 3.8

$$\dim Z = \dim \widetilde{Z} - 1 \geq \dim \widetilde{V} + \dim \widetilde{W} - (n+1) - 1 = \dim V + \dim W - n.$$

Corollary 3.10. If $V, W \subset \mathbf{P}^n(L)$ are arbitrary varieties with $\dim V + \dim W \geq n$, then $V \cap W \neq \emptyset$.

This generalizes I.5.2 and the well-known fact of projective geometry that linear varieties of complementary dimension always intersect. Note that from 3.8 and 3.9 the statements given there can also be derived at once if V and W are varieties whose irreducible components have constant dimension ("unmixed" varieties).

Since the height of an ideal $I \neq (1)$ is defined as the infimum of the heights of the prime divisors of I, it follows from the generalized Principal Ideal Theorem that an ideal $I \neq (1)$ in a Noetherian ring always has finite height

$$h(I) \leq \mu(I).$$

With a view to geometric applications we introduce the following terminology.

Definition 3.11. Let $I \neq R$ be an ideal in a Noetherian ring R.

a) I is called a *complete intersection* if $h(I) = \mu(I)$.

b) I is called a *set-theoretic complete intersection* if there are elements $a_1,\ldots,a_m \in I$ such that $\mathrm{Rad}(I) = \mathrm{Rad}(a_1,\ldots,a_m)$, where $m = h(I)$.

c) We say that I is locally a *complete intersection* if I_m is a complete intersection in R_m for all $m \in \text{Max}(R)$ with $I \subset m$.

In case c) I_p is also a complete intersection in R_p for all $p \in \text{Spec}(R)$ with $I \subset p$; for if $p \subset m$, $m \in \text{Max}(R)$, then (cf. III.4.12) $h(I_m) \leq h(I_p) \leq \mu(I_p) \leq \mu(I_m)$, and from $h(I_m) = \mu(I_m)$ it follows that $h(I_p) = \mu(I_p)$.

If I is a complete intersection, of course it is also a set-theoretic and locally a complete intersection, since $h(I_p) \geq h(I)$ for all $p \in \mathfrak{V}(I)$. In cases a) and b) of Definition 3.11 it follows at once from the generalized Principal Ideal Theorem that all minimal prime divisors of I have height m.

Definition 3.12. Let V be a d-dimensional K-variety in n-dimensional affine or projective space over L.

a) V is called an *ideal-theoretic complete intersection* if its vanishing ideal in the polynomial ring over K can be generated by $n - d$ polynomials.

b) V is called a *set-theoretic complete intersection* if V is the intersection of $n - d$ K-hypersurfaces.

c) We say that V is locally an (ideal-theoretic) *complete intersection* if for all $x \in V$ the ideal of V in the local ring O_x (defined over K) of the ambient affine or projective space (cf. III.4.14d)) is a complete intersection. (If this holds for an $x \in V$, we say that V is a *complete intersection at x*).

It is easy to see that V has one of the properties of a) or b) of 3.12 if and only if the ideal of V in the polynomial ring over K satisfies the corresponding condition in 3.11 (where, in the projective case, the a_i are chosen to be homogeneous). Just as there we find that in cases a) and b) of 3.12 all the irreducible components of V have dimension d. If V is locally a complete intersection at a point x, then all the irreducible components of V that contain x have the same dimension. In the affine case V is locally a complete intersection if and only if the ideal of V in the polynomial ring is locally a complete intersection.

It is clear that linear varieties in affine or projective space are ideal-theoretic complete intersections. We now want to consider more examples of these concepts. We take this opportunity to indicate some theorems that will be treated in the next two chapters.

Examples 3.13.

a) Affine K-varieties of dimension 0 are set-theoretic complete intersections by 1.5. A projective K-variety of dimension 0 is a set-theoretic complete intersection if it has a K-rational point (1.7).

b) Affine K-varieties of dimension 0 are ideal-theoretic complete intersections. This can easily be proved directly, but it is also a simple special case of the subsequent theorem 5.21 and the fact that 0-dimensional varieties are regular (Ch. VI).

c) There are projective varieties of dimension 0 that are not ideal-theoretic complete intersections. In fact, for any $r \in \mathbf{N}_+$ there is a finite set of points V in $\mathbf{P}^2(L)$ whose ideal $\mathfrak{I}(V)$ cannot be generated by r elements:

A homogeneous polynomial $\sum_{\nu_0+\nu_1+\nu_2=r-1} a_{\nu_0\nu_1\nu_2} Y_0^{\nu_0} Y_1^{\nu_1} Y_2^{\nu_2}$ of degree $r-1$ has $s := \binom{r+1}{2}$ coefficients $a_{\nu_0\nu_1\nu_2}$. One can find s points $P_i = \langle y_{0i}, y_{1i}, y_{2i} \rangle \in \mathbf{P}^2(L)$ such that no homogeneous polynomial of degree $r-1$ vanishes on $V := \{P_1, \ldots, P_s\}$. These must be chosen so that the determinant whose rows consist of the s products $y_{0i}^{\nu_0} y_{1i}^{\nu_1} y_{2i}^{\nu_2}$ does not vanish. For $y_{0i} = t_i^r$, $y_{1i} = t_i$, $y_{2i} = 1$ $(i = 1, \ldots, s)$ the determinant has the form

$$\det \left(\{t_i^{\nu_0 r + \nu_1}\}_{\substack{\nu_0+\nu_1 \leq r-1 \\ i=1,\ldots,s}} \right).$$

Expansion with respect to the row with $i = 1$ and induction show that the determinant does not vanish if the t_i are considered as indeterminates. By I.1.3a) it is then possible to choose special $t_i \in L$ $(i = 1, \ldots, s)$ so that the determinant also does not vanish if the indeterminates are given these special values.

A homogeneous polynomial F in $L[Y_0, Y_1, Y_2]$ of degree r has $\binom{r+2}{2} = s+r+1$ coefficients. If V is chosen as above, one sees that the homogeneous polynomials of degree r that vanish on V form a vector space over L of dimension $r+1$, since $F(P_i) = 0$ $(i = 1, \ldots, s)$ gives s linearly independent conditions on the coefficients of F.

For the homogeneous components of $\mathfrak{I}(V)$ we have: $\mathfrak{I}(V)_n = \langle 0 \rangle$ for $n = 0, \ldots, r-1, \dim_L(\mathfrak{I}(V)_r) = r+1$. Therefore, $\mathfrak{I}(V)$ cannot be generated by fewer than $r+1$ polynomials.

For more precise information about the number of generators of the vanishing ideal of a finite set of points in projective space see: A. Geramita, Remarks on the number of generators of some homogeneous ideals, *Bull. Soc. Math. France* **107**, 197–207 (1983), and the references given there.

d) It is conjectured that affine algebraic curves are always set-theoretic complete intersections. This has been proved

   α) for curves that are locally complete intersections. For curves in 3-dimensional space this was first shown by Szpiro [78]. His proof will be treated in Ch. VII. The result was then extended by Mohan Kumar [56] to curves in spaces of arbitrary dimension.

   β) For arbitrary curves, if the coordinate field has prime characteristic (Cowšik-Nori [11]). The proof uses Szpiro's and Mohan Kumar's results.

e) It is easy to give affine curves that are not local complete intersections (and so are not ideal-theoretic complete intersections either). In a classical example Macaulay [55] even showed that for any $r \in \mathbf{N}_+$ there is an irreducible space curve whose ideal requires more than r generators. (For a modern treatment of this example see Abhyankar [3] and Geyer [25]). There are also smooth space curves (definition in Ch. VI) that are not ideal-theoretic complete intersections (Abhyankar [2], Murthy [57], see also Ch. VII, §3, Exercises 4 and 5.)

f) We now treat an example investigated by Herzog [34] in which we can explicitly determine which of a certain class of affine space curves are complete intersections.

Let n_1, n_2, n_3 be natural numbers with $\gcd(n_1, n_2, n_3) = 1$. The kernel I of the K-homomorphism $\varphi : K[X_1, X_2, X_3] \to K[T]$ with $\varphi(X_i) = T^{n_i}$ ($i = 1, 2, 3$) is the vanishing ideal of the curve in $\mathsf{A}^3(L)$ with the parametrization

$$x_1 = t^{n_1}, x_2 = t^{n_2}, x_3 = t^{n_3} (t \in L).$$

We endow $K[X_1, X_2, X_3]$ with the grading in which $\deg(X_i) = n_i$ ($i = 1, 2, 3$). Let $K[T]$ have the usual grading ($\deg(T) = 1$). I is then a homogeneous prime ideal, $h(I) = 2$.

α) *Determining a minimal generating system of I*

If $F = \sum a_{\nu_1 \nu_2 \nu_3} X_1^{\nu_1} X_2^{\nu_2} X_3^{\nu_3} \in I$ is homogeneous of degree d, then $F(T^{n_1}, T^{n_2}, T^{n_3}) = (\sum a_{\nu_1 \nu_2 \nu_3}) T^d = 0$ and so $\sum a_{\nu_1 \nu_2 \nu_3} = 0$. Therefore $F = \sum a_{\nu_1 \nu_2 \nu_3} (X_1^{\nu_1} X_2^{\nu_2} X_3^{\nu_3} - X_1^{\bar{\nu}_1} X_2^{\bar{\nu}_2} X_3^{\bar{\nu}_3})$, where $X_1^{\bar{\nu}_1} X_2^{\bar{\nu}_2} X_3^{\bar{\nu}_3}$ is a fixed monomial of degree d ($\sum \bar{\nu}_i n_i = d$). Using the fact that I is a prime ideal and $X_i \notin I$ ($i = 1, 2, 3$), we get:

(∗) F is a linear combination with coefficients in $K[X_1, X_2, X_3]$ of (homogeneous) polynomials in I of the form

$$\varphi_1 = X_1^{c_1'} - X_2^{r_{12}'} X_3^{r_{13}'}, \quad \varphi_2 = X_2^{c_2'} - X_1^{r_{21}'} X_3^{r_{23}'}, \quad \varphi_3 = X_3^{c_3'} - X_1^{r_{31}'} X_2^{r_{32}'}.$$

If one of the variables X_j occurs in F to at most the power $\alpha (\alpha \in \mathsf{N})$, then we may assume that the φ_j occurring in the linear combination also have this property.

Now let $c_1 \in \mathsf{N}_+$ be the least number with $c_1 n_1 \in \mathsf{N}n_2 + \mathsf{N}n_3$; let c_2 and c_3 be defined likewise. Among the polynomials of the form φ_i there is one of least degree. Possibly after renumbering the variables we may assume that this one of least degree is of the form $F_1 = X_1^{c_1} - X_2^{r_{12}} X_3^{r_{13}}$. Here the exponent of X_1 is the number c_1 above, since otherwise there would be a polynomial of lower degree. For the same reason we have $r_{12} \leq c_2$, $r_{13} \leq c_3$. Here $r_{13} = c_3$ is equivalent to $r_{12} = 0$, and $r_{12} = c_2$ is equivalent to $r_{13} = 0$. In the last case if we change the numbering so that X_2 becomes X_3, then there are only the two following possibilities for F_1:

a) $F_1 = X_1^{c_1} - X_3^{c_3}$,

b) $F_1 = X_1^{c_1} - X_2^{r_{12}} X_3^{r_{13}}$ $(0 < r_{12} < c_2, 0 < r_{13} < c_3)$.

Since F_1 is monic in X_1, we can divide any homogeneous $F \in I$ by F_1 with a remainder R. Here R is also homogeneous, $\deg(R) = \deg(F)$, and X_1 occurs in R to at most the power $c_1 - 1$. Since $R \in I$, by (∗) R can be represented as a linear combination of polynomials of the type φ_2, φ_3. There are such polynomials, since I (according to Krull) is not a principal ideal. Let F_2 be one such of least degree.

If in case a) we have $F_2 = X_3^{c_3'} - X_1^{r_{31}'} X_2^{r_{32}'}$, then $F_2 + X_3^{c_3'-c_3} F_1 = X_3^{c_3'-c_3} X_1^{c_1} - X_1^{r_{31}'} X_2^{r_{32}'} \in I$ is of the same degree as F_2 and is not divisible by F_1. Necessarily then $r_{31}' = 0$. Hence we may assume that F_2 has the form $F_2 = X_2^{c_2'} - X_1^{r_{21}'} X_3^{r_{23}'}$ $(r_{21}' < c_1)$. Here we must have $c_2' = c_2$, since otherwise F would not be of minimal degree. We shall show that $I = (F_1, F_2)$.

If there were an $F \in I \backslash (F_1, F_2)$, then there would also be such a one of type φ_3, say $F_3 = X_3^{c_3'} - X_1^{r_{31}'} X_2^{r_{32}'}$ with $r_{3i}' < c_i$ $(i = 1, 2), c_3' \geq c_3$, since modulo (F_1, F_2) any homogeneous $F \in I$ can be reduced by $(*)$ to a polynomial that is a linear combination of polynomials of type φ_3 with $r_{3i}' < c_i$ $(i = 1, 2)$. Then $X_3^{c_3'-c_3} F_1 + F_3 = X_1^{c_1} X_3^{c_3'-c_3} - X_1^{r_{31}'} X_2^{r_{32}'} \in I$, so $X_1^{c_1-r_{31}'} X_3^{c_3'-c_3} - X_2^{r_{32}'} \in I$, contradicting $r_{32}' < c_2$.

In case b) we can (possibly after renumbering X_2 and X_3) assume that $F_2 = X_2^{c_2} - X_1^{r_{21}} X_3^{r_{23}}$ $(r_{21} < c_1)$. Here we must have $r_{23} \leq c_3$, and $r_{23} = c_3$ is equivalent to $r_{21} = 0$. If $r_{21} = 0$, then $I = (F_1, F_2)$ as in case a). Now we have to investigate I only if

$$F_2 = X_2^{c_2} - X_1^{r_{21}} X_3^{r_{23}} \qquad (0 < r_{21} < c_1, 0 < r_{23} < c_3).$$

Since no polynomial of type φ_3 is contained in (F_1, F_2) (put $X_1 = X_2 = 0$), there is a polynomial $F_3 = X_3^{c_3} - X_1^{r_{31}} X_2^{r_{32}} \in I$ with $r_{31} < c_1, r_{32} < c_2$. Necessarily also $0 < r_{3i}$ $(i = 1, 2)$. Each polynomial F of type φ_i $(i \in \{1, 2, 3, \})$ can be reduced modulo F_i to a multiple of a polynomial of type φ_j $(j \in \{1, 2, 3\})$ that has lower degree than F. By induction on the degree it then follows that $I = (F_1, F_2, F_3)$. If we consider I modulo $(X_1^{c_1}, X_2^{c_2}, X_3^{c_3})$, we easily see that in case b) too I cannot be generated by 2 elements.

The result of this discussion is that (with a suitable numbering of the indeterminates) only the two following cases can occur.

a) I has a minimal generating system of 2 polynomials

$$F_1 = X_1^{c_1} - X_3^{c_3}, F_2 = X_2^{c_2} - X_1^{r_{21}} X_3^{r_{23}} \ (0 \leq r_{21} \leq c_1).$$

b) I has a minimal generating system with 3 polynomials

$$F_1 = X_1^{c_1} - X_2^{r_{12}} X_3^{r_{13}}, F_2 = X_2^{c_2} - X_1^{r_{21}} X_3^{r_{23}},$$
$$F_3 = X_3^{c_3} - X_1^{r_{31}} X_2^{r_{32}} \text{ where } 0 < r_{ji} < c_i \ (i = 1, 2, 3, j \neq i).$$

Which of the two cases occurs can easily be decided computationally by determining the numbers c_1, c_2, c_3.

For $\{n_1, n_2, n_3\} = \{3, 4, 5\}$ we have

$$3n_1 = n_2 + n_3, \qquad 2n_2 = n_1 + n_3, \qquad 2n_3 = 2n_1 + n_2,$$
$$F_1 = X_1^3 - X_2 X_3, \qquad F_2 = X_2^2 - X_1 X_3, \qquad F_3 = X_3^2 - X_1^2 X_2.$$

The curve (t^3, t^4, t^5) is not an ideal-theoretic complete intersection.

For $\{n_1, n_2, n_3\} = \{4, 5, 6\}$ we have $3n_1 = 2n_3$, $2n_2 = n_1 + n_3$, and $I = (X_1^3 - X_3^2, X_2^2 - X_1 X_3)$ is a complete intersection.

In general one can show that I is a complete intersection if and only if the subsemigroup of $(\mathbb{N}, +)$ generated by n_1, n_2, n_3 is "symmetric" (Exercises 3–5).

β) I is always a set-theoretic complete intersection

We have to consider only case b). We put $v_1 := (-c_1, r_{12}, r_{13})$, $v_2 := (r_{21}, -c_2, r_{23})$, $v_3 := (r_{31}, r_{32}, -c_3)$, and $v := v_1 + v_2 + v_3 =: (a_1, a_2, a_3)$. Then $a_1 n_1 + a_2 n_2 + a_3 n_3 = 0$. We may assume that, say, a_2 and a_3 have the same sign. Then $a_1 = 0$ or $|a_1| = |-c_1 + r_{12} + r_{31}| \geq c_1$ by the definition of c_1. Since $0 < r_{j1} < c_1$ $(j = 2, 3)$, the second is not possible. Therefore $a_1 = 0$, and therefore also $a_2 = a_3 = 0$. In case b) we get as additional information that

$$c_1 = r_{21} + r_{31}, c_2 = r_{12} + r_{32}, c_3 = r_{13} + r_{23},$$

whence the following formula results at once:

$$X_2^{r_{32}} F_1 + X_3^{r_{13}} F_2 + X_1^{r_{21}} F_3 = 0. \tag{1}$$

$R := K[X_1, X_2, X_3]/(F_3)$ is an integral domain, since F_3 is irreducible. If ξ is the residue class of X_3 in R, then we can write

$$R = K[X_1, X_2] \oplus K[X_1, X_2]\xi \oplus \cdots \oplus K[X_1, X_2]\xi^{c_3 - 1},$$
$$\xi^{c_3} = X_1^{r_{31}} X_2^{r_{32}}. \tag{2}$$

From (1) it follows that

$$X_2^{r_{32} c_3} F_1^{c_3} \equiv (-1)^{c_3} X_3^{r_{13} c_3} F_2^{c_3} \mod (F_3);$$

therefore $X_2^{r_{32}(r_{13} + r_{23})} F_1^{c_3} \equiv (-1)^{c_3} X_1^{r_{31} r_{13}} X_2^{r_{32} r_{13}} F_2^{c_3} \mod (F_3)$ and since $X_2^{r_{32} r_{13}} \notin (F_3)$,

$$X_2^{r_{32} r_{23}} F_1^{c_3} \equiv (-1)^{c_3} X_1^{r_{31} r_{13}} F_2^{c_3} \mod (F_3).$$

Using (2) and the fact that $K[X_1, X_2]$ is a factorial ring, we find that

$$F_1^{c_3} \equiv (-1)^{c_3} X_1^{r_{31} r_{13}} P \mod (F_3), F_2^{c_3} \equiv X_2^{r_{32} r_{23}} P \mod (F_3)$$

with some $P \in K[X_1, X_2, X_3]$. Here $P \in I$, since $X_1, X_2 \notin I$. It follows that $I = \mathrm{Rad}(P, F_3)$.

Therefore, the curves $(t^{n_1}, t^{n_2}, t^{n_3})$ are always intersections of two surfaces with (quasi-)homogeneous equations; and one of the two equations can always be arbitrarily chosen from the minimal generating system of I given above.

For $\{n_1, n_2, n_3\} = \{3, 4, 5\}$ we have

$$I = \mathrm{Rad}(X_1^4 - 2X_1 X_2 X_3 + X_2^3, X_3^2 - X_1 X_2).$$

The structure of the vanishing ideal of the curves $(t^{n_1}, t^{n_2}, t^{n_3}, t^{n_4})$ in $\mathsf{A}^4(n_i \in \mathsf{N})$ is much more complicated than in the case considered above. See: H. Bresinsky, On prime ideals with generic zero $x_i = t^{n_i}$, *Proc. AMS* **47**, 329–332 (1975); Symmetric semigroups of integers generated by 4 elements, *Manuscr. math.* **17**, 205–219 (1975); and: R. Waldi, Zur Konstruktion von Weierstraßpunkten mit vorgegebener Halbgruppe, *Manuscr. math.* **30**, 257–278 (1980).

g) There are projective curves in 3-dimensional space that are not set-theoretic complete intersections. The simplest example is given by two skew lines in P^3 (Ch. VI.4.4). It is an open problem whether connected curves in P^3 are always set-theoretic complete intersections. For a result in this direction see: D. Ferrand, Set theoretical complete intersections in characteristic $p > 0$. In: Algebraic Geometry, *Springer Lect. Notes in Math.* **732** (1979).

h) There are affine surfaces that are not set-theoretic complete intersections (Hartshorne [31]). For surfaces in $\mathsf{A}^4(L)$, where L is the algebraic closure of a finite field, Murthy [58], as an analogue of Szpiro's result, has shown that they are set-theoretic complete intersections if they are local complete intersections.

i) Varieties in A^n or P^n all of whose irreducible components have dimension $n - 1$ are ideal-theoretic complete intersections, for they are just the hypersurfaces (II.3.11g) and II.4.4f)).

Exercises

1. Interpret 3.3 geometrically.

2. If a K-variety $V \subset \mathsf{A}^n(L)$ is locally a complete intersection, then $\in$s ideal in $K[X_1, \ldots, X_n]$ is generated by $n + 1$ elements. (This consequence of the Forster–Swan theorem will be sharpened in §5.)

3. A *numerical semigroup* H is a subsemigroup of $(\mathsf{N}, +)$ with $0 \in H$ and $c + \mathsf{N} \subset H$ for some $c \in \mathsf{N}$. H is called *symmetric* if there is an $m \in \mathsf{Z}$ such that for all $z \in \mathsf{Z}$, $z \in H$ if and only if $m - z \notin H$.

   a) H is symmetric if and only if the set of the $z \in \mathsf{Z} \setminus H$ with $z + h \in H$ for all $h \in H \setminus \{0\}$ consists of precisely one element.

   b) If $n_1, n_2 > 1$ are relatively prime integers, then $H := \mathsf{N} n_1 + \mathsf{N} n_2$ is a symmetric numerical semigroup.

4. In this exercise we use the notations of Example 3.13f). Let $H :=$ $\mathbb{N}n_1 + \mathbb{N}n_2 + \mathbb{N}n_3$. Then $\varphi : K[X_1, X_2, X_3] \rightarrow K[T]$ induces a surjection (!) $K[X_1, X_2, X_3]_{X_3} \rightarrow K[T]_T$ with kernel I_{X_3}. If ξ_i is the residue class of X_i modulo I_{X_3} $(i = 1, 2)$, then in case a) of the example

$$K[T]_T \cong \bigoplus_{\substack{\nu_1=0,\ldots,c_1-1 \\ \nu_2=0,\ldots,c_2-1}} K[X_3]_{X_3} \xi_1^{\nu_1} \xi_2^{\nu_2}.$$

Deduce:

a) Any $z \in \mathbb{Z}$ has a unique representation

$$z = a_1 n_1 + a_2 n_2 + a_3 n_3 \quad (0 \le a_1 < c_1, 0 \le a_2 < c_3, a_3 \in \mathbb{Z}).$$

b) $z \in H$ if and only if $a_3 \ge 0$.

c) H is symmetric (with $m := (c_1 - 1)n_1 + (c_2 - 1)n_2 - n_3$).

5. We now consider case b) in Example 3.13f). With the notations there let

$$\mu_1 := c_1 n_1 + c_2 n_2 - n_1 - n_2 - n_3 - r_{12} n_2,$$
$$\mu_2 := c_1 n_1 + c_2 n_2 - n_1 - n_2 - n_3 - r_{21} n_1.$$

Prove:

a) $\mu_1 \ne \mu_2$ and $\mu_i + h \in H$ for all $h \in H \setminus \{0\}$ $(i = 1, 2)$.

b) $\mu_i \notin H$ $(i = 1, 2)$.

c) H is not symmetric.

4. Applications of the Principal Ideal Theorem in Noetherian rings

Proposition 4.1. Any Noetherian semilocal ring has finite Krull dimension.

It is the maximum of the heights of the (finitely many) maximal ideals, which are finite by 3.4.

The following considerations will lead into a new characterization of the dimension of Noetherian local rings.

Definition 4.2. An ideal $\mathfrak{q}$ of a ring R is called *primary* if any zero divisor of $R/\mathfrak{q}$ is nilpotent.

Equivalent to this condition is the following: If $a, b \in R$ with $a \cdot b \in \mathfrak{q}$ and $a \notin \mathfrak{q}$ are given, then there is a $\rho \in \mathbb{N}$ with $b^\rho \in \mathfrak{q}$.

Remark 4.3. The radical of a primary ideal is a prime ideal.

If $\mathfrak{q}$ is a primary ideal and $\mathfrak{p} := \mathrm{Rad}(\mathfrak{q})$, then it follows from $a \cdot b \in \mathfrak{p}$ that $a^\rho \cdot b^\rho \in \mathfrak{q}$ for some $\rho \in \mathbb{N}$. If $a^\rho \notin \mathfrak{q}$, then there is a $\sigma \in \mathbb{N}$ with $(b^\rho)^\sigma \in \mathfrak{q}$, so $b \in \mathfrak{p}$.

If $\mathfrak{q}$ is primary and $\mathfrak{p} = \mathrm{Rad}(\mathfrak{q})$, then we also say that $\mathfrak{q}$ is $\mathfrak{p}$-primary.

Lemma 4.4. Let $\mathfrak{m}$ be a maximal ideal, $\mathfrak{q}$ an arbitrary ideal of a ring R. Then the following statements are equivalent.

a) $\mathfrak{q}$ is $\mathfrak{m}$-primary.

b) $\text{Rad}(\mathfrak{q}) = \mathfrak{m}$.

c) $\mathfrak{m}$ is the only minimal prime divisor of $\mathfrak{q}$.

Proof. It suffices to prove the implication c) $\rightarrow$ a). From c) it follows that $\text{Spec}(R/\mathfrak{q})$ has only one element, namely $\mathfrak{m}/\mathfrak{q}$. The elements of $\mathfrak{m}/\mathfrak{q}$ are nilpotent, those outside $\mathfrak{m}/\mathfrak{q}$ are units. The condition in 4.2 is therefore fulfilled.

In particular, the powers $\mathfrak{m}^\rho$ of a maximal ideal $\mathfrak{m}$ are always $\mathfrak{m}$-primary. If $\mathfrak{m}$ is finitely generated, then an ideal $\mathfrak{q}$ is $\mathfrak{m}$- primary if and only if $\mathfrak{m}^\rho \subset \mathfrak{q} \subset \mathfrak{m}$ for suitable $\rho \in \mathbf{N}$.

The concept of primary ideal can be viewed as a generalization of the concept of prime power. However, a power of an arbitrary prime ideal $\mathfrak{p}$ need not always be $\mathfrak{p}$-primary (Exercise 2), but the symbolic powers of $\mathfrak{p}$ are always $\mathfrak{p}$-primary (Exercise 4).

Remark 4.5. Let $(R, \mathfrak{m})$ be a Noetherian local ring, $\mathfrak{q}$ an $\mathfrak{m}$-primary ideal. Then $\mu(\mathfrak{q}) \geq \dim R$. In particular, $\mu(\mathfrak{m}) \geq \dim R$.

Since $\mathfrak{m}$ is the only minimal prime divisor of $\mathfrak{q}$, by 3.4 we have

$$\mu(\mathfrak{q}) \geq h(\mathfrak{m}) = \dim R.$$

Definition 4.6. If $(R, \mathfrak{m})$ is a Noetherian local ring, then $\mu(\mathfrak{m})$ is called the *embedding dimension* (edim R) of R.

As just shown we always have

$$\text{edim} R \geq \dim R.$$

The following lemma on avoiding prime ideals, that sharpens III.1.6, plays an important role in several later theorems.

Lemma 4.7. Let R be a Noetherian ring. Let $J \subset I$ be two ideals of R with $\mathfrak{V}(I) = \mathfrak{V}(J)$, and let $\mu(I/J) =: m$. Further, let $\mathfrak{p}_1, \ldots, \mathfrak{p}_s \in \text{Spec}(R)$ with $I \not\subset \bigcup_{j=1}^s \mathfrak{p}_j$ be given. Then one can find elements $a_1, \ldots, a_m \in I$ such that:

a) $I = (a_1, \ldots, a_m) + J$.

b) $a_i \notin \bigcup_{j=1}^s \mathfrak{p}_j$ $(i = 1, \ldots, m)$.

c) If $\mathfrak{p} \in \mathfrak{V}(a_1, \ldots, a_m)$, $\mathfrak{p} \notin \mathfrak{V}(I)$, then $h(\mathfrak{p}) \geq m$.

Proof. We construct inductively elements $a_1, \ldots, a_r \in I \setminus \bigcup_{j=1}^s \mathfrak{p}_j$ whose images $\bar{a}_1, \ldots, \bar{a}_r$ in I/J are part of a minimal generating system of this ideal and which have the property: If $\mathfrak{p} \in \mathfrak{V}(a_1, \ldots a_r)$, $\mathfrak{p} \notin \mathfrak{V}(I)$, then $h(\mathfrak{p}) \geq r$.

For $r = 0$ there is nothing to show. If $a_1, \ldots, a_r$ have already been constructed in the desired way for some r with $0 \leq r < m$, we first choose an arbitrary $a \in I$ such that $\{\bar{a}_1, \ldots, \bar{a}_r, \bar{a}\}$ is also part of a minimal generating system of I/J.

Let $q_1, \ldots, q_t$ be the minimal prime divisors of $(a_1, \ldots, a_r)$ which do not belong to $\mathfrak{V}(I)$, and let X be the set of maximal elements (with respect to inclusion) of $\{q_1, \ldots, q_t, \mathfrak{p}_1, \ldots, \mathfrak{p}_s\}$. Then $X = X_1 \cup X_2$, where X_1 consists of the $\mathfrak{p} \in X$ with $a \in \mathfrak{p}$ and X_2 of the $\mathfrak{p} \in X$ with $a \notin \mathfrak{p}$.

Since $\mathfrak{V}(J) = \mathfrak{V}(I)$, $J \not\subset \bigcup_{\mathfrak{p} \in X} \mathfrak{p}$, so there is $b \in J$ such that $b \notin \mathfrak{p}$ for all $\mathfrak{p} \in X$. Further, by III.1.6 there is $\lambda \in \bigcap_{\mathfrak{p} \in X_2} \mathfrak{p}$ with $\lambda \notin \bigcup_{\mathfrak{p} \in X_1} \mathfrak{p}$. If we now put $a_{r+1} := a + \lambda b$, then $a_{r+1} \notin \mathfrak{p}$ for all $\mathfrak{p} \in X$, so in particular $a_{r+1} \notin \mathfrak{p}_j$ $(j = 1, \ldots, s)$. Since $a_{r+1} \equiv a \bmod J$, the images of $a_1, \ldots, a_{r+1}$ in I/J are part of a minimal generating system of this ideal.

If $\mathfrak{p} \in \mathfrak{V}(a_1, \ldots, a_{r+1}) \setminus \mathfrak{V}(I)$, then $h(\mathfrak{p}) \geq r + 1$, since $\mathfrak{p}$ contains one of the q_i $(i = 1, \ldots, t)$ and by hypothesis $h(q_i) \geq r$, but $a_{r+1} \notin q_i$ $(i = 1, \ldots, t)$. This proves the lemma.

As a first application we have a converse of the generalized Principal Ideal Theorem.

Proposition 4.8. Let R be a Noetherian ring. If $\mathfrak{p} \in \mathrm{Spec}(R)$ has height m, then there are elements $a_1, \ldots, a_m \in \mathfrak{p}$ such that $\mathfrak{p}$ is a minimal prime divisor of $(a_1, \ldots, a_m)$.

Proof. We put $I := \mathfrak{p}$ and $J := \mathfrak{p}^2$. Since $\mu(\mathfrak{p}/\mathfrak{p}^2) \geq \mu_\mathfrak{p}(\mathfrak{p}/\mathfrak{p}^2) = \mathrm{edim} R_\mathfrak{p} \geq \dim R_\mathfrak{p} = h(\mathfrak{p}) = m$, there are, as shown in the proof of 4.7, elements $a_1, \ldots, a_m \in \mathfrak{p}$ such that $h(\mathfrak{p}') \geq m$ for all $\mathfrak{p}' \in \mathrm{Spec}(R)$ with $(a_1, \ldots, a_m) \subset \mathfrak{p}', \mathfrak{p} \not\subset \mathfrak{p}'$. Then $\mathfrak{p}$ is certainly a minimal prime divisor of $(a_1, \ldots, a_m)$.

For the Krull dimension of Noetherian local rings we get the following description.

Corollary 4.9. In any Noetherian local ring $(R, \mathfrak{m})$ there is an $\mathfrak{m}$-primary ideal q which is a complete intersection: $\mu(q) = \dim R$. We have

$$\dim R = \mathrm{Min}\{\mu(q) \,|\, q \text{ is } \mathfrak{m}\text{-primary}\}.$$

Proof. Let $m := \dim R$. By 4.8 there are elements $a_1, \ldots, a_m \in \mathfrak{m}$ such that $\mathfrak{m}$ is the only minimal prime divisor of $q := (a_1, \ldots, a_m)$. By 4.4 q is $\mathfrak{m}$-primary and hence is also a complete intersection. Since $\mu(q') \geq \dim R$ for any $\mathfrak{m}$-primary ideal q' (4.5), the dimension formula also follows.

Definition 4.10. A set $\{a_1, \ldots, a_d\}$ of elements of a d-dimensional Noetherian local ring $(R, \mathfrak{m})$ is called a *system of parameters* of R if it generates an $\mathfrak{m}$-primary ideal.

By 4.9 such systems always exist.

Proposition 4.11. Let $(R, \mathfrak{m})$ be a Noetherian local ring, $a_1, \ldots, a_m$ a system of elements of $\mathfrak{m}$. Then

a) $\dim R \geq \dim R/(a_1, \ldots, a_m) \geq \dim R - m$.

b) $\dim R/(a_1,\ldots,a_m) = \dim R - m$ if and only if $\{a_1,\ldots,a_m\}$ can be extended to a system of parameters of R.

Proof.

a) Let $\delta := \dim R/(a_1,\ldots,a_m)$ and let $\{b_1,\ldots,b_\delta\}$ be a system of elements of $\mathfrak{m}$ whose images in $R/(a_1,\ldots,a_m)$ form a system of parameters of this ring. Then $(a_1,\ldots,a_m,b_1,\ldots,b_\delta)$ is an $\mathfrak{m}$-primary ideal, so $m + \delta \geq \dim R$ by 4.5.

b) If $m + \delta = \dim R$, then $\{a_1,\ldots,a_m,b_1,\ldots,b_\delta\}$ is a system of parameters of R. Conversely, if $\{a_1,\ldots,a_m\}$ can be extended to a parameter system of R by adjoining elements $b_1,\ldots,b_t \in \mathfrak{m}$, then $m + t = \dim R$ and the residue classes of $b_1,\ldots,b_t$ in $R/(a_1,\ldots,a_m)$ generate a primary ideal belonging to the maximal ideal of this ring.

Then $t \geq \dim R/(a_1,\ldots,a_m) \geq \dim R - m = t$, so $\dim R - m = \dim R/(a_1,\ldots,a_m)$.

Corollary 4.12. Let the ideal $I \subset \mathfrak{m}$ be a complete intersection; $I = (a_1,\ldots,a_m)$ with $h(I) = m$. Then $\{a_1,\ldots,a_m\}$ is a subsystem of a parameter system of R, so $\dim R/I = \dim R - m$. In particular, if $a \in \mathfrak{m}$ is a non-zerodivisor of R, then $\dim R/(a) = \dim R - 1$.

Proof. For any minimal prime divisor $\mathfrak{p}$ of I we have $h(\mathfrak{p}) = m$. Let $\mathfrak{p}$ be chosen so that $\dim R/I = \dim R/\mathfrak{p}$. By 4.11a) we have $\dim R - m \leq \dim R/I = \dim R/\mathfrak{p} \leq \dim R - h(\mathfrak{p}) = \dim R - m$ and so $\dim R/(a_1,\ldots,a_m) = \dim R - m$. By 4.11b), $\{a_1,\ldots,a_m\}$ is a subsystem of a parameter system of R. If $a \in \mathfrak{m}$ is not a zero divisor, then $h(a) = 1$ by 3.1. Therefore, (a) is a complete intersection, and $\dim R/(a) = \dim R - 1$.

Next we shall derive a characterization of the complete intersections of global Noetherian rings.

Definition 4.13. A system of elements $\{a_1,\ldots,a_m\}$ $(m \geq 0)$ of a ring R is called *independent* if the following conditions hold:

a) $(a_1,\ldots,a_m) \neq R$.

b) If $F \in R[X_1,\ldots,X_m]$ is a homogeneous polynomial† with $F(a_1,\ldots,a_m) = 0$, then all the coefficients of F are contained in $\mathrm{Rad}(a_1,\ldots,a_m)$.

b) is equivalent to the following statement, which is often more convenient. If $F \in R[X_1,\ldots,X_m]$ is homogeneous of degree d and $b := F(a_1,\ldots,a_m) \in (a_1,\ldots,a_m)^{d+1}$, then all the coefficients of F belong to $\mathrm{Rad}(a_1,\ldots,a_m)$.

In fact, $b \in (a_1,\ldots,a_m)^{d+1}$ can be written in the form $b = G(a_1,\ldots,a_m)$, where G is homogeneous of degree of d with coefficients in $(a_1,\ldots,a_m)$. Now apply 4.13b) to $F - G$.

---

† In the sequel when we speak of polynomials it is always assumed that the polynomial ring is endowed with its canonical grading.

Theorem 4.14. In a Noetherian ring let an ideal $I = (a_1, \ldots, a_m) \neq R$ be given. Then $h(I) = m$ (and so I is a complete intersection) if and only if $\{a_1, \ldots, a_m\}$ is independent.

The proof of the theorem uses a technique going back to E. Davis ([12], [13]). We prepare the way with some results of independent interest.

Lemma 4.15. Let $R[X]$ be the polynomial ring over a Noetherian ring R. For any ideal I of R with $I \neq R$, we have

$$h(IR[X]) = h(I), \quad h((I,X)R[X]) = h(I) + 1.$$

If $\dim R < \infty$, then

$$\dim R[X] = \dim R + 1.$$

Proof. For any $\mathfrak{p} \in \operatorname{Spec}(R)$, $\mathfrak{p}R[X]$ is a prime ideal of the ring $R[X]$, since $R[X]/\mathfrak{p}R[X] \cong R/\mathfrak{p}[X]$. Further, $\mathfrak{P} := (\mathfrak{p}, X)R[X] \in \operatorname{Spec}(R[X]))$, and we have $\mathfrak{p}R[X] \cap R = \mathfrak{p}, \mathfrak{P} \cap R = \mathfrak{p}$.

If $\mathfrak{p}$ is a minimal prime divisor of I, then $\mathfrak{p}R[X]$ is a minimal prime divisor of $IR[X]$ and $\mathfrak{P}$ is a minimal prime divisor of $(I,X)R[X]$. Conversely, if $\mathfrak{Q}$ is a minimal prime divisor of $IR[X]$ (resp. of $(I,X)R[X]$) and $\mathfrak{q} := \mathfrak{Q} \cap R$, then $\mathfrak{q}$ is a minimal prime divisor of I and $\mathfrak{Q} = \mathfrak{q}R[X]$ (resp. $\mathfrak{Q} = (\mathfrak{q}, X)R[X]$). Hence it suffices to show the height formulas for prime ideals $I = \mathfrak{p}$.

For any prime ideal chain $\mathfrak{p}_0 \subset \cdots \subset \mathfrak{p}_t = \mathfrak{p}$ in R we have that $\mathfrak{p}_0 R[X] \subset \cdots \subset \mathfrak{p}_t R[X] = \mathfrak{p}R[X]$ is a prime ideal chain in $R[X]$ and so $h(\mathfrak{p}R[X]) \geq h(\mathfrak{p})$. Further, $\mathfrak{p}R[X] \subsetneq \mathfrak{P} = (\mathfrak{p}, X)R[X]$. Since X is not a zero divisor of $R[X]_{\mathfrak{P}}$, by 4.12 it follows that

$$h(\mathfrak{P}) = \dim R[X]_{\mathfrak{P}} = \dim(R[X]_{\mathfrak{P}}/(X)) + 1 = \dim R_{\mathfrak{p}} + 1 = h(\mathfrak{p}) + 1.$$

Hence it follows that $h(\mathfrak{p}R[X]) \leq h(\mathfrak{p})$, but then also $h(\mathfrak{p}R[X]) = h(\mathfrak{p})$.

It has been shown that $\dim R[X] \geq \dim R + 1$. Now let $\mathfrak{P} \in \operatorname{Spec}(R[X])$ be given and $\mathfrak{p} := \mathfrak{P} \cap R$. Then $R[X]_{\mathfrak{P}}/\mathfrak{p}R[X]_{\mathfrak{P}} \cong (R_{\mathfrak{p}}/\mathfrak{p}R_{\mathfrak{p}})[X]_{\mathfrak{P}^*}$ with some $\mathfrak{P}^* \in \operatorname{Spec}(R_{\mathfrak{p}}/\mathfrak{p}R_{\mathfrak{p}}[X])$. Since this ring is a principal ideal ring, there is an $f \in \mathfrak{P}$ such that $\mathfrak{P}R_{\mathfrak{p}}[X]_{\mathfrak{P}} = (\mathfrak{p}, f)R_{\mathfrak{p}}[X]_{\mathfrak{P}}$. Let $\dim R_{\mathfrak{p}} =: d$ and let $\{a_1, \ldots, a_d\}$ be a parameter system of $R_{\mathfrak{p}}$. Then $\mathfrak{p}R_{\mathfrak{p}} = \operatorname{Rad}((a_1, \ldots, a_d)R_{\mathfrak{p}})$, and it follows that $\mathfrak{P}R_{\mathfrak{p}}[X]_{\mathfrak{P}} = \operatorname{Rad}(a_1, \ldots, a_d, f)$. This implies that

$$h(\mathfrak{P}) = \dim R_{\mathfrak{p}}[X]_{\mathfrak{P}} \leq d + 1 \leq \dim R + 1.$$

Since this is correct for all $\mathfrak{P} \in \operatorname{Spec}(R[X])$, it follows that $\dim R[X] \leq \dim R + 1$ and hence equality also results.

Lemma 4.16. Let $R[X]$ be the polynomial ring over a Noetherian ring R. Let $\{a_1, \ldots, a_m\}$ be a system of elements of R with $(a_1, \ldots, a_m) \neq R$. $\{a_1, \ldots, a_m\}$ is independent in R if and only if $\{a_1, \ldots, a_m, X\}$ is independent in $R[X]$.

Proof. If $\{a_1, \ldots, a_m, X\}$ is independent in $R[X]$, then $\{a_1, \ldots, a_m\}$ is independent in R, since $\operatorname{Rad}((a_1, \ldots, a_m, X) \cdot R[X]) \cap R = \operatorname{Rad}(a_1, \ldots, a_m)$.

Conversely, let the elements $\{a_1, \ldots, a_m\}$ be independent in R and let $F \in R[X][X_1, \ldots, X_m, T]$ be a homogeneous polynomial (in the variables $X_1, \ldots, X_m, T$) with $F(a_1, \ldots, a_m, X) = 0$. We write

$$F = \sum_{\nu_1 + \cdots + \nu_m + \nu = d} \rho_{\nu_1 \ldots \nu_m \nu} X_1^{\nu_1} \ldots X_m^{\nu_m} T^\nu \qquad (d := \deg F, \ \rho_{\nu_1 \ldots \nu_m \nu} \in R[X])$$

and put $I := (a_1, \ldots, a_m)$, $J := (I, X)R[X]$. Then

$$\sum_{\nu_1 + \cdots + \nu_m + \nu = d} \rho_{\nu_1 \ldots \nu_m \nu}(0) a_1^{\nu_1} \ldots a_m^{\nu_m} X^\nu$$

$$\in J^{d+1} = I^{d+1} + I^d X + \cdots + I X^d + X^{d+1} R[X].$$

Comparing coefficients shows that

$$\sum_{\nu_1 + \cdots + \nu_m = d - \nu} \rho_{\nu_1 \ldots \nu_m \nu}(0) a_1^{\nu_1} \ldots a_m^{\nu_m} \in I^{d - \nu + 1} \qquad (\nu = 0, \ldots, d).$$

Since $\{a_1, \ldots, a_m\}$ is independent, $\rho_{\nu_1 \ldots \nu_m v}(0) \in \mathrm{Rad}(I)$ for all $(\nu_1, \ldots, \nu_m, \nu)$ and therefore also $\rho_{\nu_1 \ldots \nu_m \nu} \in \mathrm{Rad}(J)$, q. e. d.

We now consider a system of elements $\{a_1, \ldots, a_m\}$ of a ring R, where $(a_1, \ldots, a_m) \neq R$ and a_1 is not a zero divisor of R. In the full ring of fractions of R we can then form the subring

$$R' := R[a_2/a_1, \ldots, a_m/a_1].$$

The passage from R to R' is an example of a "monoidal transformation." We shall not systematically investigate this concept but only use some of its properties.

Let

$$\alpha : R[Y_2, \ldots, Y_m] \to R[a_2/a_1, \ldots, a_m/a_1]$$

be the R-epimorphism with $\alpha(Y_i) = a_i/a_1$ $(i = 2, \ldots, m)$. Its kernel $\mathfrak{a}$ contains $\mathfrak{a}^* := (a_1 Y_2 - a_2, \ldots, a_1 Y_m - a_m)$, and for any $F \in \mathfrak{a}$ there is a $\rho \in \mathbb{N}$ with $a_1^\rho \cdot F \in \mathfrak{a}^*$, as one sees, for example, by passing over to the ring of fractions R_{a_1}. For a homogeneous polynomial $F \in R[X_1, \ldots, X_m]$ we have $F(a_1, \ldots, a_m) = 0$ if and only if $F(1, Y_2, \ldots, Y_m) \in \mathfrak{a}$. This immediately gives us a criterion for the independence of $\{a_1, \ldots, a_m\}$:

Lemma 4.17. $\{a_1, \ldots, a_m\}$ is independent if and only if for any $\mathfrak{p} \in \mathrm{Spec}(R)$ with $(a_1, \ldots, a_m) \subset \mathfrak{p}$ we have

$$\mathfrak{a} \subset \mathfrak{p} R[Y_2, \ldots, Y_m].$$

Proof. Let $\{a_1, \ldots, a_m\}$ be independent and let $\varphi \in \mathfrak{a}$ be given. Further, let $F \in R[X_1, \ldots, X_n]$ be the homogenization of φ. Then $F(a_1, \ldots, a_m) = 0$; therefore, all the coefficients of F and hence also those of φ lie in $\mathrm{Rad}(a_1, \ldots, a_m)$, and so in each $\mathfrak{p} \in \mathrm{Spec}(R)$ with $(a_1, \ldots, a_m) \subset \mathfrak{p}$.

Conversely, if $\mathfrak{a} \subset \mathfrak{p} R[Y_2, \ldots Y_m]$ for all such $\mathfrak{p}$ and $F(a_1, \ldots, a_m) = 0$ for a homogeneous polynomial $F \in R[X_1, \ldots, X_m]$, it follows that all the coefficients of F belong to $\mathrm{Rad}(a_1, \ldots a_m)$, so $\{a_1, \ldots, a_m\}$ is independent.

Corollary 4.18. If $\{a_1, \ldots, a_m\}$ is independent in R and $m \geq 2$, then $\{a_1, a_3, \ldots, a_m\}$ is independent in $R_1 := R[a_2/a_1]$.

Proof. The kernel $\mathfrak{b}$ of the R_1-epimorphism $\beta : R_1[Y_3, \ldots, Y_m] \to R'$ with $\beta(Y_i) = a_i/a_1$ is the image of $\mathfrak{a}$ in $R_1[Y_3, \ldots, Y_m]$. If $(a_1, a_3, \ldots, a_m) \subset \mathfrak{P}$ for some $\mathfrak{P} \in \mathrm{Spec}(R_1)$ and $\mathfrak{p} := \mathfrak{P} \cap R$, then also $(a_1, a_2, \ldots, a_m) \subset \mathfrak{p}$, since $a_2 = a_1 \cdot (a_2/a_1)$. By 4.17 $\mathfrak{a} \subset \mathfrak{p} R[Y_2, \ldots, Y_m]$ and hence $\mathfrak{b} \subset \mathfrak{p} R_1[Y_3, \ldots, Y_m] \subset \mathfrak{P} R_1[Y_3, \ldots, Y_m]$. Again by 4.17 it follows that $\{a_1, a_3, \ldots, a_m\}$ is independent in R_1.

Proof of 4.14. Let $I = (a_1, \ldots, a_m)$. For $J = (I, X) \cdot R[X]$ we have $h(J) = h(I) + 1$ by 4.15; and by 4.16 $\{a_1, \ldots, a_m\}$ is independent in R if and only if $\{a_1, \ldots, a_m, X\}$ is independent in $R[X]$. Hence we may from the start assume that a_1 is not a zero divisor in R, since if necessary we can pass over to $R[X]$ and adjoin X to the system of elements.

a) Let $h(I) = m$ and $\mathfrak{a}^* = (a_1 Y_2 - a_2, \ldots, a_1 Y_m - a_m)$. For any minimal prime divisor $\mathfrak{p}$ of I we have $\mathfrak{a}^* \subset \mathfrak{p} R[Y_2, \ldots, Y_m]$. Since $\mathfrak{a}^*$ is generated by $m-1$ elements, we have $h(\mathfrak{P}) \leq m - 1$ for any minimal prime divisor $\mathfrak{P}$ of $\mathfrak{a}^*$ with $\mathfrak{P} \subset \mathfrak{p} R[Y_2, \ldots, Y_m]$. We have $a_1 \notin \mathfrak{P}$, since otherwise we would have $IR[Y_2, \ldots, Y_m] \subset \mathfrak{P}$; but this is impossible because $h(IR[Y_2, \ldots, Y_m]) = h(I) = m$ (4.15). Since $a_1^\rho \cdot \mathfrak{a} \subset \mathfrak{a}^* \subset \mathfrak{P}$ for some $\rho \in \mathbb{N}$, it follows that $\mathfrak{a} \subset \mathfrak{P} \subset \mathfrak{p} R[Y_2, \ldots, Y_m]$. By 4.17 $\{a_1, \ldots, a_m\}$ is independent in R.

b) Let $\{a_1, \ldots a_m\}$ be independent in R. For any minimal prime divisor $\mathfrak{p}$ of I we have $h(\mathfrak{p}) \leq m$. Hence it suffices to prove that $h(\mathfrak{p}) \geq m$. For $m = 1$ certainly $h(\mathfrak{p}) = 1$, since a_1 is not a zero divisor of R. Hence we assume that $m > 1$ and that the statement has already been shown for $m-1$ independent elements.

$\mathfrak{p} R[Y_2]$ is a minimal prime divisor of $IR[Y_2]$, and by 4.17 $\mathfrak{p} R[Y_2]$ contains the kernel of $\beta : R[Y_2] \to R_1$ ($Y_2 \mapsto a_2/a_1$). Hence $\mathfrak{p} R_1$ is a minimal prime divisor of $IR_1 = (a_1, a_3, \ldots, a_m) R_1$. By 4.18 $\{a_1, a_3, \ldots a_m\}$ is independent in R_1 and hence $h(\mathfrak{p} R_1) \geq m - 1$.

We have $a_1 Y_2 - a_2 \in \mathrm{Ker} \beta$ and $a_1 Y_2 - a_2$ is not a zero divisor of $R[Y_2]$ since a_1 is not a zero divisor of R. Therefore, $h(\mathfrak{p}) = h(\mathfrak{p} R[Y_2]) > h(\mathfrak{p} R_1) \geq m - 1$, q. e. d.

Corollary 4.19. Let $a_1, \ldots, a_d$ be elements in the maximal ideal of a Noetherian local ring $(R, \mathfrak{m})$ of dimension d. $\{a_1, \ldots, a_d\}$ is a parameter system of R if and only if $\{a_1, \ldots, a_d\}$ is independent. Here this means: If $F(a_1, \ldots, a_d) = 0$ with some homogeneous polynomial $F \in R[X_1, \ldots, X_d]$, then $F \in \mathfrak{m} R[X_1, \ldots, X_d]$.

Exercises

1. In a Noetherian local ring $(R, \mathfrak{m})$ an ideal $\mathfrak{q} \neq R$ is $\mathfrak{m}$-primary if and only if $R/\mathfrak{q}$ is of finite length.

2. In the polynomial ring $K[X_1, X_2, X_3]$ over a field,

$$\mathfrak{p} := (X_1^3 - X_2 X_3, X_2^2 - X_1 X_3, X_3^2 - X_1^2 X_2)$$

is a prime ideal for which all $\mathfrak{p}^\nu$ ($\nu \geq 2$) are not primary. (One can show that for $\mathfrak{p} \in \operatorname{Spec}(K[X_1, X_2, X_3])$, $\mathfrak{p}^2$ is primary if and only if $\mathfrak{p}$ is a local complete intersection [35]).

3. For a prime ideal $\mathfrak{p}$ of a ring R the following statements are equivalent.

    a) $\mathfrak{p}^n$ is $\mathfrak{p}$-primary for $n = 1, \ldots, m$.

    b) The $R/\mathfrak{p}$-modules $\mathfrak{p}^n/\mathfrak{p}^{n+1}$ are torsion-free for $n = 1, \ldots, m - 1$.

4. Let S be a multiplicatively closed subset of a ring R and $i : R \to R_S$ the canonical mapping into the ring of fractions.

    a) An ideal J in R_S is primary if and only if $I := i^{-1}(J)$ is a primary ideal in R.

    b) If $S = R \setminus \mathfrak{p}$ with some $\mathfrak{p} \in \operatorname{Spec}(R)$ and $\mathfrak{p}^{(n)} := S(\mathfrak{p}^n)$ is the S-component of $\mathfrak{p}^n$, then $\mathfrak{p}^{(n)}$ is $\mathfrak{p}$-primary.

5. Let $(R, \mathfrak{m})$ and $(S, \mathfrak{n})$ be local rings, $\phi : R \to S$ a local homomorphism (that is, $\phi(\mathfrak{m}) \subset \mathfrak{n}$). If R and S are Noetherian, then for such a homomorphism

$$\dim S \leq \dim R + \dim S/\phi(\mathfrak{m})S.$$

(Hint: Apply parameter systems of R and $S/\phi(\mathfrak{m})S$.)

6. Let V and W be irreducible affine algebraic varieties and $\phi : V \to W$ a dominant morphism (III.§2, Exercise 6). Let Z' be an irreducible subvariety of W and Z an irreducible component of $\phi^{-1}(Z')$ whose image in Z' is dense. Then

$$\dim \phi^{-1}(Z') \geq \dim Z \geq \dim V - \dim W + \dim Z'.$$

(Suggestion: Apply results from the exercises of Ch. III, §2, and the preceding Exercise 5.)

7. Let R be a ring. For $f = r_0 + r_1 X + \cdots + r_m X^m \in R[X]$ with $r_m \neq 0$ put $\lambda(f) := r_m$; further, let $\lambda(0) := 0$. For an ideal I of $R[X]$ let $\lambda(I) := \{\lambda(f) | f \in I\}$.

    a) $\lambda(I)$ is an ideal in R and $\lambda(I) \subset \lambda(\operatorname{Rad}(I)) \subset \operatorname{Rad}(\lambda(I)), I \cap R \subset \lambda(I)$.

    b) If $I \cap R = (0)$, then $\lambda(I) = R$ if and only if $R[X]/I$ is integral over R.

    c) If I_1, I_2 are two ideals in $R[X]$, then $\lambda(I_1) \cdot \lambda(I_2) \subset \lambda(I_1, I_2)$.

    d) If R is Noetherian and $\lambda(I) \neq R$, then $h(\lambda(I)) \geq h(I)$. (Show this, with the aid of 4.15, first when I is a prime ideal and then reduce the general case to this case by means of a) and c)).

8. Let e be an idempotent element of a ring R. Show that $(e, X) \cdot R[X]$ is a principal ideal of $R[X]$.

5. The graded ring and the conormal module of an ideal

For an ideal I of a ring R the R/I-module I/I^2 is called the *conormal module* of I. We have $\mu(I) \geq \mu(I/I^2)$. From the study of the conormal module we may hope to get information about I itself, in particular about $\mu(I)$.

More generally, for all $n \in \mathbb{N}$ we can consider the quotients $gr_I^n(R) := I^n/I^{n+1}$ $(gr_I^0(R) := R/I)$. Put $gr_I(R) := \bigoplus_{n \in \mathbb{N}} gr_I^n(R)$ (direct sum of R/I-modules) and make $gr_I(R)$ into a graded ring as follows.

If $x = a + I^{m+1} \in gr_I^m(R)$ and $y = b + I^{n+1} \in gr_I^n(R)$, where $a \in I^m, b \in I^n$, then

$$x \cdot y := ab + I^{m+n+1} \in gr_I^{n+m}(R).$$

Evidently this result is independent of the choice of the representatives a, b of x, y. This defines the product of homogeneous elements of $gr_I(R)$. For arbitrary elements we define the product so that the distributive law holds.

Definition 5.1. $gr_I(R)$ is called the *graded ring* (or *form-ring*) of I. If $a \in I^n, a \notin I^{n+1}$, then

$$L_I(a) := a + I^{n+1} \in gr_I^n(R)$$

is called the *leading form* of a with respect to I. For $a \in \bigcap_{n \in \mathbb{N}} I^n$ put $L_I(a) := 0$. The degree $v_I(a)$ of $L_I(a)$ is also called the *degree of a with respect to I*.

Example. If $R = P[X_1, \ldots, X_n]$ is the polynomial ring over a ring P and $I := (X_1, \ldots, X_n)$, then it easily follows that $gr_I(R) \cong R$ (as graded rings). For $F \in R \setminus \{0\}, L_I(F)$ is the homogeneous component of F of lowest degree, and the degree of F with respect to I is the degree of this component.

The graded ring of an ideal I in a ring R was introduced by Krull [46] among others so that, using I in R, degree considerations and coefficient comparisons might be made as in polynomial rings. On the geometric meaning of the form-ring we shall have something to say in Ch. VI (cf. VI.1.1).

In the following let R be any ring, I an ideal of R.

Lemma 5.2.

a) For $a, b \in R$,

$$L_I(a) \cdot L_I(b) = L_I(a \cdot b) \qquad \text{or} \qquad L_I(a) \cdot L_I(b) = 0.$$

b) If $I = (a_1, \ldots, a_m)$ and $x_i := a_i + I^2$ $(i = 1, \ldots, m)$, then

$$gr_I(R) = R/I[x_1, \ldots, x_m].$$

Proof.

a) If $a \in I^m \setminus I^{m+1}, b \in I^n \setminus I^{n+1}$, then $L_I(a) \cdot L_I(b) = ab + I^{m+n+1}$. If $ab \notin I^{m+n+1}$, then $L_I(a) \cdot L_I(b) = L_I(a \cdot b)$; otherwise, $L_I(a) \cdot L_I(b) = 0$.

b) follows from the definition of $gr_I(R)$, since I^ρ is generated by the products $a_1^{v_1} \cdot \ldots \cdot a_m^{v_m}$ with $\sum_{i=1}^m v_i = \rho$.

Now let there be given another ring S, an ideal $J \subset S$ and a ring homomorphism $h : R \to S$ with $h(I) \subset J$. h induces a ring homomorphism

$$gr(h) : gr_I(R) \to gr_J(S)$$

as follows. If $x = a + I^{m+1} \in gr_I^m(R)$ is given, then we take $gr(h)(x) = h(a) + J^{m+1} \in gr_J^m(S)$. Because $h(I^\rho) \subset J^\rho$, this is independent of the choice of the representative a of x. Through linear continuation this assignment extends to all the elements of $gr_I(R)$. By the definition, for all $a \in R$ we have the formula

$$gr(h)(L_I(a)) = \begin{cases} L_J(h(a)), & \text{if } \nu_J(h(a)) = \nu_I(a), \\ 0, & \text{otherwise.} \end{cases} \qquad (1)$$

$gr(h)$ is a homogeneous homomorphism; that is, homogeneous elements of $gr_I(R)$ are mapped to homogeneous elements of the same degree in $gr_J(S)$.

Lemma 5.3. Let h be surjective and $h(I) = J$. If $\mathfrak{a} := \text{Ker}(h)$, then

$$\text{Ker}(gr(h)) = gr_I(\mathfrak{a}) := (\{L_I(a)\}_{a \in \mathfrak{a}}),$$

the ideal in $gr_I(R)$ generated by all the leading forms of the $a \in \mathfrak{a}$. In particular, we have a canonical isomorphism of graded rings

$$gr_J(S) \cong gr_I(R)/gr_I(\mathfrak{a}).$$

Proof. By (1) we have $gr_I(\mathfrak{a}) \subset \text{Ker}(gr(h))$ and further $\text{Ker}(gr(h))$ is a homogeneous ideal. If $x = L_I(a)$ for some $a \in I^m \setminus I^{m+1}$ and $gr(h)(x) = 0$, then $h(a) \in J^{m+1} = h(I^{m+1})$. Then there is $b \in I^{m+1}$ with $h(b) = h(a)$. $a^* := a - b$ belongs to $\mathfrak{a}$ and $x = a^* + I^{m+1} = L_I(a^*)$.

Warning. If $\mathfrak{a} = (a_1, \ldots, a_m)$, then $gr_I(\mathfrak{a})$ need not be generated by $L_I(a_1), \ldots, L_I(a_m)$. Yet under quite general assumptions the converse does hold (Exercise 1).

Lemma 5.4. If under the assumptions of 5.3 $\mathfrak{a} = (a)$ is a principal ideal and $L_I(a)$ is not a zero divisor of $gr_I(R)$, then $gr_I((a)) = (L_I(a))$ is the principal ideal generated by the leading form of a.

By 5.2a) we have $L_I(ra) = L_I(r) \cdot L_I(a)$ for all $r \in R$, because $L_I(a)$ is not a zero divisor of $gr_I(R)$.

We are interested in the question of when, for every $a \in R \setminus \{0\}$, we have $L_I(a) \neq 0$. About this we get some information from

Theorem 5.5. (Krull's Intersection Theorem) Let R be a Noetherian ring, I and ideal of R, M a finitely generated R-module and $\widetilde{M} := \bigcap_{n \in \mathbb{N}} I^n M$. Then

$$\widetilde{M} = I \cdot \widetilde{M}.$$

We shall prove this with the aid of

Lemma 5.6. (Artin–Rees): Under the assumptions of 5.5 let $U \subset M$ be a submodule. Then there is a $k \in \mathbb{N}$ such that for all $n \in \mathbb{N}$

$$I^{n+k} M \cap U = I^n \cdot (I^k M \cap U).$$

For the proof we form the graded ring $\mathfrak{R}_I(R) = \bigoplus_{n \in \mathbb{N}} I^n$ (with the obvious multiplication), which is also called the *Rees ring* of R with respect to I. Further, consider the $\mathfrak{R}_I(R)$-module $\mathfrak{R}_I(M) := \bigoplus_{n \in \mathbb{N}} I^n M$. This is a "graded module over the graded ring $\mathfrak{R}_I(R)$": the elements of $\mathfrak{R}_I^n(M) = I^n M$ are called homogeneous elements of degree n of $\mathfrak{R}_I(M)$. If such an element is multiplied by a homogeneous element of degree m in $\mathfrak{R}_I(R)$, one gets an element of $\mathfrak{R}_I(M)$ of degree $m + n$.

Since I and M are finitely generated, $\mathfrak{R}_I(R)$ is finitely generated as an algebra over $I^0 = R$; therefore, it is Noetherian, and $\mathfrak{R}_I(M)$ is a finitely generated $\mathfrak{R}_I(R)$-module. If we put $U_n = I^n M \cap U$ and $\overline{U} := \bigoplus_{n \in \mathbb{N}} U_n$, then $\overline{U}$ is a submodule of the $\mathfrak{R}_I(R)$- module $\mathfrak{R}_I(M)$. By Hilbert's Basis Theorem for modules, $\overline{U}$ is generated by finitely many elements $v_1, \ldots, v_s$. Since for all $u \in \overline{U}$, all the homogeneous components of u belong to $\overline{U}$, we can choose the v_i to be homogeneous elements. Then, if $m_i := \deg(v_i)$ $(i = 1, \ldots, s)$ and $k := \mathrm{Max}_{i=1,\ldots,s}\{m_i\}$, we have $U_{n+k} = I^n U_k$ for all $n \in \mathbb{N}$. In fact, obviously $I^n U_k \subset U_{n+k}$; conversely, if $u \in U_{n+k}$ is given, then there is a representation $u = \sum_{i=1}^{s} \rho_i v_i$, where $\rho_i \in \mathfrak{R}_I(R)$ is homogeneous of degree $n+k-m_i$; therefore $\rho_i \in I^{n+k-m_i}$ $(i = 1, \ldots, s)$, and it follows that $u \in I^n U_k$.

Krull's Intersection Theorem follows by applying 5.6 to the submodule $U = \widetilde{M}$ of M. Indeed, $I^n M \cap \widetilde{M} = \widetilde{M}$ for all $n \in \mathbb{N}$.

Corollary 5.7. Under the assumptions of 5.5, let I be contained in the intersection of all the maximal ideals of R. For any submodule U of M,

$$\bigcap_{n \in \mathbb{N}} (U + I^n M) = U.$$

In particular, $\bigcap_{n \in \mathbb{N}} I^n M = \langle 0 \rangle$.

Proof. If we put $M' := M/U$, then from 5.5 and Nakayama's Lemma it follows that $\bigcap_{n \in \mathbb{N}} I^n M' = \langle 0 \rangle$. Considering inverse images in M, the assertion follows.

Typical applications of the corollary are the following. If $(R, \mathfrak{m})$ is a Noetherian local ring and if $\mathfrak{a}, I$ are ideals of R with $I \subset \mathfrak{m}$, then

$$\bigcap_{n \in \mathbb{N}} (\mathfrak{a} + I^n) = \mathfrak{a}.$$

If R is any Noetherian ring, $I \subset R$ is an ideal, and if there is a $\mathfrak{p} \in \mathrm{Spec}(R)$ with $I \subset \mathfrak{p}$ such that the canonical mapping $R \to R_\mathfrak{p}$ is injective, then $\bigcap_{n \in \mathbb{N}} I^n = (0)$.

We now return to characterizing complete intersections. If $I = (a_1, \ldots, a_m)$ is an ideal of a ring R, then there is an epimorphism of graded R/I-algebras

$$\alpha : R/I[X_1, \ldots, X_m] \to gr_I(R) \qquad (\alpha(X_i) = a_i + I^2).$$

Remark 5.8. $\{a_1, \ldots, a_m\}$ is independent if and only if $I \neq R$ and $\operatorname{Ker}(\alpha) \subset \operatorname{Rad}(I) \cdot R/I[X_1, \ldots X_m]$.

This follows immediately from the remark following Definition 4.13. If $\operatorname{Rad}(I) = I$, then the independence of $\{a_1, \ldots, a_m\}$ is equivalent to the facts that $I \neq R$ and α is an isomorphism. We will now deal with the last condition, which is of particular interest. It is connected with the concept of a regular sequence.

Definition 5.9. Let M be an R-module. $a \in R$ is called an M-regular element (or a non-zerodivisor of M) if $ax = 0$ with $x \in M$ implies $x = 0$. A sequence $\{a_1, \ldots, a_m\}$ $(m \geq 0)$ of elements of R is called an M-regular sequence if

a) $M \neq (a_1, \ldots, a_m) \cdot M$.

b) For $i = 0, \ldots m - 1$, a_{i+1} is not a zero divisor of $M/(a_1, \ldots, a_i)M$.

If we put $M_i := M/(a_1, \ldots, a_i)M$, then b) is equivalent to the condition that the multiplication mapping

$$\mu_{a_{i+1}} : M_i \to M_i \qquad (x \mapsto a_{i+1}x)$$

is injective for $i = 0, \ldots, m-1$. In particular, a_1 is not a zero divisor of $M_0 = M$.

If $\{a_1, \ldots, a_m\}$ is an M-regular sequence and if M is finitely generated, then it is also an $M_\mathfrak{p}$-regular sequence for all $\mathfrak{p} \in \operatorname{Supp}(M) \cap V(I)$, because for these $\mathfrak{p}$ we also have $M_\mathfrak{p} \neq IM_\mathfrak{p}$ and $(\mu_{a_{i+1}})_\mathfrak{p}$ is injective for all $i = 0, \ldots, n-1$. Further, for an M-regular sequence $\{a_1, \ldots, a_m\}$, $\{a_1^{\nu_1}, \ldots, a_m^{\nu_m}\}$ is also M-regular for any $\nu_i \in \mathbb{N}_+$, and $\{a_{i+1}, \ldots, a_m\}$ is an M_i-regular sequence $(i = 0, \ldots, m - 1)$.

A simple example of an R-regular sequence is $\{X_1, \ldots, X_m\}$ if $R = P[X_1, \ldots, X_m]$ is a polynomial ring over a ring P.

Proposition 5.10. Let R be a ring, $\{a_0, \ldots, a_m\}$ $(m \geq 0)$ an R-regular sequence and $I := (a_0, \ldots, a_m)$. Then:

a) The R/I-epimorphism

$$\alpha : R/I[X_0, \ldots X_m] \to gr_I(R) \qquad (X_i \mapsto a_i + I^2)$$

is an isomorphism (in particular, $\{a_0, \ldots, a_m\}$ is independent).

b) The R-epimorphism

$$\beta : R[Y_1, \ldots, Y_m] \to R[a_1/a_0, \ldots, a_m/a_0] \qquad (Y_i \mapsto a_i/a_0)$$

has kernel $(a_0 Y_1 - a_1, \ldots, a_0 Y_m - a_m)$. (The a_i/a_0 are here considered as elements of $Q(R)$.)

Proof (after Davis [12]).

a) Statement a) is equivalent to the following: For any homogeneous polynomial $F \in R[X_0, \ldots X_m]$ of degree d with $F(a_0, \ldots, a_m) \in I^{d+1}$ we have $F \in IR[X_0, \ldots X_m]$. This is the case if and only if for each homogeneous polynomial F with $F(a_0, \ldots, a_m) = 0$, $F \in IR[X_0, \ldots, X_m]$. If such an F with $\deg F := d$ is given, then

$$F(1, a_1/a_0, \ldots, a_m/a_0) = (1/a_0^d)F(a_0, \ldots, a_m) = 0,$$

and so $F(1, Y_1, \ldots, Y_m) \in \operatorname{Ker}\beta$. Once b) has been proved, it will follow that all the coefficients of F lie in I, because the $a_0 Y_i - a_i$ have this property.

b) It is clear that $a_0 Y_i - a_i \in J := \operatorname{Ker}(\beta)$ for $i = 1, \ldots, m$. We first consider the case $m = 1$. For $F(Y_1) \subset J$ we see, by using the division algorithm, that there exist $d \in \mathbf{N}$, $\varphi(Y_1) \in R[Y_1]$, and $r \in R$ such that

$$a_0^d F(Y_1) = \varphi(Y_1)(a_0 Y_1 - a_1) + r.$$

From $F(a_1/a_0) = 0$ follows $r = 0$. If we now consider the equation modulo a_0^d, it easily follows, because $\{a_0^d, a_1\}$ is also an R- regular sequence, that all the coefficients of φ are divisible by a_0^d. But then $F \in (a_0 Y_1 - a_1)$.

In case $m > 1$, β is the composition

$$R[Y_1, \ldots, Y_m] \xrightarrow{\beta_1} R[a_1/a_0][Y_2, \ldots, Y_m] \xrightarrow{\beta_2} R[a_1/a_0, \ldots, a_m/a_0],$$

where, as already shown, $\operatorname{Ker}(\beta_1) = (a_0 Y_1 - a_1) \cdot R[Y_1, \ldots, Y_m]$. If we put $R' := R[a_1/a_0]$, then $R'/a_0 R' = R'/(a_0, a_1)R' \cong R/(a_0, a_1)[Y_1]$, and from this we see that $\{a_0, a_2, \ldots, a_m\}$ is also an R'-regular sequence, since $\{a_2, \ldots, a_m\}$ is an $R/(a_0, a_1)$-regular sequence. Hence, by the induction hypothesis, $\operatorname{Ker}(\beta_2) = (a_0 Y_2 - a_2, \ldots, a_0 Y_m - a_m)$ and assertion b) follows.

Corollary 5.11. If an ideal I of a ring R is generated by a regular sequence, then the R/I-modules $gr_I^m(R) = I^m/I^{m+1}$ are free for all $m \in \mathbf{N}$. If R is Noetherian, then I is a complete intersection in R.

The following proposition contains a partial converse of 5.10a).

Proposition 5.12. Let $\{a_1, \ldots, a_m\}$ be a generating system of an ideal I in a ring R. Let $R/I \neq \{0\}$ and for $R_i := R/(a_1, \ldots, a_i)$ suppose that $\bigcap_{n \in \mathbf{N}} I^n R_i = (0)$ $(i = 0, \ldots, m)$.

a) If the epimorphism

$$\alpha : R/I[X_1, \ldots, X_m] \to gr_I(R) \qquad (\alpha(X_i) = a_i + I^2)$$

is an isomorphism, then $\{a_1, \ldots, a_m\}$ is an R-regular sequence.

b) If moreover R is Noetherian and I is a prime ideal, then R is an integral domain and R_I is integrally closed in its field of fractions.

Proof. For $m = 0$ the statements of the proposition are trivial, so let $m > 0$.

a) a_1 is not a zero divisor in R. Suppose $ra_1 = 0$ for some $r \in R \setminus (0)$. Then there is $n \in \mathbb{N}$ with $r \in I^n \setminus I^{n+1}$, so $L_I(r) \neq 0$. Since $L_I(a_1) = a_1 + I^2$ is not a zero divisor in $gr_I(R)$, it follows that $L_I(ra_1) = L_I(r) \cdot L_I(a_1)$, contradicting $ra_1 = 0$.

If we put $I_1 := I/(a_1)$, then 5.4 shows that

$$gr_{I_1}(R_1) \cong gr_I(R)/(L_I(a_1)) \cong R_1/I_1[X_2, \dots, X_m].$$

By induction we can assume that $\{a_2, \dots, a_m\}$ is an R_1-regular sequence. Then $\{a_1, \dots, a_m\}$ is an R-regular sequence.

b) For $r, s \in R \setminus (0)$ we have $L_I(r) \cdot L_I(s) \neq (0)$, since $gr_I(R)$ is an integral domain, and hence by 5.2 $L_I(rs) \neq 0$, so $rs \neq 0$. We now put $S := R_I, M = IR_I$. Then $gr_M(S) \cong gr_I(R)_I \cong S/M[X_1, \dots, X_m]$, a polynomial ring over the field S/M. Let $x \in Q(R)$ be integral over S, $x = r/s$ with $r, s \in S$. Then there is an $n \in \mathbb{N}$ such that $S[x] = S + Sx + \dots + Sx^{n-1}$, and so $s^{n-1}x^m \in S$ for all $m \in \mathbb{N}$. For $m \geq n$ it follows that $r^m \in s^{m-n+1}S$, and so $L_M(r)^m \in L_M(s)^{m-n+1}gr_M(S)$. Since $gr_M(S)$ is a factorial ring, $L_M(s)$ must therefore be a divisor of $L_M(r)$. Then there is $r_0 \in S$ with

$$v_M(r - r_0 s) > v_M(r).$$

$x_1 := (r - r_0 s)/s = x - r_0$ is also integral over S. Hence there is also $r_1 \in S$ with $v_M(r - r_1 s) > v_M(r - r_0 s)$ etc. It follows that $r \in \bigcap_{n \in \mathbb{N}}((s) + M^n) = (s)$, and therefore $x = r/s \in S$, q. e. d.

The intersection conditions in 5.12 are fulfilled, for example, if R is Noetherian and I is contained in the intersection of all the maximal ideals of R, hence in particular for ideals $I \neq R$ of a Noetherian local ring.

Corollary 5.13. Let $(R, \mathfrak{m})$ be a Noetherian local ring, $\{a_1, \dots, a_m\}$ $(m \geq 0)$ a sequence of elements in $\mathfrak{m}, I := (a_1, \dots, a_m)$. Then the following statements are equivalent.

a) $\{a_1, \dots, a_m\}$ is an R-regular sequence.

b) $\alpha : R/I[X_1, \dots, X_m] \to gr_I(R)$ $(\alpha(X_i) = a_i + I^2)$ is an isomorphism.

If $\text{Rad}(I) = I$, then a) and b) are equivalent to

c) $\{a_1, \dots, a_m\}$ is independent in R; that is, I is a complete intersection.

To prove the last part of the corollary apply 5.8 and 4.14.

Corollary 5.14. Under the assumptions of 5.13, if $\{a_1, \dots, a_m\}$ is an R-regular sequence, then so is $\{a_{\pi(1)}, \dots, a_{\pi(n)}\}$ for any permutation π of $\{1, \dots, m\}$.

In arbitrary Noetherian rings this is not always so (Exercise 4).

A sequence of elements $\{a_1, \ldots, a_m\}$ of a ring R with $I := (a_1, \ldots, a_m) \neq R$ is called *quasi-regular*, if it is a regular sequence in R_M for all $M \in \mathrm{Max}(R)$ with $I \subset M$. If R is Noetherian by 5.13 and the local–global principle, this is the case if and only if the epimorphism α in 5.13b) is an isomorphism. Therefore any permutation of a quasi-regular sequence is again quasi-regular. For this reason quasi-regular sequences are easier to handle than regular sequences. Of course, any permutation of a regular sequence is quasi-regular and in Noetherian local rings the two notions agree.

Corollary 5.15. Let $(R, \mathfrak{m})$ be a Noetherian local ring, $\mathfrak{p} \in \mathrm{Spec}(R)$ a complete intersection: $\mathfrak{p} = (a_1, \ldots, a_m)$ with $m = h(\mathfrak{p})$. Then:

a) R is an integral domain and $R_\mathfrak{p}$ is integrally closed in $Q(R)$.

b) For any subset $\{i_1, \ldots, i_t\}$ of $\{1, \ldots, m\}$, $(a_{i_1}, \ldots, a_{i_t})$ is a prime ideal and a complete intersection.

c) $\{a_1, \ldots, a_m\}$ is an R-regular sequence.

Proof. By 4.14 $\{a_1, \ldots, a_m\}$ is independent in R. Since $\mathrm{Rad}(\mathfrak{p}) = \mathfrak{p}$, this means that $\alpha : R/\mathfrak{p}[X_1, \ldots, X_m] \to gr_\mathfrak{p}(R)$ is an isomorphism. By 5.13 $\{a_1, \ldots a_m\}$ is an R-regular sequence; by 5.12b) R is an integral domain and $R_\mathfrak{p}$ is integrally closed in $Q(R)$.

In $R/(a_1)$, $\mathfrak{p}/(a_1)$ is a prime ideal of height $m - 1$ generated by $m - 1$ elements, so it is a complete intersection. Then $R/(a_1)$ is also an integral domain and hence (a_1) is a prime ideal of R. Induction now proves assertion b) of the corollary if 5.14 is applied again.

5.15b) can be interpreted geometrically in the following way. If $V \subset \mathbb{A}^n$ is an irreducible d-dimensional variety that is locally a complete intersection at $x \in V$, then there are $n - d$ irreducible hypersurfaces such that V in the neighborhood of x, is the intersection of these hypersurfaces and the intersection of any $\delta \leq n - d$ of them is also irreducible in the neighborhood of x, i.e. x lies on only one component of this intersection.

We now want to discuss what connection there is between the minimal number of generators of an ideal and that of its conormal module.

Lemma 5.16. For an ideal I of a Noetherian ring R the following statements are equivalent.

a) $I = I^n$ for all $n \geq 1$.

b) $gr_I^n(R) = \langle 0 \rangle$ for all $n \geq 1$.

c) I is a principal ideal, generated by an idempotent element of R.

Proof. It is clear that a) and b) are equivalent and that a) follows from c). To derive c) from a) we choose a system of generators $a_1, \ldots, a_m$ of I and write

$$a_i = \sum_{k=1}^m r_{ik} a_k \quad (i = 1, \ldots, m; r_{ik} \in I),$$

which is possible because $I = I^2$. Then

$$\det(E - (r_{ik})) \cdot a_k = 0 \qquad (k = 1, \ldots, m), \tag{2}$$

where E is the m-rowed unit matrix. Expanding the determinant shows that $\det(E - (r_{ik})) = 1 - a$ with some $a \in I$. Since a is a linear combination of the a_k it follows from (2) that $(1 - a)a = 0$, so $a^2 = a$. From $(1 - a)a_k = 0$ we get $a_k = a_k \cdot a \ (k = 1, \ldots, m)$, and so $I = (a)$.

Proposition 5.17. For an ideal I of a Noetherian ring R,

$$h(I) \leq \mu(I/I^2) \leq \mu(I) \leq \mu(I/I^2) + 1.$$

If $\mu(I/I^2) > \dim R$, then $\mu(I) = \mu(I/I^2)$.

Proof. Since $\mu(I_\mathfrak{p}/I_\mathfrak{p}^2) = \mu(I_\mathfrak{p})$ for all $\mathfrak{p} \in \mathfrak{V}(I)$ by Nakayama, and $h(I_\mathfrak{p}) \leq \mu(I_\mathfrak{p})$ by Krull, it follows that $h(I) \leq \mu(I/I^2)$. To show that $\mu(I) \leq \mu(I/I^2) + 1$, we consider elements $a_1, \ldots, a_m \in I$ whose residue classes in I/I^2 form a minimal generating system of this R/I-module. If we put $\overline{R} := R/(a_1, \ldots, a_m)$ and $\overline{I} := I/(a_1, \ldots, a_m)$, then $\overline{I} = \overline{I}^2$. Since by 5.16 $\overline{I}$ is a principal ideal, I is generated by $m + 1$ elements.

Now let $m := \mu(I/I^2) > \dim R$. By 4.7 we can choose elements $a_1, \ldots, a_m \in I$ so that every $\mathfrak{p} \in \mathfrak{V}(a_1, \ldots, a_m)$ with $\mathfrak{p} \notin \mathfrak{V}(I)$ has height $\geq m$. In our case $\mathfrak{V}(a_1, \ldots, a_m) \setminus \mathfrak{V}(I) = \emptyset$; that is, $\overline{I} := I/(a_1, \ldots, a_m)$ is a nilpotent ideal of $\overline{R} := R/(a_1, \ldots, a_m)$. Since it is also generated by an idempotent element, necessarily $\overline{I} = (0)$ and $\mu(I) = \mu(I/I^2)$.

For ideals in polynomial rings we can sharpen 5.17. In the proof Quillen's and Suslin's results on projective modules are used.

Theorem 5.18. (Mohan Kumar [56]) Let $R = P[X]$ be the polynomial ring over a Noetherian ring P of finite Krull dimension. Let the ideal I of R contain a monic polynomial, and let $m := \mu(I/I^2) \geq \dim R/I + 2$. Then there is a finitely generated projective P-module M of rank m such that I is a homomorphic image of the extension module $M[X]$. In particular, if all finitely generated projective P-modules are free, then†

$$\mu(I) = \mu(I/I^2)$$

Proof.

1. As a first step we show that there are elements $f, g \in P$ with $D(f) \cup D(g) = \mathrm{Spec}(R)$ such that I_f is generated as an R_f-module by m elements and $I_g = R_g$.

---

† It can be shown that this formula holds without the assumption that finitely generated projective P-modules be free, see: S. Mandal, On efficient generation of ideals, *Inv. Math.* **75**, 59–67 (1984). For some other cases in which the formula is true, see: B. Nashier, Efficient generation of ideals in polynomial rings, *J. Alg.* **85**, 287–302 (1983).

Let $a_1 \in I$ be an element whose residue class in I/I^2 is contained in a minimal generating system of this R/I-module. We can assume that a_1 is a monic polynomial, since if necessary we can add-to a_1 a sufficiently high power of a monic polynomial in I and thus get an element of the desired form.

Then $S := R/(a_1)$ is a finitely generated P-module. If $J := I \cap P$ and I^* is the image of I in S, then $I^* \cap P = J = JS \cap P$ and S/I^*, as well as S/JS, are integral ring extensions of P/J; therefore, $\dim R/I = \dim S/I^* = \dim P/J = \dim S/JS = \dim S/J^2 S$ (II.2.13). If $\overline{I}$ is the image of I in $S/J^2 S$, then

$$\mu(\overline{I}/\overline{I}^2) = \mu(I/I^2) - 1 > \dim R/I = \dim S/J^2 S;$$

by 5.17 the ideal $\overline{I}$ is generated by $m - 1$ elements. Then there are also elements $a_2^*, \ldots, a_m^* \in I$ with

$$I^* = (a_2^*, \ldots, a_m^*) + JS. \tag{3}$$

We now localize S at $N := 1 + J$ to be able to apply Nakayama's Lemma. To show that JS_N is contained in all maximal ideals of S_N, it suffices to show that JP_N lies in all maximal ideals of P_N (II.2.10), since S_N is integral over P_N. As is checked at once, $1 + JP_N$ consists only of units of P_N. If, for some $\mathfrak{m} \in \mathrm{Max}(P_N)$, there were some $x \in JP_N$, $x \notin \mathfrak{m}$, then we would have an equation $1 = \rho_1 x + \rho_2$ with $\rho_1 \in P_N, \rho_2 \in \mathfrak{m}$, and we would have $\rho_2 = 1 - \rho_1 x \in \mathfrak{m} \cap (1 + JP_N)$, a contradiction.

From (3) it follows by Nakayama's Lemma that $I_N^* = (a_2^*, \ldots, a_m^*)S_N$. If $a_2, \ldots, a_m \in I$ are representatives of the a_i^*, then $I_N = (a_1, \ldots, a_m)R_N$. Therefore, there also exists an $f \in N$ with

$$I_f = (a_1, \ldots, a_m)R_f.$$

If we put $g := 1 - f$, then $g \in J$ and hence $I_g = R_g$. Further, $D(f) \cup D(g) = \mathrm{Spec}(R)$ since $1 = f + g$.

2 . To construct the sought projective P-module M we consider the presentation of the R_f-module I_f belonging to $\{a_1, \ldots, a_m\}$:

$$0 \to K \to R_f^m \to I_f \to 0. \tag{4}$$

Since I_g is a free R_g-module, localizing at g shows that K_g is a projective module over $R_{fg} = P_{fg}[X]$, since (4) splits after localizing.

From the exact sequence

$$0 \to (K_g)_{a_1} \to (R_{fg})_{a_1}^m \to (I_{fg})_{a_1} \to 0,$$

we see, because $(I_{fg})_{a_1} = (R_{fg})_{a_1} \cdot a_1$, that $(K_g)_{a_1}$ is a free module over $(R_{fg})_{a_1} = P_{fg}[X]_{a_1}$. Since a_1 is monic, it follows from the theorem of Quillen and Suslin (IV.3.14) that K_g is a free R_{fg}-module.

Thus the assumptions hold under which the following lemma can be applied.

Lemma 5.19. Let I be a finitely generated module over a ring R. Let there be given a number $m \in \mathbf{N}_+$ and elements $f, g \in R$ such that:

a) $D(f) \cup D(g) = \operatorname{Spec}(R)$,

b) I_g is free R_g-module of rank $\leq m$.

c) There exists an exact sequence of R_f-modules

$$0 \to K \to R_f^m \xrightarrow{\beta_1} I_f \to 0,$$

where K_g is free as an R_{fg}-module.

Then I is a homomorphic image of a projective R-module of rank m.

Proof. Let $F_1 := R_f^m, F_2 := R_g^m$. By b) there is an epimorphism $\beta_2 : F_2 \to I_g$ with free kernel. Since K_g is also a free R_{fg}-module, there is an isomorphism $\alpha : (F_1)_g \xrightarrow{\sim} (F_2)_f$ of R_{fg}-modules such that the diagram

$$
\begin{array}{c}
(F_1)_g \xrightarrow{(\beta_1)_g} \\
\alpha \downarrow \qquad\qquad I_{fg} \\
(F_2)_f \quad (\beta_2)_f
\end{array}
$$

commutes.

Now let F be the fiber product of F_1 and F_2 over $(F_2)_f$ with respect to $F_1 \to (F_1)_g \xrightarrow{\alpha} (F_2)_f$ and $F_2 \to (F_2)_f$. Since I is the fiber product of I_f and I_g over I_{fg} (IV.1.7), on the basis of the universal property a mapping $l : F \to I$ is induced such that the diagram

$$
\begin{array}{c}
F_1 \xrightarrow{\beta_1} I_f \\
F \dashrightarrow[l]{} I \qquad I_{fg} \\
F_2 \xrightarrow[\beta_2]{} I_g
\end{array}
$$

commutes. By IV.1.6 l is surjective, since l_f can be identified with β_1 and l_g with β_2; by hypothesis these mappings are surjective and we have $\operatorname{Spec}(R) = D(f) \cup D(g)$.

Since $F_f \cong F_1$ and $F_g \cong F_2$, by IV.1.14 F is finitely presentable. Since F is also locally free of rank m, F is thus a finitely generated projective R-module of rank m, q. e. d.

If we now apply the above construction of F in the situation of Theorem 5.18, we see that, because f, g were chosen in P, F is locally extended. By Quillen's Theorem (IV.1.20) F is then globally extended too: $F \cong M[X]$ with $M \cong F/XF$. Since F is a projective R-module of rank m, M is a projective P-module of rank m. Since I is a homomorphic image of $M[X]$, Theorem 5.18 is thus proved.

Corollary 5.20. Let K be a field and I an ideal of $K[X_1,\ldots,X_n]$ with $\mu(I/I^2) \geq \dim K[X_1,\ldots,X_n]/I + 2$. Then

$$\mu(I) = \mu(I/I^2).$$

Proof. Since $I \neq (0)$, we may, possibly after tranforming the variables, assume that I contains a monic polynomial in X_n with coefficients in $P :=$ $K[X_1,\ldots,X_{n-1}]$ (II.3.2). The assertion follows then by 5.18, since over P all finitely generated projective modules are free.

The corollary also holds more generally, if K is a principal ideal domain (cf. Exercise 9). The method applied in its proof—to determine $\mu(I)$ by first mapping a projective module to I and then using its freeness—will later be applied again.

Theorem 5.21. Let I be an ideal of the polynomial ring $K[X_1,\ldots,X_n]$ over a field K. If I is locally a complete intersection then I can be generated by n elements.

Proof. Put $R := K[X_1,\ldots,X_n]$ and assume $I \neq (0)$. Since $I_\mathfrak{p}$ is a complete intersection in $R_\mathfrak{p}$ for all $\mathfrak{p} \in \mathfrak{V}(I)$ the $R_\mathfrak{p}/I_\mathfrak{p}$-module $I_\mathfrak{p}/I_\mathfrak{p}^2$ is generated by $h(I_\mathfrak{p}) \leq h(\mathfrak{p})$ elements, hence we have $\mu_\mathfrak{p}(I/I^2) + \dim R/\mathfrak{p} \leq n$. Since $\mathfrak{V}(I) =$ $\mathrm{Supp}(I/I^2)$ the Forster-Swan Theorem (IV.2.14) tells us that $\mu(I/I^2) \leq n$.

Let $d := \dim R/I$ and consider first the case $d \leq n - 2$. If in this case $\mu(I/I^2) \leq d+1$ then $\mu(I) \leq d+2 \leq n$ by 5.17; if $\mu(I/I^2) \geq d+2$ then 5.20 can be applied and it follows that $\mu(I) = \mu(I/I^2) \leq n$.

Since for $d = n$ we have $I = (0)$ it remains only to consider the case $d = n-1$. We then have $h(I) = 1$. Because I is contained in a prime ideal of height 1 of the factorial ring R the elements of I have a greatest common divisor F with $(F) \neq R$. It follows that $I = (F) \cdot I'$ with some ideal I' of R. The elements of I' now have greatest common divisor 1, therefore I' is an ideal of height > 1, hence $\dim R/I' \leq n - 2$.

Along with I also I' is locally a complete intersection, since $\mu_\mathfrak{p}(I') =$ $\mu_\mathfrak{p}(I) = h(I_\mathfrak{p}) \leq h(I'_\mathfrak{p})$ for all $\mathfrak{p} \in \mathfrak{V}(I')$ and therefore $\mu_\mathfrak{p}(I') = h(I'_\mathfrak{p})$. Since $\dim R/I' \leq n-2$ and since the theorem was proved above in this case I' can be generated by n elements, hence also $I = (F) \cdot I'$.

Corollary 5.22. If an affine variety in n-space is locally a complete intersection then its ideal in the polynomial ring can be generated by n elements.

The corollary holds, in particular, for nonsingular varieties, because these are locally complete intersections as will be shown in the next chapter (see VI.1.11). It was conjectured in this case by Forster [24].

Exercises

1. a) Let I and J be ideals of a ring R and suppose $\bigcap_{n \in \mathbb{N}} I^n = (0)$. Assume there are $a_1,\ldots,a_m \in J$ with $gr_I(J) = (L_I(a_1),\ldots,L_I(a_m))$ and $\bigcap_{n \in \mathbb{N}}(I^n + J') = J'$, where $J' := (a_1,\ldots,a_m)$. Then $J = J'$.

b) With the same method prove the Hilbert Basis Theorem for power series rings: If R is Noetherian ring, then so is $R[[X_1,\ldots,X_n]]$, the ring of formal power series in $X_1,\ldots,X_n$ over R. (Suggestion: Put $I := (X_1,\ldots,X_n)$ and proceed as in a).)

2. Let R be a Noetherian ring. For $\mathfrak{p} \in \operatorname{Spec}(R)$ the following statements are equivalent.

   a) $gr_{\mathfrak{p}}(R)$ is an integral domain.

   b) $gr_{\mathfrak{p}R_{\mathfrak{p}}}(R_{\mathfrak{p}})$ is an integral domain and the canonical mapping $gr_{\mathfrak{p}}(R) \to gr_{\mathfrak{p}R_{\mathfrak{p}}}(R_{\mathfrak{p}})$ is injective.

   c) $gr_{\mathfrak{p}R_{\mathfrak{p}}}(R_{\mathfrak{p}})$ is an integral domain and $\mathfrak{p}^n = \mathfrak{p}^{(n)}$ for all $n \in \mathbb{N}_+$.

3. Let I be a prime ideal of a Noetherian ring R. For any $\mathfrak{P} \in \mathfrak{V}(I)$ assume that $I_{\mathfrak{P}}$ is generated by an $R_{\mathfrak{P}}$-regular sequence. Then I^n is I-primary for all $n \in \mathbb{N}_+$.

4. In the polynomial ring $K[X,Y,Z]$ over a field K, $\{X(X-1),XY-1,XZ\}$ is a regular sequence, but not $\{X(X-1),XZ,XY-1\}$. The sequences generate a maximal ideal.

5. Let R be a Noetherian ring, $I \supset J$ two ideals. The following statements are equivalent.

   a) $\mathfrak{R}_I(R) = \bigoplus_{n\in\mathbb{N}} I^n$ is finitely generated as an $\mathfrak{R}_J(R)$-module.

   b) There is an $n \in \mathbb{N}$ with $JI^n = I^{n+1}$.

   If a) or b) holds then $\operatorname{Rad}(I) = \operatorname{Rad}(J)$.

6. Let R be a positively graded Noetherian ring, where $R_0 =: K$ is a field. Let $I \supset J$ be homogeneous ideals of R and $\mathfrak{m} := \bigoplus_{i>0} R_i$. Statements a), b) of Exercise 5 are then equivalent to c) $gr_I(R)/\mathfrak{m}gr_I(R) = \mathfrak{R}_I(R)/\mathfrak{m}\mathfrak{R}_I(R)$ is a finitely generated module over the K-subalgebra generated by the elements $x + \mathfrak{m}I \in I/\mathfrak{m}I$ with $x \in J$. (This can also be shown for ideals $I \supset J$ in the maximal ideal $\mathfrak{m}$ of a Noetherian local ring R.)

7. Under the assumptions of Exercises 6, J is called (according to Northcott–Rees [61]) a *reduction* of I if one of the conditions a)–c) of Exercise 6 is fulfilled.

   a) For any reduction J of I

   $$\mu(J) \geq \dim gr_I(R)/\mathfrak{m}gr_I(R).$$

   b) If K is an infinite field, using the Noether Normalization Theorem show that any (homogeneous) ideal I has a reduction J with

   $$\mu(J) = \dim gr_I(R)/\mathfrak{m}gr_I(R).$$

   (In this case $\dim gr_I(R)/\mathfrak{m}gr_I(R)$ is an upper bound on the minimal number m for which there are (homogeneous) elements $a_1,\ldots,a_m \in R$ with $\operatorname{Rad}(I) = \operatorname{Rad}(a_1,\ldots,a_m)$.)

8. Let $(R, \mathfrak{m})$ be a Noetherian local ring, $I \subset \mathfrak{m}$ an ideal generated by n elements that has a minimal prime divisor of height n. Then $gr_I(R)/\mathfrak{m}gr_I(R)$ is isomorphic to the polynomial ring in n variables over $R/\mathfrak{m}$.

9. Let R be a Noetherian ring of dimension d, $I \subset R[X_1, \ldots, X_n]$ an ideal with $h(I) > d$. Then there are elements $Y_1, \ldots, Y_n \in R[X_1, \ldots, X_n]$ with
   • $R[X_1, \ldots, X_n] = R[Y_1, \ldots, Y_n]$ such that I contains a monic polynomial in Y_n. (Put $S := R[X_1, \ldots, X_{n-1}]$, $X := X_n$. We may assume that $Y_1, \ldots, Y_{n-1}$ have already been found such that the ideal $\lambda(I)$ from §4, Exercise 7 contains a polynomial monic in Y_{n-1}. *Suslin's variable exchange trick:* $Z_{n-1} := X - Y_{n-1}^\rho$, $Z_n := Y_{n-1}$, $Z_i := Y_i$ $(i = 1, \ldots, n-2)$ with sufficiently large $\rho \in \mathbb{N}$ then does the job.)

10. Let R be a ring, $\mathfrak{m} \in \mathrm{Max}(R[X])$ contains a monic polynomial if and only if $\mathfrak{m} \cap R \in \mathrm{Max}(R)$.

11. Let R be the subring of the formal power series ring $K[\|T\|]$ over a field K consisting of all the series $\sum a_v T^v$ with $a_1 = 0$.

    a) The R-homomorphism $R[X] \to Q(R)$ with $X \mapsto 1/T$ is surjective, so its kernel is a maximal ideal $\mathfrak{m}$.

    b) $\mu(\mathfrak{m}) = 2$ and $\mu(\mathfrak{m}/\mathfrak{m}^2) = 1$.

12. Let I be a homogeneous ideal of a Noetherian positively graded ring R for which R_0 is a field. Then $\mu(I) = \mu(I/I^2)$.

13. Let an affine algebra A over a field K have two representations

$$A = K[X_1, \ldots, X_n]/I = K[Y_1, \ldots, Y_m]/J$$

as homomorphic images of the polynomial rings $R := K[X_1, \ldots, X_n]$ and $S := K[Y_1, \ldots, Y_m]$. Then there is an isomorphism of graded A-algebras $gr_I(R)[Y_1, \ldots, Y_m] \cong gr_J(S)[X_1, \ldots, X_n]$, where X_i and Y_j are all of degree 1. In particular, we have an A-module isomorphism

$$I/I^2 \oplus A^m \cong J/J^2 \oplus A^n.$$

(Hint: It suffices to show that $gr_J(S) \cong gr_I(R)[Y_1, \ldots, Y_m]$ for $S = K[X_1, \ldots, X_n, Y_1, \ldots, Y_m]$ and $J = IS + (Y_1, \ldots, Y_m)S$.)

14. In the situation of Exercise 13 deduce that:

    a) If $I \in \mathrm{Spec}(R)$ and I^v is primary for $v = 1, \ldots, r$, then J^v is also primary for $v = 1, \ldots, r$.

    b) For an $\mathfrak{m} \in \mathrm{Max}(R)$, $\mathfrak{m} \supset I$, let $\mathfrak{n} \in \mathrm{Max}(S)$ be the inverse image of $\mathfrak{m}/I$. If $I_\mathfrak{m}$ is a complete intersection in $R_\mathfrak{m}$, then $J_\mathfrak{n}$ is a complete intersection in $S_\mathfrak{n}$.

    c) If $\dim A = 1$ and I/I^2 is a free A-module, then J/J^2 is also a free A-module. (Apply IV. §3, Exercise 10.)

References

The Principal Ideal Theorem, which is fundamental to the whole chapter, was proved by Krull in [44]. There are many generalizations that refer to the height of determinantal ideals in Noetherian rings (for example, see Eagon–Northcott [16], Eisenbud–Evans [18] and W. Bruns, The Eisenbud-Evans generalized principal ideal theorem and determinantal ideals, *Proc. AMS* **83**, 19–24 (1981)).

The theorem of Storch and Eisenbud–Evans treated in §1 was first proved for 3-dimensional space by M. Kneser [41]. Similarly, Abhyankar [2] first gave a geometric proof of Theorem 5.21 in the case of nonsingular curves in A^3 (see also Abhyankar–Sathaye [4]), which Murthy [57] later supported with a brief homological proof for local complete intersections in A^3. Sathaye [71] then proved it for varieties of any dimension over an infinite ground field, before Mohan Kumar [56] solved the general case.

The generation of maximal ideals in polynomial rings $R[X]$ over a Noetherian ring R is dealt with by Davis–Germita [15] and S. M. Bhatwadekar, Projective generation of maximal ideals in polynomial rings, *J. Alg.* (to appear).

In special cases Davis [14] proves that an ideal of a Noetherian ring that is a prime ideal and a complete intersection is generated by a regular sequence every subset of which generates a prime ideal (this generalizes 5.15). But Heitmann [33] has also given an example where this fails.

For information on ideal-theoretic complete intersections in projective space see: Szpiro [78] and G. Faltings, Ein Kriterium für vollständige Durchschnitte, *Inv. math.* **62**, 393–401 (1981).

Chapter VI
Regular and singular points of algebraic varieties

The points of algebraic varieties can be divided into those at which the variety is "smooth" and the "singularities" of the variety. This chapter deals with various characterizations of this concept, mainly through properties of the local rings at the points of the variety. It is shown how investigating the ideal theory of local rings makes possible inferences about the nature of the singularity set. The study of regular sequences, begun in Chapter V, is continued. We get more invariants of Noetherian local rings and a classification of these rings (complete intersections, Gorenstein rings, Cohen–Macaulay rings). This corresponds to a (rough) classification of singularities. In this chapter we are again especially interested in statements that have applications to complete intersections.

1. Regular points of algebraic varieties. Regular local rings

The algebraic definition of regular and singular points of algebraic varieties can best be made intuitive with the example of affine hypersurfaces. Let L be an algebraically closed field, $H \subset A^n(L)$ a hypersurface. To investigate H at one of its points x we may, after a translation, assume that $x = (0, \ldots, 0)$, the origin. Suppose the ideal of H in $L[X_1, \ldots, X_n]$ is generated by F, where $F = F_m + F_{m+1} + \cdots + F_d$, F_i homogeneous of degree i, $F_m \neq 0$, $F_d \neq 0$, $m \leq d$. Then $0 < m$, since $F(0, \ldots, 0) = 0$.

For any line $g = \{t(\xi_1, \ldots, \xi_n) \mid t \in L\}$ with some $(\xi_1, \ldots, \xi_n) \in L^n \setminus \{0\}$ we get the points of $g \cap H$ by solving the equation

$$F(t\xi_1, \ldots, t\xi_n) = t^m(F_m(\xi_1, \ldots, \xi_n) + \cdots + t^{d-m}F_d(\xi_1, \ldots, \xi_n)) = 0$$

for t. For $t = 0$ we get x itself, in fact as an "at least m-fold" intersection point of g and H. x is an "exactly m-fold" intersection point if $F_m(\xi_1, \ldots, \xi_n) \neq 0$.

Therefore, the number m is the minimum of the "intersection multiplicities" of H with an arbitrary line through x. We call m the "multiplicity" of x on H. A line that cuts H at x with multiplicity $> m$ is called a *tangent to H at* x. The direction vectors $(\xi_1, \ldots, \xi_n)$ of the tangents are given by the equation $F_m(\xi_1, \ldots, \xi_n) = 0$. The union of all tangents to H at x is the cone with equation $L(F) = 0$, where $L(F) = F_m$ is the leading form of F. This is called the *geometric tangent cone* of H at x. x is called a *regular* or *simple* point of H if $m = 1$, otherwise a *singular* point or *singularity*. At a simple point the tangent cone is the hyperplane with equation $\sum_{i=1}^n a_i X_i = 0$, $a_i := \frac{\partial F}{\partial X_i}(0)$, which is called the *tangent hyperplane* of H at x.

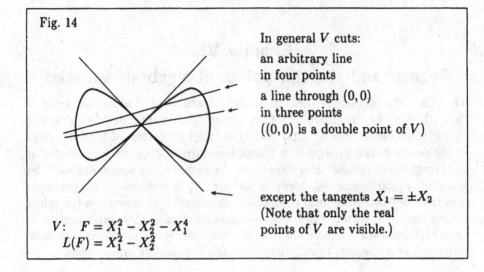

Fig. 14

In general V cuts:

an arbitrary line
in four points

a line through $(0,0)$
in three points
($(0,0)$ is a double point of V)

except the tangents $X_1 = \pm X_2$
(Note that only the real
points of V are visible.)

$$V: \quad F = X_1^2 - X_2^2 - X_1^4$$
$$L(F) = X_1^2 - X_2^2$$

Now let there be given an arbitrary variety $V \subset \mathsf{A}^n(L)$, which we shall take to be an L-variety; that is, all the concepts occurring in what follows refer to the ground field $K = L$.

For $x \in V$ the local ring O_x with maximal ideal $\mathfrak{m}_x$ has been defined (III.§2).

Definition 1.1. $\mathrm{Spec}(gr_{\mathfrak{m}_x}(O_x))$ is called the *tangent cone of V at x*.

In what sense this generalizes the concept of tangent cone considered above will be shown by the following discussion. After a translation we may again assume that $x = (0, \ldots, 0)$. $R := L[X_1, \ldots, X_n]_{(X_1, \ldots, X_n)}$ is then the local ring of x in $\mathsf{A}^n(L)$ and by III.4.18 we have

$$O_x \cong R/\mathfrak{I}(V)R,$$

if $\mathfrak{I}(V)$ is the ideal of V in $L[X_1, \ldots, X_n]$. If we denote the maximal ideal of R by $\mathfrak{m}$, then by V.5.3

$$gr_{\mathfrak{m}_x}(O_x) = gr_{\mathfrak{m}}(R)/gr_{\mathfrak{m}}(\mathfrak{I}(V)R). \tag{1}$$

Since $\mathfrak{m} = (X_1, \ldots, X_n)$ and $\{X_1, \ldots, X_n\}$ is an R-regular sequence, the canonical epimorphism of L-algebras

$$L[X_1, \ldots, X_n] \to gr_{\mathfrak{m}}(R) \qquad (X_i \mapsto X_i + \mathfrak{m}^2)$$

is an isomorphism by V.5.10. If we identify these two rings, $gr_{\mathfrak{m}}(\mathfrak{I}(V)R)$ is mapped to the ideal generated by the $L(F), F \in \mathfrak{I}(V)$. In fact, for $F/G \in R$ ($F, G \in L[X_1, \ldots, X_n]$, $G(0, \ldots, 0) \neq 0$), we have

$$L_{\mathfrak{m}}(F/G) = L(F)/G(0, \cdots, 0)$$

with the usual leading form $L(F)$ of the polynomial F, since

$$L_{\mathfrak{m}}(F) = L_{\mathfrak{m}}(G \cdot (F/G)) = L_{\mathfrak{m}}(G) \cdot L_{\mathfrak{m}}(F/G)$$

by V.5.2 and $L_{\mathfrak{m}}(F) = L(F)$, $L_{\mathfrak{m}}(G) = G(0, \ldots, 0)$.

We therefore get

Proposition 1.2. There is an isomorphism of graded L-algebras

$$gr_{\mathfrak{m}_x}(\mathcal{O}_x) \cong L[X_1, \ldots, X_n]/(\{L(F)\}_{F \in \mathfrak{I}(V)}).$$

The zero set $\mathfrak{V}(\{L(F)\}_{F \in \mathfrak{I}(V)})$ in $\mathbf{A}^n(L)$ is called the *geometric tangent cone* of V at x. The lines through x that belong to the geometric tangent cone are precisely the lines tangent at x to all the hypersurfaces containing V.

Example. If V is a hypersurface with $\mathfrak{I}(V) = (F)$, then by V.5.4

$$gr_{\mathfrak{m}_x}(\mathcal{O}_x) \cong L[X_1, \ldots, X_n]/(L(F)).$$

In general, the tangent cone contains more information than the geometric tangent cone. For example, if $V \subset \mathbf{A}^3(L)$ is given by $X_2^2 - X_1^2 X_3 = 0$ (Fig. 15), then $gr_{\mathfrak{m}_x}(\mathcal{O}_x) \cong L[X_1, X_2, X_3]/(X_2^2)$, the plane $X_2 = 0$ "counted twice," while the geometric tangent cone is the plane $X_2 = 0$ (Fig. 16). At any other point of the X_3-axis the tangent cone is a pair of planes.

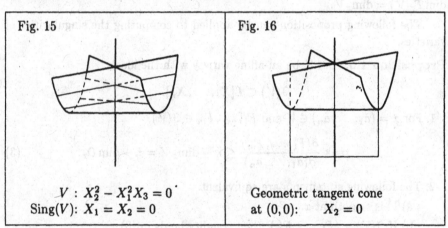

Fig. 15	Fig. 16
$V : X_2^2 - X_1^2 X_3 = 0$ $\mathrm{Sing}(V):\ X_1 = X_2 = 0$	Geometric tangent cone at $(0,0)$: $\quad X_2 = 0$

Let $\mathfrak{I}_1(V)$ denote the ideal spanned by the homogeneous components of degree 1 of the $F \in \mathfrak{I}(V)$ in $L[X_1, \ldots, X_n]$:

$$\mathfrak{I}_1(V) = (\{dF\}_{F \in \mathfrak{I}(V)}), \qquad \text{where} \qquad dF := \sum_{i=1}^n \frac{\partial F}{\partial X_i}(0) \cdot X_i.$$

By restricting to the homogeneous degree 1 components we immediately get from (1) an exact sequence of L-vector spaces

$$0 \to \mathfrak{I}_1(V)/\mathfrak{I}_1(V) \cap \mathfrak{m}^2 \to \mathfrak{m}/\mathfrak{m}^2 \to \mathfrak{m}_x/\mathfrak{m}_x^2 \to 0, \qquad (2)$$

where $\mathfrak{I}_1(V)/\mathfrak{I}_1(V) \cap \mathfrak{m}^2$ is isomorphic to the L-vector space spanned by the linear forms dF ($F \in \mathfrak{I}(V)$).

Definition 1.3. $T_x(V) := \mathfrak{V}(\mathfrak{I}_1(V))$ is called the *tangent space* of V at x.

$T_x(V)$ is a linear variety in $\mathsf{A}^n(L)$ that contains the tangent cone of V at x, since $\mathfrak{I}_1(V) \subset (\{L(F)\}_{F \in \mathfrak{I}(V)})$. We have

$$\dim T_x(V) = n - \dim_L(\mathfrak{I}_1(V)/\mathfrak{I}_1(V) \cap \mathfrak{m}^2)$$
$$= \dim_L \mathfrak{m}_x/\mathfrak{m}_x^2 = \mathrm{edim}\, O_x.$$

As is easily seen, $T_x(V)$ is independent of the coordinates for all $x \in V$. In the projective case we can introduce $T_x(V)$ as the projective closure of the affine tangent space.

Definition 1.4. For an affine or projective variety V a point $x \in V$ is called a *regular* (or *simple*) point (or we say that V is *regular* or *smooth* at x) if

$$\mathrm{edim}\, O_x = \dim O_x.$$

If V is not smooth at x, then x is called a *singularity* of V. A variety that has no singularities is called *nonsingular* or *smooth*.

Since $\dim O_x =: \dim_x V$ is the maximum of the dimensions of the irreducible components of V that contain x (III.4.14d), V is smooth at x if and only if $\dim T_x(V) = \dim_x V$.

The following proposition can be applied to computing the singularities of varieties.

Proposition 1.5. Let V be an affine variety with the ideal

$$\mathfrak{I}(V) \subset L[X_1, \dots, X_n].$$

1. For $x = (a_1, \dots, a_n) \in V$ and $F_1, \dots, F_m \in \mathfrak{I}(V)$,

$$\mathrm{rank}\, \frac{\partial(F_1, \dots, F_m)}{\partial(a_1, \dots, a_n)} \leq n - \dim_x V = n - \dim O_x. \qquad (3)$$

2. The following statements are equivalent.
   a) V is smooth at x.
   b) If $\mathfrak{I}(V) = (F_1, \dots, F_m)$, then equality holds in (3).
   c) There are polynomials $F_1, \dots, F_m \in \mathfrak{I}(V)$ such that equality holds in (3).

(Naturally $\partial(F_1, \dots, F_m)/\partial(a_1, \dots, a_n)$ is the Jacobian matrix gotten by replacing $(X_1, \dots, X_n)$ by $(a_1, \dots, a_n)$ in the formal partial derivatives.)

The proof follows immediately from the fact that $T_x(V)$ is described by the system of linear equations

$$\sum_{i=1}^{n} \frac{\partial F_i}{\partial X_k}(a_1, \dots, a_n)(X_k - a_k) = 0 \qquad (\mathfrak{I}(V) = (F_1, \dots, F_m)),$$

and $\dim T_x(V) = \mathrm{edim}\, O_x \geq \dim O_x = \dim_x V$.

Definition 1.6. A Noetherian local ring is called *regular* if edim $R = \dim R$ (in other words, if the maximal ideal of R is generated by $\dim R$ elements).

In this terminology x is regular on V if and only if $\mathcal{O}_x$ is a regular local ring. If V is defined over a subfield $K \subset L$ and $\mathcal{O}_x$ is the local ring of V at x formed with respect to K, then we call x a *K-regular* point if $\mathcal{O}_x$ is a regular local ring. In general this concept depends of the field of definition K (Exercise 10). In the classical terminology the regular points of V in the sense of definition 1.4 are called "absolutely regular."

Examples of regular local rings are fields, and the local rings $R_{(\pi)}$, where R is a factorial ring and π is a prime element of R, since $\dim R_{(\pi)} = 1$ and the maximal ideal of $R_{(\pi)}$ is generated by π. In particular, the local rings $\mathbf{Z}_{(p)}$ (p a prime number) are regular.

More examples are provided by

Proposition 1.7. If $(R, \mathfrak{m})$ is a regular local ring, then so is $R[X]_{\mathfrak{P}}$ for all $\mathfrak{P} \in \operatorname{Spec}(R[X])$ with $\mathfrak{P} \cap R = \mathfrak{m}$.

Proof. Since $\mathfrak{m}R[X] \subset \mathfrak{P}$, by V.4.15 we have the formula: $\dim R = h(\mathfrak{m}) = h(\mathfrak{m}R[X]) \leq h(\mathfrak{P}) = \dim R[X]_{\mathfrak{P}}$. If $\mathfrak{m}R[X] = \mathfrak{P}$, then $\dim R = \dim R[X]_{\mathfrak{P}}$ and $\mathfrak{P}$ is generated by $\dim R$ elements, so $R[X]_{\mathfrak{P}}$ is regular. If $\mathfrak{m}R[X] \neq \mathfrak{P}$, then $\dim R[X]_{\mathfrak{P}} \geq \dim R + 1$. Since $R[X]/\mathfrak{m}R[X]$ is a principal ideal ring, the image of $\mathfrak{P}$ in this ring is generated by one element, and so $\mathfrak{P}$ is generated by $\dim R + 1$ elements. It follows that $\operatorname{edim}R[X]_{\mathfrak{P}} \leq \dim R + 1 \leq \dim R[X]_{\mathfrak{P}}$; and since we always have $\operatorname{edim}R[X]_{\mathfrak{P}} \geq \dim R[X]_{\mathfrak{P}}$, we find that $R[X]_{\mathfrak{P}}$ is regular.

Corollary 1.8. If K is a principal ideal ring without zero divisors, then $K[X_1, \ldots, X_n]_{\mathfrak{P}}$ is regular for any $\mathfrak{P} \in \operatorname{Spec}(K[X_1, \ldots, X_n])$.

Proof. If $\mathfrak{p} := \mathfrak{P} \cap K[X_1, \ldots, X_{n-1}]$, then

$$K[X_1, \ldots, X_n]_{\mathfrak{P}} \cong K[X_1, \ldots, X_{n-1}]_{\mathfrak{p}}[X_n]_{\mathfrak{P}}$$

and the maximal ideal of $K[X_1, \ldots, X_n]_{\mathfrak{P}}$ lies over that of $K[X_1, \ldots, X_{n-1}]_{\mathfrak{p}}$. The assertion now follows from 1.7 by induction on n, if (at the start of the induction) one uses the fact that $K_{\mathfrak{p}}$ is regular for all $\mathfrak{p} \in \operatorname{Spec}(K)$.

Since the local rings defined over K of the points of affine or projective space are of the form given in 1.8, all such points are K-regular. The same holds for the points of 0-dimensional varieties, since their local rings are fields.

In what follows we study the properties of regular local rings. In particular, from the properties of the rings $\mathcal{O}_x$ we shall then get statements on regular and singular points.

By definition a Noetherian local ring $(R, \mathfrak{m})$ is regular if and only if $\mathfrak{m}$ is a complete intersection in R. Therefore, the propositions of V.§5 immediately provide the following statements about regular local rings.

1. If $\dim R = d$, then R is regular if and only if $gr_{\mathfrak{m}}(R)$ is isomorphic, as a graded $R/\mathfrak{m}$-algebra, to the polynomial algebra in d variables over $R/\mathfrak{m}$ (V.5.13). (From this it follows in particular that the geometric tangent cone at a regular point of an affine variety coincides with the tangent space.)

2. Regular local rings are integral domains and integrally closed in their fields of fractions (V.5.15a)).

3. If $(R, \mathfrak{m})$ is regular, then any minimal system of generators $\{a_1, \ldots, a_d\}$ of $\mathfrak{m}$ is a parameter system of R and an R-regular sequence. Any subsystem $\{a_{i_1}, \ldots, a_{i_\delta}\}$ generates a prime ideal of R (V.5.15b) and c)).

Definition 1.9. If $(R, \mathfrak{m})$ is a regular local ring, then any minimal system of generators of $\mathfrak{m}$ is called a *regular system of parameters* of R.

Statement 3 above is extended by

Proposition 1.10. Let $(R, \mathfrak{m})$ be a regular local ring, $I \subset \mathfrak{m}$ an ideal. The following statements are equivalent.

a) R/I is also a regular local ring.

b) I is generated by a subsystem of a regular parameter system of R.

Proof.

b)$\rightarrow$a). Let $\{a_1, \ldots, a_d\}$ be a regular system of parameters of R and let I be generated by $\{a_{\delta+1}, \ldots, a_d\}$ with some $\delta \in [0, d]$. Then $\dim R/I = \delta$ by V.4.11, and the maximal ideal of R/I is generated by the images of $a_1, \ldots, a_\delta$. Therefore R/I is regular.

a)$\rightarrow$b). Let R/I be regular of dimension δ and let $a_1, \ldots, a_\delta \in \mathfrak{m}$ be a system of representatives of a regular system of parameters of R/I. Then $\mathfrak{m} = (a_1, \ldots, a_\delta) + I$. If $\overline{\mathfrak{m}}$ is the maximal ideal of R/I, then we have an exact sequence of vector spaces over $R/\mathfrak{m}: 0 \rightarrow I/I \cap \mathfrak{m}^2 \rightarrow \mathfrak{m}/\mathfrak{m}^2 \rightarrow \overline{\mathfrak{m}}/\overline{\mathfrak{m}}^2 \rightarrow 0$. By Nakayama's Lemma it follows that $\{a_1, \ldots, a_\delta\}$ can be extended to a minimal generating system of $\mathfrak{m}$ by adjoining suitable elements $a_{\delta+1}, \ldots, a_d \in I$.

Let $I' := (a_{\delta+1}, \ldots, a_d)$. R/I' (as shown above) is a regular local ring of dimension δ, and R/I is a homomorphic image of R/I'. Since R/I' is an integral domain and $\dim R/I' = \dim R/I$, we must have $R/I' = R/I$ and so $I = (a_{\delta+1}, \ldots, a_d)$, q. e. d.

Corollary 1.11. At any K-regular point a K-variety is a local complete intersection.

This follows from 1.8 and 1.10, since the local rings of the variety are of the form $K[X_1, \ldots, X_n]_{\mathfrak{p}}/I$.

To derive some geometric statements we now consider an L-variety $V \subset \mathbb{A}^n(L)$ and assume that $x \in V$ is the origin, $\mathcal{O}_x = R/\mathfrak{J}(V)R$ with $R = L[X_1, \ldots, X_n]_{(X_1, \ldots, X_n)}$. A system $\{F_1, \ldots, F_n\}$ in $L[X_1, \ldots, X_n]$ is a regular parameter system of R if and only if the leading forms $L(F_i)$ are of degree 1 $(i = 1, \ldots, n)$ and are linearly independent over L.

For we have $\mathfrak{m} = (F_1, \ldots, F_n)R$ if and only if the residue classes $F_i + \mathfrak{m}^2 \in \mathfrak{m}/\mathfrak{m}^2$ $(i = 1, \ldots, n)$ span the $R/\mathfrak{m}$-vector space $\mathfrak{m}/\mathfrak{m}^2$, which is equivalent to the statement above. On the other hand, the statement is also equivalent to the following: the hypersurfaces $H_i := \mathfrak{V}(F_i)$ are smooth at x $(i = 1, \ldots, n)$ and the tangent hyperplanes $T_x(H_i) = \mathfrak{V}(L(F_i))$ are linearly independent (i.e. the $L(F_i)$ are linearly independent over L).

If a regular systems of parameters $\{F_1, \ldots, F_n\}$ of R is given with $F_i \in L[X_1, \ldots, X_n]$ $(i = 1, \ldots, n)$, then the $H_i := \mathfrak{V}(F_i)$ are something like the "coordinate hypersurfaces" of a local coordinate system of $\mathsf{A}^n(L)$ at x. That V is smooth at x is by 1.10 equivalent to the existence of a regular parameter system $\{F_1, \ldots, F_n\}$ of R such that $\mathfrak{I}(V)R = (F_{d+1}, \ldots, F_n)$, where d is the dimension of V at x. Here, without loss of generality, it may be assumed that $F_1, \ldots, F_n \in L[X_1, \ldots, X_n]$. In the neighborhood of x, V equals the intersection $H_{d+1} \cap \cdots \cap H_n$ of $n - d$ coordinate hypersurfaces. The other d hypersurfaces $H_1, \ldots, H_d$ correspond to a regular parameter system $\{\phi_1, \ldots, \phi_d\}$ of O_x, $\phi_i := \bar{F}_i + \mathfrak{I}(V)R$ $(i = 1, \ldots, d)$. They define a "local coordinate system" on V at x.

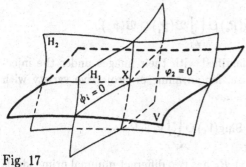

Fig. 17

Now we can also say what it means geometrically for a smooth affine variety to be an ideal-theoretic complete intersection.

Proposition 1.12. Let $V \subset \mathsf{A}^n(L)$ be a nonsingular L-variety of dimension d. V is an ideal-theoretic complete intersection if and only if there are $n - d$ L-hypersurfaces $H_1, \ldots, H_{n-d} \subset \mathsf{A}^n(L)$ such that:

1. $V = H_1 \cap \cdots \cap H_{n-d}$.

2. For all $x \in V$, the H_i are smooth at x $(i = 1, \ldots, n - d)$.

3. The tangent hyperplanes $T_x(H_i)$ $(i = 1, \ldots, n - d)$ are linearly independent for all $x \in V$.

Proof. It remains only to prove that the conditions are sufficient. If hypersurfaces $H_1, \ldots, H_{n-d}$ are given satisfying 1–3 and if $\mathfrak{I}(H_i) = (F_i)$, then $\mathfrak{I}(V) = \mathrm{Rad}(F_1, \ldots, F_{n-d})$, and it suffices to prove that $\mathfrak{I}(V)_\mathfrak{m} = (F_1, \ldots, F_{n-d})_\mathfrak{m}$ for any maximal ideal $\mathfrak{m} \supset \mathfrak{I}(V)$. After a coordinate transformation we may assume that $\mathfrak{m} = (X_1, \ldots, X_n)$ is the ideal of the origin. Since $F_1, \ldots, F_{n-d} \in \mathfrak{I}(V)$

and since the leading forms of the F_i are of degree 1 and linearly independent over L, the local discussion above shows that $\mathfrak{I}(V)_m$ is in fact generated by $F_1, \ldots, F_{n-d}$.

Remark. If V is not smooth, then it is more difficult to give a geometric criterion for V to be an ideal-theoretic complete intersection. This can be done in the framework of the theory of the intersection-multiplicities of algebraic varieties. Instead of 2. and 3. one must require that the hypersurfaces "intersect along V with multiplicity 1."

For a Noetherian ring R

$$\mathrm{Reg}(R) := \{ \mathfrak{p} \in \mathrm{Spec}(R) \mid R_{\mathfrak{p}} \text{ is a regular local ring } \}$$

is called the *regular locus* of R, and $\mathrm{Sing}(R) := \mathrm{Spec}(R) \setminus \mathrm{Reg}(R)$ the *singular locus*. Likewise, for an affine or projective variety V the set $\mathrm{Reg}(V)$ of points x at which V is smooth is called the *regular locus* of V, and $\mathrm{Sing}(V) := V \setminus \mathrm{Reg}(V)$ the *singular locus* of V.

Proposition 1.13. Let $R \neq \{0\}$ be a reduced Noetherian ring, $\{\mathfrak{p}_1, \ldots, \mathfrak{p}_s\}$ the set of minimal prime ideals of R, and $R_i := R/\mathfrak{p}_i$ $(i = 1, \ldots, s)$. Then

$$\mathrm{Sing}(R) = \bigcup_{i=1}^{s} \mathrm{Sing}(R_i) \cup \bigcup_{i \neq j} \mathfrak{V}(\mathfrak{p}_i) \cap \mathfrak{V}(\mathfrak{p}_j).$$

(Here the prime ideals of R_i are identified with their images under the injection $\mathrm{Spec}(R_i) \to \mathrm{Spec}(R)$). Likewise, for an affine or projective variety with irreducible components $V_1, \ldots, V_s$ we have

$$\mathrm{Sing}(V) = \bigcup_{i=1}^{s} \mathrm{Sing}(V_i) \cup \bigcup_{i \neq j} V_i \cap V_j.$$

Proof. For $\mathfrak{p} \in \mathfrak{V}(\mathfrak{p}_i) \cap \mathfrak{V}(\mathfrak{p}_j)$ $(i \neq j)$, $R_{\mathfrak{p}}$ has two different minimal prime ideals and so is not an integral domain and therefore is not regular. On the other hand, if $\mathfrak{p}$ contains exactly one of the $\mathfrak{p}_i$, then $\mathfrak{p}_i R_{\mathfrak{p}} = (0)$, since R is reduced. Further, $R_{\mathfrak{p}} = R_{\mathfrak{p}}/\mathfrak{p}_i R_{\mathfrak{p}} = (R/\mathfrak{p}_i)_{\mathfrak{p}} = (R_i)_{\mathfrak{p}}$. Hence we have $\mathfrak{p} \in \mathrm{Reg}(R)$ if and only if its image in R_i lies in $\mathrm{Reg}(R_i)$.

For affine and projective varieties the proof proceeds in a completely analogous way.

Corollary 1.14. Nonsingular hypersurfaces in $\mathbf{P}^n(L)$ $(n \geq 2)$ are irreducible.

Since the irreducible components of hypersurfaces are themselves hypersurfaces and since two hypersurfaces in $\mathbf{P}^n$ $(n \geq 2)$ always intersect (I.5.2), 1.14 follows from 1.13.

1.14 gives a sufficient *irreducibility criterion* for polynomials in $n \geq 2$ variables. For such a polynomial $F \in K[X_1, \ldots, X_n]$ (K a field), consider the hypersurfaces H in $\mathbf{P}^n(L)$ belonging to the homogenization $F^* \in K[Y_0, \ldots, Y_n]$ where L is the algebraic closure of K. If H is nonsingular, then F is (up to a unit) a power of an irreducible polynomial. But if we know that F can have no multiple factors, then F is irreducible.

Example. The polynomial $F = a_1 X_1^m + a_2 X_2^m + \cdots + a_n X_n^m + a_{n+1}$ $(a_i \neq 0)$ is irreducible over any field K whose characteristic does not divide m. Indeed, $\partial F/\partial X_i = m a_i X_i^{m-1}$. If F had a multiple factor, then F and one of its partial derivatives would have a common divisor, which is evidently not the case. The affine hypersurface H in A^n defined by F is smooth by 1.5, since the Jacobian matrix $(\frac{\partial F}{\partial X_i}(x), \ldots, \frac{\partial F}{\partial X_n}(x))$ has rank 1 at any point $x \in H$. The corresponding statement holds for the projective closure of H, since if one first homogenizes F and then dehomogenizes with respect to one of the variables $Y_1, \ldots, Y_n$, one gets back a polynomial like F. (That F is irreducible can also be shown by Eisenstein's irreducibility criterion).

The next theorem will generalize the previous Proposition 1.5. Let $A = K[X_1, \ldots, X_n]/I$ be an affine algebra over a field K and $I = (F_1, \ldots, F_m)$. For $\mathfrak{p} \in \mathrm{Spec}(A)$ we then have $A_\mathfrak{p}/\mathfrak{p}A_\mathfrak{p} = Q(A/\mathfrak{p}) = K(\xi_1, \ldots, \xi_n)$, where ξ_i is the image of X_i in $A_\mathfrak{p}/\mathfrak{p}A_\mathfrak{p}$. If t is the transcendence degree of $K(\xi_1, \ldots, \xi_n)$ over K, then $t = \dim A/\mathfrak{p}$ by II.3.6a).

$$J(\mathfrak{p}) := \frac{\partial(F_1, \ldots, F_m)'}{\partial(\xi_1, \ldots, \xi_n)} = (\frac{\partial F_i}{\partial X_k}(\xi_1, \ldots, \xi_n))_{i=1,\ldots,m, k=1,\ldots,n}$$

is called the *Jacobian matrix at the place* $\mathfrak{p}$ (with respect to the given presentation of A).

Theorem 1.15. (Jacobian criterion for regular local rings) Let K be a perfect field and A an integral domain with $\dim A =: d$. Then

$$\mathrm{rank}(J(\mathfrak{p})) = n - d - (\mathrm{edim}A_\mathfrak{p} - \dim A_\mathfrak{p}).$$

Hence $\mathfrak{p} \in \mathrm{Reg}(A)$ if and only if $\mathrm{rank}(J(\mathfrak{p})) = n - d$.

Proof. Let $L := K(\xi_1, \ldots, \xi_n)$. We shall use the following proposition of field theory: Since K is perfect, from $\{\xi_1, \ldots, \xi_n\}$ one can choose a transcendence basis of L/K—say $\{\xi_1, \ldots, \xi_t\}$ with a suitable numbering—such that L is separably algebraic over $K(\xi_1, \ldots, \xi_t)$.

If $\mathfrak{P}$ is the inverse image of $\mathfrak{p}$ in $R := K[X_1, \ldots, X_n]$, then

$$\mathfrak{P} \cap K[X_1, \ldots, X_t] = (0),$$

since $\{\xi_1, \ldots, \xi_t\}$ is algebraically independent over K. Hence $K(X_1, \ldots, X_t) \subset R_\mathfrak{P}$ and $R_\mathfrak{P} = S_\mathfrak{M}$ with

$$S := K(X_1, \ldots, X_t)[X_{t+1}, \ldots, X_n], \qquad \mathfrak{M} := \mathfrak{P}S.$$

. By 1.8 $S_\mathfrak{M}$ is a regular local ring and $\dim S_\mathfrak{M} = \dim R_\mathfrak{P} = n - \dim R/\mathfrak{P} = n - \dim A/\mathfrak{p} = n - t$. Further, $A_\mathfrak{p} \cong S_\mathfrak{M}/IS_\mathfrak{M}$. Hence we have an exact sequence of vector spaces over L

$$0 \to \Lambda \to \mathrm{gr}^1_{\mathfrak{M}S_\mathfrak{M}}(S_\mathfrak{M}) \to \mathrm{gr}^1_{\mathfrak{p}A_\mathfrak{p}}(A_\mathfrak{p}) \to 0,$$

where Λ is spanned by the elements $F_i + \mathfrak{M}^2 S_{\mathfrak{M}} \in gr^1_{\mathfrak{M} S_{\mathfrak{M}}}(S_{\mathfrak{M}})$ $(i = 1, \ldots, m)$. Here $\dim_L(\Lambda) = n - t - \mathrm{edim} A_{\mathfrak{p}} = n - d - (\mathrm{edim} A_{\mathfrak{p}} - \dim A_{\mathfrak{p}})$, since $d = \dim A_{\mathfrak{p}} + t$ (II.3.6). It is now a matter of proving that $\dim_L(\Lambda)$ coincides with the rank of $J(\mathfrak{p})$.

To do this we first want to get a minimal generating system of $\mathfrak{M}$. Since $L \cong S/\mathfrak{M}$ is algebraic over $K(\xi_1, \ldots, \xi_t) \cong K(X_1, \ldots, X_t)$, we can get a generating system of $\mathfrak{M}$ as follows. For $i = 1, \ldots, n - t$ let $g_i(X)$ denote the minimal polynomial of ξ_{t+i} over $K(\xi_1, \ldots, \xi_t)[\xi_{t+1}, \ldots, \xi_{t+i-1}]$. Let $G_i(X_1, \ldots, X_{t+i}) \in K(X_1, \ldots, X_t)[X_{t+1}, \ldots, X_{t+i}]$ be a polynomial gotten from $g_i(X)$ by choosing representatives in $K(X_1, \ldots, X_t)[X_{t+1}, \ldots, X_{t+i-1}]$ for the nonzero coefficients of $g_i(X)$ and replacing X by X_{t+i}. It is clear that $(G_1, \ldots, G_{n-t}) \subset \mathfrak{M}$. Further, $S/(G_1, \ldots, G_{n-t})$ as a vector space over $K(X_1, \ldots, X_t)$ has dimension $\prod_{i=1}^{n-t} d_i$, if d_i is the degree of $g_i(X)$ $(i = 1, \ldots, n - t)$. Since also $[L : K(X_1, \ldots, X_t)] = \prod_{i=1}^{n-t} d_i$, we must have $(G_1, \ldots, G_{n-t}) = \mathfrak{M}$. In particular, $\{G_1, \ldots, G_{n-t}\}$ is then a regular parameter system of $S_{\mathfrak{M}}$ and $L_{\mathfrak{M} S_{\mathfrak{M}}}(G_i)$ $(i = 1, \ldots, n - t)$ is a basis of $gr^1_{\mathfrak{M} S_{\mathfrak{M}}}(S_{\mathfrak{M}})$.

Because $L/K(\xi_1, \ldots, \xi_t)$ is separable, we have

$$\frac{\partial G_i}{\partial X_{t+i}}(\xi_1, \ldots, \xi_n) = g_i'(\xi_{t+i}) \neq 0 \qquad (i = 1, \ldots, n - t),$$

and since only the indeterminates $X_1, \ldots, X_{t+i}$ occur in G_i, the matrix $\partial(G_1, \ldots, G_{n-t})/\partial(\xi_1, \ldots, \xi_n)$ has rank $n - t$. We now write

$$F_i = \sum_{k=1}^{n-t} \sigma_{ik} G_k \qquad (i = 1, \ldots, m; \sigma_{ik} \in S_{\mathfrak{M}})$$

and, by the product rule for differentiation, get

$$\frac{\partial(F_1, \ldots, F_m)}{\partial(\xi_1, \ldots, \xi_n)} = (\sigma_{ik}(\xi_1, \ldots, \xi_n)) \cdot \frac{\partial(G_1, \ldots, G_{n-t})}{\partial(\xi_1, \ldots, \xi_n)}, \qquad (4)$$

where $\sigma_{ik}(\xi_1, \ldots, \xi_n)$ is the image of σ_{ik} in L. On the other hand,

$$F_i + \mathfrak{M}^2 S_{\mathfrak{M}} = \sum_{k=1}^{n-t} \sigma_{ik}(\xi_1, \ldots, \xi_n) \cdot L_{\mathfrak{M} S_{\mathfrak{M}}}(G_k).$$

Since the $L_{\mathfrak{M} S_{\mathfrak{M}}}(G_k)$ form a basis of $gr^1_{\mathfrak{M} S_{\mathfrak{M}}}(S_{\mathfrak{M}})$ and the $F_i + \mathfrak{M}^2 S_{\mathfrak{M}}$ span the L-vector space Λ, we have $\dim_L(\Lambda) = \mathrm{rank}(\sigma_{ik}(\xi_1, \ldots, \xi_n))$. Since $\partial(G_1, \ldots, G_{n-t})/\partial(\xi_1, \ldots, \xi_n)$ has (maximal) rank $n - t$, it follows from (4) that also rank $\partial(F_1, \ldots, F_m)/\partial(\xi_1, \ldots, \xi_n) = \dim_L(\Lambda)$, q. e. d.

Corollary 1.16. For any reduced affine algebra $A \neq \{0\}$ over a perfect field K, $\mathrm{Reg}(A)$ is open in $\mathrm{Spec}(A)$ and $\mathrm{Reg}(A) \cap \mathrm{Max}(A) \neq \emptyset$.

Proof. By 1.13 it suffices to consider the case where A is an integral domain: $A = K[X_1, \ldots, X_n]/(F_1, \ldots, F_m)$. Let x_k denote the image of X_k in A and $\partial F_i/\partial x_k$ that of the partial derivative $\partial F_i/\partial X_k$. Let $\mathfrak{d}(A)$ be the ideal of A generated by the $(n-d)$-rowed minors of the Jacobian matrix $\partial(F_1, \ldots, F_m)/\partial(x_1, \ldots, x_n)$, where $d := \dim A$. For $\mathfrak{p} \in \mathrm{Spec}(A)$ we have $\mathfrak{p} \in \mathrm{Sing}(A)$ if and only if $\mathfrak{d}(A) \subset \mathfrak{p}$, for this condition is equivalent to the vanishing of all the $(n-d)$-rowed minors of $J(\mathfrak{p}) = \partial(F_1, \ldots, F_m)/\partial(\xi_1, \ldots, \xi_n)$, where ξ_k is the image of X_k in $A_\mathfrak{p}/\mathfrak{p}A_\mathfrak{p}$. Since always $\mathrm{rank}(J(\mathfrak{p})) \leq n - d$, by 1.15 this condition is equivalent with $\mathrm{rank}(J(\mathfrak{p})) < n - d$, hence with $\mathfrak{p} \in \mathrm{Sing}(A)$.

Therefore, $\mathrm{Sing}(A) = \mathfrak{V}(\mathfrak{d}(A))$ is a closed subset of $\mathrm{Spec}(A)$, and hence $\mathrm{Reg}(A)$ is open. Since the zero ideal belongs to $\mathrm{Reg}(A), \mathrm{Reg}(A) \neq \emptyset$. Then there is also an $f \in A \setminus \{0\}$ with $D(f) \subset \mathrm{Reg}(A)$. Since the intersection of all maximal ideals of A is the zero ideal, there is at least one $\mathfrak{m} \in \mathrm{Max}(R)$ with $f \notin \mathfrak{m}$. Therefore, $\mathrm{Reg}(A) \cap \mathrm{Max}(A) \neq \emptyset$.

Remark. More generally, one can show that 1.16 also holds for an arbitrary field K, but the proof is more complicated. On the other hand, there are Noetherian rings R for which $\mathrm{Reg}(R)$ is not open in $\mathrm{Spec}(R)$. It can also be shown that the ideal $\mathfrak{d}(A)$ in the proof of 1.16 is an invariant of A, i.e. is independent of the representation $A = K[X_1, \ldots, X_n]/(F_1, \ldots, F_m)$ ($\mathfrak{d}(A)$ is called the Jacobian ideal of A).

Corollary 1.17. The regular locus of any nonempty affine or projective variety is open and not empty. Algebraic curves have only finitely many singularities.

Proof. It suffices to prove this in the affine case. If A is the coordinate ring of an affine variety V, then $\mathrm{Reg}(V) \neq \emptyset$ and is open, since $\mathrm{Reg}(A) \cap \mathrm{Max}(A) \neq \emptyset$ and $\mathrm{Reg}(A)$ is open in $\mathrm{Spec}(A)$.

If V is a curve, by 1.13 $\mathrm{Sing}(V)$ contains no irreducible components of V and so is 0-dimensional, so is a finite point set.

Note that in the above proof of $\mathrm{Reg}(V) \neq \emptyset$ it was useful to prove the corresponding proposition for $\mathrm{Spec}(A)$ first, and thus not to consider just the maximal ideals of A. This shows one of the advantages of working with spectra instead of varieties.

Corollary 1.18. Under the assumptions of 1.16 let $\mathfrak{p} \in \mathrm{Reg}(A)$. Then there is an $\mathfrak{m} \in \mathrm{Max}(A) \cap \mathrm{Reg}(A)$ with $\mathfrak{p} \subset \mathfrak{m}$. Further, $\mathrm{Spec}(A_\mathfrak{p}) = \mathrm{Reg}(A_\mathfrak{p})$; that is, $(A_\mathfrak{p})_\mathfrak{P}$ is a regular local ring for any $\mathfrak{P} \in \mathrm{Spec}(A_\mathfrak{p})$.

Proof. We choose $f \in A$ with $\mathfrak{p} \in D(f) \subset \mathrm{Reg}(A)$. Since $\mathfrak{p}$ is the intersection of all maximal ideals of A that contain $\mathfrak{p}$, there is an $\mathfrak{m} \in \mathrm{Max}(A)$ with $\mathfrak{p} \subset \mathfrak{m}$ and $f \notin \mathfrak{m}$, therefore $\mathfrak{m} \in D(f) \subset \mathrm{Reg}(A)$. This proves the first assertion.

Any $\mathfrak{P} \in \mathrm{Spec}(A_\mathfrak{p})$ is of the form $\mathfrak{P} = \mathfrak{q}A_\mathfrak{p}$ with $\mathfrak{q} \in \mathrm{Spec}(A)$, $\mathfrak{q} \subset \mathfrak{p}$, and $(A_\mathfrak{p})_\mathfrak{P} = A_\mathfrak{q}$. If we had $\mathfrak{q} \in \mathrm{Sing}(A)$, then we would also have $\mathfrak{p} \in \mathrm{Sing}(A)$ because $\mathrm{Sing}(A)$ is closed. Therefore $(A_\mathfrak{p})_\mathfrak{P}$ is regular.

In Ch. VII it will be shown that, more generally, for any regular local ring R and any $\mathfrak{P} \in \mathrm{Spec}(R), R_{\mathfrak{P}}$ is also a regular local ring (VII.2.6).

An irreducible subvariety W of an affine or projective variety V is called *regular on V* if the local ring $\mathcal{O}_{V,W}$ is regular. The geometric meaning of this concept is gotten from

Corollary 1.19. $\mathcal{O}_{V,W}$ is a regular local ring if and only if W contains a regular point of V (thus is not entirely contained in $\mathrm{Sing}(V)$).

Fig. 18

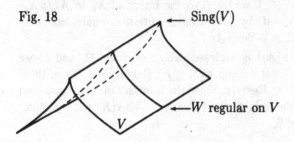

Sing(V)

W regular on V

V

Proof. It suffices to prove the statement in the affine case. Let A be the coordinate ring of V and $\mathfrak{p}$ the prime ideal of W in A. Then $\mathcal{O}_{V,W} \cong A_{\mathfrak{p}}$. For all $\mathfrak{m} \in \mathrm{Max}(A)$ with $\mathfrak{p} \subset \mathfrak{m}, A_{\mathfrak{p}}$ is a localization of $A_{\mathfrak{m}}$. By 1.18 $A_{\mathfrak{p}}$ is regular if and only if such an $\mathfrak{m}$ can be found in $\mathrm{Reg}(A)$. This happens if and only if $\mathrm{Reg}(V) \cap W \neq \emptyset$.

Definition 1.20. A Noetherian ring R is called *regular* if $\mathrm{Max}(R) \subset \mathrm{Reg}(R)$. (Then $\mathrm{Reg}(R) = \mathrm{Spec}(R)$, as will be shown in VII.2.6.)

We already know many examples of Noetherian regular rings: all fields, all Dedekind domains, all polynomial rings over principal ideal domains, all regular local rings, and all coordinate rings of smooth affine varieties. A direct product of finitely many regular Noetherian rings is also one. Regular rings can therefore have zero divisors (in contrast to regular local rings).

Exercises

1. Determine the singularities and the tangent cones at the singular points of the surfaces with the equations given in I.§1, Exercise 2.

2. The ring $K[[X_1, \ldots, X_n]]$ of formal power series over a field is a regular local ring.

3. In a regular local ring $(R, \mathfrak{m})$ of dimension 1, for any ideal $I \neq (0)$ there is a unique $n \in \mathbb{N}$ with $I = \mathfrak{m}^n$.

4. A *discrete valuation* v of a field K is a mapping $v : K \to \mathbb{Z} \cup \{\infty\}$ with the following properties. For all $a, b \in K$:

   a) $v(a \cdot b) = v(a) + v(b)$,

   b) $v(a + b) \geq \text{Min}\{v(a), v(b)\}$,

   c) $v(a) = \infty$ if and only if $a = 0$.

   The *valuation ring* of v is the set of all $a \in K$ with $v(a) \geq 0$. v is called trivial if $v(a) = 0$ for all $a \neq 0$. Show: the valuation ring of a nontrivial discrete valuation is a regular local ring of dimension 1, and any such ring is the valuation ring of a discrete valuation of its field of fractions.

5. Let $(R, \mathfrak{m})$ be a Noetherian local integral domain.

   a) If $\mathfrak{m}$ is an invertible ideal (Ch. IV, §3, Exercise 6), then R is regular and $\dim R = 1$.

   b) If $\dim R = 1$, then for any $x \in R \setminus \{0\}$ there is $y \notin (x)$ with $\mathfrak{m}y \subset (x)$. (Hint: $R/(x)$ is of finite length.) Further, $\mathfrak{m}^{-1} := \{a \in Q(R) \mid \mathfrak{m}a \subset R\} \neq R$.

   c) If R is integrally closed in $Q(R)$, then from $\mathfrak{m} \cdot \mathfrak{m}^{-1} = \mathfrak{m}$ it follows that $\mathfrak{m}^{-1} = R$. (Hint: For $x \in \mathfrak{m}^{-1}$ we have $\mathfrak{m}R[x] \subset \mathfrak{m}$. Deduce that $R[x]$ is finitely generated as an R-module.)

   Using a)–c) show: A Noetherian local integral domain of dimension 1 is regular if and only if it is integrally closed in its field of fractions.

In Exercises 6 through 9 we consider only varieties whose fields of definition equal their coordinate fields.

6. An algebraic variety V is called *regular in codimension* 1 if for any irreducible subvariety $W \subset V$ of codimension 1 the local ring $\mathcal{O}_{V,W}$ is regular. This condition is equivalent to: $\text{Sing}(V)$ has codimension at least 2 in V.

7. A variety V is called *normal* if $\mathcal{O}_{V,x}$ is an integrally closed ring for any $x \in V$.

   a) Normal varieties are regular in codimension 1.

   b) An algebraic curve is normal if and only if it is smooth.

8. Let $r \neq 0$ be a rational function on an irreducible normal variety V. Show that the zero set and the pole set of r are finite unions of irreducible subvarieties W of codimension 1 in V. (If v_W is the discrete valuation belonging to $\mathcal{O}_{V,W}$ (Exercise 4), then $v_W(r)$ is called the order of r at W (the order of the zero or pole), and the mapping given by $W \mapsto v_W(r)$ from the set of irreducible subvarieties of codimension 1 into $\mathbb{Z}$ is called the divisor of the function r.)

9. Let $V \subset \mathbf{P}^n(L)$ be an algebraic variety. By the duality principle of projective geometry the hyperplanes $H_{\langle a_0, \dots, a_n \rangle}$ (with equation $\sum_{i=0}^n a_i X_i = 0$) correspond bijectively to the points $\langle a_0, \dots, a_n \rangle \in \mathbf{P}^n(L)$; under this the hyperplanes through a point $x \in \mathbf{P}^n(L)$ correspond bijectively to the points of a hyperplane $H_x \subset \mathbf{P}^n(L)$. Show that if x is a regular point of V, then

the set of $\langle a_0, \ldots, a_n \rangle \subset \mathbf{P}^n(L)$ such that x is regular on $V \cap H_{\langle a_0, \ldots, a_n \rangle}$ is an open nonempty subset of H_x.

10. For a prime number $p > 2$ let $K := \mathbf{F}_p(u)$ be the field of rational functions in one indeterminate u over the field with p elements, and let L be the algebraic closure of K. Let $u' \in L$ be the element with $u'^p = u$.

   a) $K[X, Y]/(X^2 + Y^p - u)$ is an integral domain which is integrally closed in its field of fractions (and hence is a regular ring).

   b) $L[X, Y]/(X^2 + Y^p - u)$ is an integral domain but is not integrally closed in its field of fractions.

   c) For the point $(0, u')$ on the curve with equation $X^2 + Y^p - u = 0$, the local ring over K is regular, but the one over L is not.

11. Let p be a prime number, R a ring containing a field of characteristic p, and R^p the subring of all p-th powers in R. A system of elements $\{x_1, \ldots, x_n\}$ in R is called a (finite) p-basis of R if $R = R^p[x_1, \ldots, x_n]$ and if the products $x_1^{\alpha_1} \cdot \ldots \cdot x_n^{\alpha_n}$ $(0 \le \alpha_i \le p - 1)$ are linearly independent over R^p.

   a) Any field K of characteristic p with $[K : K^p] < \infty$ has p-basis. It can be chosen from any system of ring generators of K over K^p.

   b) If R has a p-basis, so does any ring of fractions R_S and any polynomial ring $R[X_1, \ldots, X_m]$.

   c) If $\{x_1, \ldots, x_n\}$ is a p-basis of R and x_1 is not a unit, then $R/(x_1)$ has p-basis $\{\overline{x}_2, \ldots, \overline{x}_n\}$, where $\overline{x}_i$ is the image of x_i in $R/(x_1)$.

   d) If R has a p-basis, then for any $\mathfrak{m} \in \mathrm{Max}(R)$ there is also a p-basis $\{x_1, \ldots, x_d, y_1, \ldots, y_n\}$, where the y_i represent a p-basis of $R/\mathfrak{m}$ and $x_1, \ldots, x_d \in \mathfrak{m}$.

   e) If a reduced Noetherian ring has a p-basis, it is regular.

   f) If R is a regular local ring that is finitely generated as an R^p-module, then R has a p-basis. (For generalizations see [48]).

2.   The zero divisors of a ring or module. Primary decomposition

As we have often seen, the study of the zero divisors of a ring is indispensable to many questions of ring theory. In this section we collect more facts on the zero divisors of a ring or module and establish the connection with the primary decomposition of ideals and modules, which represents a farreaching generalization of the prime decomposition in factorial rings.

   Let M be a module over a ring R.

Definition 2.1. $\mathfrak{p} \in \mathrm{Spec}(R)$ is said to be associated to M if there is an $m \in M$ such that $\mathfrak{p} = \mathrm{Ann}(m)$.

   The set of associated prime ideals of M will be denoted by $\mathrm{Ass}(M)$. If $\mathfrak{p} = \mathrm{Ann}(m)$ for $m \in M$, then $Rm \cong R/\mathfrak{p}$. Hence $\mathrm{Ass}(M)$ is also the set of $\mathfrak{p} \in \mathrm{Spec}(R)$ for which there is a submodule of M isomorphic to $R/\mathfrak{p}$. We have $\mathrm{Ann}(M) \subset \mathfrak{p}$ for all $\mathfrak{p} \in \mathrm{Ass}(M)$.

Lemma 2.2. For all $\mathfrak{p} \in \mathrm{Spec}(R)$ we have $\mathrm{Ass}(R/\mathfrak{p}) = \{\mathfrak{p}\}$, and $\mathfrak{p}$ is the annihilator of any $x \neq 0$ in $R/\mathfrak{p}$.

Indeed, if $x = r + \mathfrak{p}$ with $r \in R \setminus \mathfrak{p}$, then $r' \cdot (r + \mathfrak{p}) = 0$ for $r' \in R$ if and only if $r' \in \mathfrak{p}$.

Lemma 2.3. For any submodule $U \subset M$,

$$\mathrm{Ass}(U) \subset \mathrm{Ass}(M) \subset \mathrm{Ass}(U) \cup \mathrm{Ass}(M/U).$$

Proof. It is clear that $\mathrm{Ass}(U) \subset \mathrm{Ass}(M)$. If $\mathfrak{p} \in \mathrm{Ass}(M) \setminus \mathrm{Ass}(U)$ and $\mathfrak{p} = \mathrm{Ann}(m)$ for $m \in M$, then $Rm \cong R/\mathfrak{p}$ and $Rm \cap U = \langle 0 \rangle$, since otherwise by 2.2 we would have $\mathfrak{p} \in \mathrm{Ass}(U)$. Under the canonical epimorphism $M \to M/U$, Rm is mapped to a submodule of M/U isomorphic to $R/\mathfrak{p}$; that is, $\mathfrak{p} \in \mathrm{Ass}(M/U)$.

Proposition 2.4. If R is Noetherian and $M \neq \langle 0 \rangle$, then $\mathrm{Ass}(M) \neq \emptyset$.

Proof. The set of ideals that occur as annihilators of elements $m \neq 0$ in M is not empty and hence contains a maximal element $\mathfrak{p}$ because R is Noetherian. We shall show that $\mathfrak{p}$ is a prime ideal.

Let $\mathfrak{p} = \mathrm{Ann}(m)$. Further, let a, b be elements of R with $ab \in \mathfrak{p}, b \notin \mathfrak{p}$. Then $bm \neq 0$. Moreover, $\mathfrak{p} \subset \mathrm{Ann}(bm)$; and because of the maximalilty of $\mathfrak{p}$ we have $\mathfrak{p} = \mathrm{Ann}(bm)$. From $abm = 0$ it follows that $a \in \mathfrak{p}$.

Proposition 2.5. If R is Noetherian, then $\bigcup_{\mathfrak{p} \in \mathrm{Ass}(M)} \mathfrak{p}$ is the set of zero divisors of M.

Proof. The elements of the $\mathfrak{p} \in \mathrm{Ass}(M)$ are obviously zero divisors of M. Conversely, if $rm = 0$ for some $r \in R, m \in M, m \neq 0$, then $\mathrm{Ass}(Rm) \neq \emptyset$ by 2.4 and so there are $\mathfrak{p} \in \mathrm{Spec}(R)$ and $r' \in R$ with $\mathfrak{p} = \mathrm{Ann}(r'm)$. Since $rr'm = 0$, it follows that $r \in \mathfrak{p}$.

In particular, we get the important fact that the set of zero divisors of a Noetherian ring is the union of the associated prime ideals of the ring. (Compare also with I.4.10).

Proposition 2.6. Let M be a finitely generated module over a Noetherian ring R. Then there is a chain of submodules

$$M = M_0 \supset M_1 \supset \cdots \supset M_n = \langle 0 \rangle$$

such that $M_i/M_{i+1} \cong R/\mathfrak{p}_i$ for some $\mathfrak{p}_i \in \mathrm{Spec}(R)$ $(i = 0, \ldots, n-1)$.

Proof. Let $M \neq \langle 0 \rangle$. The set A of submodules $\neq \langle 0 \rangle$ of M for which the proposition is correct is not empty, since by 2.4 M contains a submodule isomorphic to $R/\mathfrak{p}$ for some $\mathfrak{p} \in \mathrm{Spec}(R)$. Since M is a Noetherian module, there exists a maximal element N in A.

If we had $M \neq N$, then M/N would contain a submodule isomorphic to $R/\mathfrak{q}$ for some $\mathfrak{q} \in \mathrm{Spec}(R)$. If N' is the inverse image of this submodule under the canonical epimorphism $M \to M/N$, then $N'/N = R/\mathfrak{q}$, contradicting the maximal property of N. It follows that $N = M$, and the proposition is proved.

Corollary 2.7. For a finitely generated module M over a Noetherian ring R, $\mathrm{Ass}(M)$ is a finite set. In particular, $\mathrm{Ass}(R)$ is finite.

Proof. From 2.2 and 2.3 it follows that $\mathrm{Ass}(M) \subset \{\mathfrak{p}_0, \ldots, \mathfrak{p}_{n-1}\}$ if the $\mathfrak{p}_i$ are the prime ideals of 2.6.

Corollary 2.8. Let M be a finitely generated module over a Noetherian ring R. If an ideal I of R consists of zero divisors of M alone, then there is an $m \in M, m \neq 0$, such that $Im = \langle 0 \rangle$.

Proof. From $I \subset \bigcup_{\mathfrak{p} \in \mathrm{Ass}(M)} \mathfrak{p}$ we see, because $\mathrm{Ass}(M)$ is finite, that $I \subset \mathfrak{p}$ for some $\mathfrak{p} \in \mathrm{Ass}(M)$; the assertion follows.

Between the associated prime ideals and the support of a module there is the following relation.

Proposition 2.9. Let R be a Noetherian ring, $M \neq \langle 0 \rangle$ a finitely generated R-module. $\mathrm{Supp}(M)$ is the set of all $\mathfrak{p} \in \mathrm{Spec}(R)$ that contain a prime ideal associated to M. In particular, the minimal prime ideals of a Noetherian ring $R \neq \{0\}$ belong to $\mathrm{Ass}(R)$.

Proof. If $\mathfrak{p} \in \mathrm{Ass}(M)$, then $\mathfrak{p} \supset \mathrm{Ann}(M)$ and therefore by III.4.6 $\mathfrak{p} \in \mathrm{Supp}(M)$. All elements of $\mathrm{Spec}(R)$ that contain $\mathfrak{p}$ therefore belong to $\mathrm{Supp}(M)$.

Conversely, let $\mathfrak{p} \in \mathrm{Supp}(M)$, so $M_\mathfrak{p} \neq 0$. By 2.4 there is a prime ideal $\mathfrak{q}R_\mathfrak{q}, \mathfrak{q} \in \mathrm{Spec}(R), \mathfrak{q} \subset \mathfrak{p}$, associated to $M_\mathfrak{p}$. It suffices to show that $\mathfrak{q} \in \mathrm{Ass}(M)$. This follows from

Lemma 2.10. Let M be a module over a Noetherian ring R, $S \subset R$ a multiplicatively closed subset. Then

$$\mathrm{Ass}(M_S) = \{\mathfrak{p}_S \mid \mathfrak{p} \in \mathrm{Ass}(M), \mathfrak{p} \cap S = \emptyset\}.$$

Proof. For $\mathfrak{p} \in \mathrm{Ass}(M)$ with $\mathfrak{p} \cap S = \emptyset$, there is an $m \in M$ such that $\mathfrak{p} = \mathrm{Ann}(m)$. Then $\mathfrak{p}_S = \mathrm{Ann}(m/1)$, so $\mathfrak{p}_S \in \mathrm{Ass}(M_S)$.

Conversely, let $\mathfrak{p}_S \in \mathrm{Ass}(M_S), \mathfrak{p}_S = \mathrm{Ann}(m/s)$. Let $\{r_1, \ldots, r_n\}$ be a system of generators of $\mathfrak{p}$. From $(r_i/1) \cdot (m/s) = 0$ $(i = 1, \ldots, n)$ it follows that there are elements $s_i \in S$ with $s_i r_i m = 0$ $(i = 1, \ldots, n)$. With $s' := \prod_{i=1}^n s_i$ and $m' := s'm$ it follows that $\mathfrak{p} \subset \mathrm{Ann}(m')$. If $rm' = 0$ for $r \in R$, it follows that $(rs'm)/s = 0$, and so $(rs')/1 \in \mathrm{Ann}(m/s) = \mathfrak{p}_S$; therefore $r \in \mathfrak{p}$ since $s' \notin \mathfrak{p}$. Hence $\mathfrak{p} = \mathrm{Ann}(m')$, and so $\mathfrak{p} \in \mathrm{Ass}(M)$.

We shall now discuss the connection between associated prime ideals and primary decomposition in rings and modules, which historically was the starting point of the whole theory (Lasker [51]).

An element $r \in R$ is called *nilpotent* for M if there is a $\rho \in \mathbb{N}$ with $r^\rho M = \langle 0 \rangle$. If M is finitely generated, this is equivalent to

$$r \in \mathrm{Rad}(\mathrm{Ann}(M)) = \bigcap_{\mathfrak{p} \in \mathrm{Supp}(M)} \mathfrak{p}.$$

Definition 2.11. A submodule $P \subset M$ is called *primary* if $\mathrm{Ass}(M/P)$ consists of a single element. If $\mathfrak{p}$ is this prime ideal, then P is also called $\mathfrak{p}$-*primary*.

In the Noetherian case this definition generalizes the previous definition of a primary ideal (V.4.2).

Lemma 2.12. Let R be Noetherian, M finitely generated, and $P \subset M$ a submodule. The following statements are equivalent.

a) P is primary.

b) Any zero divisor of M/P is nilpotent for M/P.

Proof. If P is $\mathfrak{p}$-primary, then $\mathfrak{p}$ is the set of all zero divisors of M/P. On the other hand, $\mathrm{Rad}(\mathrm{Ann}(M/P)) = \mathfrak{p}$. Hence b) follows.

Conversely, by 2.5 and 2.9 it follows from b) that $\bigcup_{\mathfrak{p} \in \mathrm{Ass}(M/P)} \mathfrak{p} = \bigcap_{\mathfrak{p} \in \mathrm{Ass}(M/P)} \mathfrak{p}$. But this can happen only if $\mathrm{Ass}(M/P)$ consists of just one element.

Lemma 2.13. Let R be Noetherian. The intersection of finitely many $\mathfrak{p}$-primary submodules of M is $\mathfrak{p}$-primary.

Proof. It suffices to show this for two $\mathfrak{p}$-primary modules $P_1, P_2 \subset M$. We have an exact sequence

$$0 \to P_1/P_1 \cap P_2 \to M/P_1 \cap P_2 \to M/P_1 \to 0,$$

where $P_1/P_1 \cap P_2 \cong P_1 + P_2/P_2$. By 2.3 and 2.4 $\mathrm{Ass}(M/P_1 \cap P_2) \neq \emptyset$ and $\mathrm{Ass}(M/P_1 \cap P_2) \subset \mathrm{Ass}(M/P_1) \cup \mathrm{Ass}(P_1 + P_2/P_2) \subset \mathrm{Ass}(M/P_1) \cup \mathrm{Ass}(M/P_2) = \{\mathfrak{p}\}$, and so $\mathrm{Ass}(M/P_1 \cap P_2) = \{\mathfrak{p}\}$.

Definition 2.14. A submodule $Q \subset M$ is called *irreducible* (in M) if the following condition is satisfied: If $Q = U_1 \cap U_2$ with two submodules $U_i \subset M$ $(i = 1, 2)$, then $Q = U_1$ or $Q = U_2$.

In particular, this also defines the concept of an irreducible ideal in a ring.

We see at once that $Q \subset M$ is irreducible if and only if in M/Q the zero module is irreducible.

Proposition 2.15. Let R be Noetherian, $Q \neq M$ an irreducible submodule. Then Q is primary.

Proof. If $\mathrm{Ass}(M/Q)$ contained two distinct prime ideals $\mathfrak{p}_1$ and $\mathfrak{p}_2$, then M/Q would contain two submodules $U_i \cong R/\mathfrak{p}_i$ $(i = 1, 2)$, and we would have $U_1 \cap U_2 = \langle 0 \rangle$, since $x \in U_i, x \neq 0$ has annihilator $\mathfrak{p}_i$ $(i = 1, 2)$. Since in M/Q the zero module is irreducible, $U_1 = \langle 0 \rangle$ or $U_2 = \langle 0 \rangle$, a contradiction. On the other hand, $\mathrm{Ass}(M/Q) \neq \emptyset$ by 2.4. Hence $\mathrm{Ass}(M/Q)$ consists of precisely one prime ideal.

Definition 2.16. A submodule $U \subset M$ has a *primary decomposition* if there are primary submodules $P_1, \ldots, P_s$ $(s \geq 1)$ of M such that

$$U = P_1 \cap \cdots \cap P_s. \tag{1}$$

The primary decomposition (1) is called reduced if the following holds:

a) If P_i is $\mathfrak{p}_i$-primary $(i = 1, \ldots, s)$, then $\mathfrak{p}_i \neq \mathfrak{p}_j$ for $i \neq j$ $(i, j = 1, \ldots, s)$.

b) $\bigcap_{j \neq i} P_j \not\subset P_i$ for $i = 1, \ldots, s$.

The P_i occurring in a reduced primary decomposition are also called the *primary components of U*.

If R is Noetherian and U has a primary decomposition, then it also has a reduced one, since by 2.13 one can collect together the primary modules with the same prime ideal, and one gets b) by then omitting the superfluous modules.

Theorem 2.17. (Existence of a primary decomposition) Let R be a Noetherian ring, M a finitely generated R-module. Any submodule $U \neq M$ has a reduced primary decomposition.

Proof. By 2.15 it suffices to show that U is the intersection of finitely many irreducible submodules of M. If this were not so, there would, by the maximal condition, be a largest submodule $U \neq M$ of M for which this statement is false. Since U is not irreducible, there would be submodules $U_i \neq U$ of M $(i = 1, 2)$ with $U = U_1 \cap U_2$. Since the U_i properly contain the module U, they are representable as intersections of finitely many irreducible submodules of M, hence so is U, a contradiction. Therefore the theorem is true.

In particular, any ideal I in a Noetherian ring has a primary decomposition. This theorem generalizes the theorem on factorization in $\mathbf{Z}$, to which it is easily seen to specialize (cf. also Exercise 1). As in that case the question arises of the uniqueness of the primary decomposition.

Theorem 2.18. (First Uniqueness Theorem) Let R be a Noetherian ring, $U \subset M$ a submodule with a reduced primary decomposition $U = P_1 \cap \cdots \cap P_s$, where P_i is $\mathfrak{p}_i$-primary $(i = 1, \ldots, s)$. Then $\{\mathfrak{p}_1, \ldots, \mathfrak{p}_s\} = \mathrm{Ass}(M/U)$. The primary ideals occurring in a reduced primary decomposition of U are therefore uniquely determined by M and U.

Proof. Let $U_i := \bigcap_{j \neq i} P_j$, so $U = U_i \cap P_i$, $U \neq U_i$ $(i = 1, \ldots, s)$. From $U_i/U \cong U_i + P_i/P_i \subset M/P_i$ follows $\emptyset \neq \mathrm{Ass}(U_i/U) \subset \mathrm{Ass}(M/P_i) = \{\mathfrak{p}_i\}$, and from $\mathrm{Ass}(U_i/U) \subset \mathrm{Ass}(M/U)$ follows $\{\mathfrak{p}_1, \ldots, \mathfrak{p}_s\} \subset \mathrm{Ass}(M/U)$.

The opposite inclusion follows by induction on s. For $s = 1$ there is nothing to show. Therefore, let $s > 1$ and suppose the proposition proved for primary decompositions with $s - 1$ components. $U_i = \bigcap_{j \neq i} P_j$ is reduced primary decomposition, so $\mathrm{Ass}(M/U_i) = \{\mathfrak{p}_j \mid j \neq i\}$. From $M/U_i \cong M/U/U_i/U$ by 2.3 it follows that

$$\mathrm{Ass}(M/U) \subset \mathrm{Ass}(U_i/U) \cup \mathrm{Ass}(M/U_i) \subset \{\mathfrak{p}_1, \ldots, \mathfrak{p}_s\}, \quad \text{q. e. d.}$$

Before proving another uniqueness theorem, we investigate the behavior of a primary decomposition under localization.

Proposition 2.19. Let $S \subset R$ be multiplicatively closed.

a) If $P \subset M$ is a $\mathfrak{p}$-primary submodule of M and $\mathfrak{p} \cap S = \emptyset$, then P_S is a $\mathfrak{p}_S$-primary submodule of M_S and $S(P) = P$. If $\mathfrak{p} \cap S \neq \emptyset$, then $P_S = M_S$. ($S(P)$) denotes the S-component of P (III.4.9).)

b) If $U = \bigcap_{i=1}^{s} P_i$ is a reduced primary decomposition of a submodule $U \subset M$ where P_i is $\mathfrak{p}_i$-primary $(i = 1, \ldots, s)$, then

$$U_S = \bigcap_{\mathfrak{p}_i \cap S = \emptyset} (P_i)_S$$

is a reduced primary decomposition of U_S.

Proof. The first statement in a) follows from 2.10 and the definition of a primary module. According to the definition, $S(P) = \{m \in M \mid \exists_{s \in S} sm \in P\}$. Since $\mathfrak{p}$ is the set of zero divisors of M/P and $\mathfrak{p} \cap S = \emptyset$, we get $S(P) = P$. On the other hand, if $\mathfrak{p} \cap S \neq \emptyset$, then $S(P) = M$ and $P_S = M_S$. b) follows from a) because localization and intersection are permutable operations (for finite intersections).

Theorem 2.20. (Second Uniqueness Theorem) Let $U \subset M$ be a submodule of M with a reduced primary decomposition $U = P_1 \cap \cdots \cap P_s$, where P_i is $\mathfrak{p}_i$-primary $(i = 1, \ldots, s)$. If $\mathfrak{p}_i$ is a minimal element of the set $\{\mathfrak{p}_1, \ldots, \mathfrak{p}_s\}$ and $S_i := R \setminus \mathfrak{p}_i$, then

$$P_i = S_i(U).$$

Therefore, the primary component of U belonging to the minimal elements of $\mathrm{Ass}(M/U)$ are uniquely determined by M and U.

Proof. Since $S_i \cap \mathfrak{p}_j \neq \emptyset$ for $j \neq i$, it follows from 2.19 that $U_{S_i} = (P_i)_{S_i}$ and $S_i(U) = S_i(P_i) = P_i$, q. e. d.

In particular, the primary decomposition of U is certainly unique if in $\mathrm{Ass}(M/U)$ no prime ideal is contained in another.

If P_i is a primary component of U with prime ideal $\mathfrak{p}_i$ that is not minimal in $\mathrm{Ass}(M/U)$, then P_i is called an *embedded primary component* of U. In general the embedded primary components are not unique.

Example 2.21. In the polynomial ring $K[X, Y]$ over a field K

$$(X^2, XY) = (X) \cap (X^2, Y).$$

Here (X) is a prime ideal and (X^2, Y) is a primary ideal with prime ideal (X, Y). Therefore, this is a primary decomposition. On the other hand, we also have

$$(X^2, XY) = (X) \cap (X^2, X + Y),$$

where $(X^2, X + Y)$ is also primary with prime ideal (X, Y) but is different from (X^2, Y).

If $I = \mathfrak{q}_1 \cap \cdots \cap \mathfrak{q}_s$ is a primary decomposition of an ideal I in a Noetherian ring R, where $\mathfrak{q}_i$ is a $\mathfrak{p}_i$-primary ideal, then

$$\mathrm{Rad}(I) = \mathfrak{p}_1 \cap \cdots \cap \mathfrak{p}_\sigma, \tag{2}$$

if $\mathfrak{p}_1, \ldots, \mathfrak{p}_\sigma$ are the minimal elements of the set $\{\mathfrak{p}_1, \ldots, \mathfrak{p}_s\}$. This follows because, by 2.9, $I \subset \mathfrak{p}$ if and only if $\mathfrak{p}_i \subset \mathfrak{p}$ for some $\mathfrak{p}_i \in \mathrm{Ass}(R/I)$. Then $\mathfrak{p}_1, \ldots, \mathfrak{p}_\sigma$ are just the minimal prime divisors of I. Of course (2) is the primary decomposition of $\mathrm{Rad}(I)$.

A Noetherian ring R is reduced if and only if in the primary decomposition of its zero ideal the primary components are all prime ideals, namely the minimal prime ideals of R.

Exercises

1. Let R be a Dedekind ring.

   a) For all $\mathfrak{p} \in \mathrm{Spec}(R)$, $\mathfrak{p} \neq (0)$, the $\mathfrak{p}$-primary ideals are just the powers of $\mathfrak{p}$.

   b) An ideal I of R, $I \neq (0)$, $I \neq R$, has a unique representation as a product of powers of prime ideals.

2. Let R be a 0-dimensional Noetherian ring and let $(0) = \mathfrak{q}_1 \cap \cdots \cap \mathfrak{q}_s$ be the reduced primary decomposition of the zero ideals of R. Then we have the length formula

$$l(R) = \sum_{i=1}^{s} l(R/\mathfrak{q}_i).$$

3. Let S/R be a ring extension, where R is Noetherian and S is finitely generated as an R-module. Let $\mu_R(S)$ be the length of a shortest system of generators of the R-module S.

   a) For $\mathfrak{p} \in \mathrm{Spec}(R)$, the extension ideal $\mathfrak{p}S_\mathfrak{p}$ has no embedded primary components. ($S_\mathfrak{p}$ is the ring of fractions of S with denominator set $R \setminus \mathfrak{p}$.)

   b) If $\mathfrak{P}_1, \ldots, \mathfrak{P}_s$ are the prime ideals of S lying over $\mathfrak{p}$, then we have the degree formula

$$\mu_R(S) \geq \sum_{i=1}^{s} [S_{\mathfrak{P}_i}/\mathfrak{P}_i S_{\mathfrak{P}_i} : R_\mathfrak{p}/\mathfrak{p}R_\mathfrak{p}] \cdot l_{S_{\mathfrak{P}_i}}(S_{\mathfrak{P}_i}/\mathfrak{p}S_{\mathfrak{P}_i}),$$

   where $l_{S_{\mathfrak{P}_i}}$ denotes the length of an $S_{\mathfrak{P}_i}$-module. If S is a free R-module, then equality holds in this formula.

4. A morphism $\varphi : V \to W$ of affine varieties is called *finite* if it is dominant and $K[V]$ is finitely generated as a $K[W]$-module. Let $\mu_{V/W}$ be the length of a shortest generating system of this module. For any irreducible subvariety Z of W, $\varphi^{-1}(Z)$ has at most $\mu_{V/W}$ irreducible components.

5. If an ideal of a Noetherian ring contains a non-zerodivisor, then it is generated by its non-zerodivisors. Is this true for any ring?

3. Regular sequences. Cohen–Macaulay modules and rings

Having learned more about the zero divisors of a ring or module in the last section, we now continue the study of regular sequences that was begun in V. §5.

Let M be a finitely generated module over a Noetherian ring R, I an ideal of R with $IM \neq M$. For any M-regular sequence $\{a_1, \ldots, a_m\}$ we have $(a_1, \ldots, a_i)M \neq (a_1, \ldots, a_{i+1})M$ for $i = 0, \ldots, m-1$. Since M is a Noetherian module, it follows at once that any M-regular sequence $\{a_1, \ldots, a_m\}$ with elements $a_i \in I$ can be lengthened to a maximal such sequence, i.e. to an M-regular sequence $\{a_1, \ldots, a_n\} \subset I$ $(n \geq m)$ such that any $a \in I$ is a zero divisor of $M/(a_1, \ldots, a_n)M$. We have the nonobvious

Proposition 3.1. *Any two maximal M-regular sequences in I have the same number of elements.*

The proof (according to Northcott–Rees [62]) uses an exchange process for regular sequences. We first show:

Lemma 3.2. *If $\{a, b\}$ is an M-regular sequence and b is not a zero divisor of M, then $\{b, a\}$ is also an M-regular sequence.*

Proof. If a were a zero divisor of M/bM, there would be an $m \in M, m \notin bM$ with $am = bm'$ $(m' \in M)$. Since $\{a, b\}$ is an M-regular sequence, we must have $m' \in aM$, hence $m' = am''$ with $m'' \in M$; then $m = bm''$, since a is not a zero divisor of M. But this contradicts $m \notin bM$.

Proof of 3.1. Among all the maximal M-regular sequences in I there is one with the least number of elements n. We argue by induction on n. If $n = 0$, then I consists of only zero divisors of M and there is nothing to show. Hence let $n > 0$, let $\{a_1, \ldots, a_n\}$ be a maximal M-regular sequence in I, and let $\{b_1, \ldots, b_n\}$ be another M-regular sequence in I. We must show that I consists only of zero divisors of $M/(b_1, \ldots, b_n)M$.

If $n = 1$, then I consists of only zero divisors of M/a_1M. By 2.8 there is therefore an $m \in M, m \notin a_1M$ with $Im \subset a_1M$. In particular, $b_1m = a_1m'$ with some $m' \in M$. If we had $m' \in b_1M$, then we would have $m \in a_1M$, so $m' \notin b_1M$. From $a_1Im' = Ib_1m \subset a_1b_1M$ it follows that $Im' \subset b_1M$, and therefore I consists only of zero divisors of M/b_1M.

If $n > 1$, put $M_i := M/(a_1, \ldots, a_i)M$, $M_i' := M/(b_1, \ldots, b_i)M$ for $i = 0, \ldots, n-1$, and choose $c \in I$ that is not a zero divisor of M_i and M_i' for all $i = 0, \ldots, n-1$. This is possible, since the sets of zero divisors of M_i and M_i' are finite unions of prime ideals (2.5 and 2.7) and I is contained in none of these sets.

By applying 3.2 repeatedly it follows that both $\{c, a_1, \ldots, a_{n-1}\}$ and $\{c, b_1, \ldots, b_{n-1}\}$ are M-regular sequences in I, where $\{c, a_1, \ldots, a_{n-1}\}$ is maximal, for $\{a_1, \ldots, a_{n-1}, c\}$ is maximal on the basis of the case $n = 1$ (applied to M_{n-1}) already treated. Then $\{a_1, \ldots, a_{n-1}\}$ and $\{b_1, \ldots, b_{n-1}\}$ are M/cM-regular sequences in I; the first is maximal, so by the induction hypothesis the second is too. But if $\{b_1, \ldots, b_{n-1}, c\}$ is a maximal M-regular sequence, then so is $\{b_1, \ldots, b_n\}$, again by the case $n = 1$. This proves Proposition 3.1.

Definition 3.3. Under the hypotheses of 3.1 the number of elements of a maximal M-regular sequence in I is called the I-*depth* of M (written $d(I,M)$) or the *grade of M with respect to I*. If R is local and I is the maximal ideal of R, then we call $d(I,M)$ simply the *depth* of M and write $d(M)$. In particular, this also defines $d(R)$.

The equation $d(I,M) = 0$ holds if and only if I consists only of zero divisors of M. If $(R, \mathfrak{m})$ is a Noetherian local ring, then $d(M) = 0$ is equivalent to $\mathfrak{m} \in \mathrm{Ass}(M)$. From 3.1 also immediately follows

Corollary 3.4. If $\{a_1, \ldots, a_m\}$ is any M-regular sequence in I, then

$$d(I, M/(a_1, \ldots, a_m)M) = d(I, M) - m.$$

On the connection between the depth and the number of generators of an ideal we have the following statements

Proposition 3.5. Under the hypotheses of 3.1, let the ideal I be generated by n elements, and put $d(I,M) =: m$. Then $m \leq n$ and there is a generating system $\{a_1, \ldots, a_n\}$ of I for which $\{a_1, \ldots, a_m\}$ is an M-regular sequence.

Proof. Let $\{a_1, \ldots, a_n\}$ be any generating system of I and let $(a_1, \ldots, a_k) \subset \bigcup_{\mathfrak{p} \in \mathrm{Ass}(M)} \mathfrak{p}, (a_1, \ldots, a_{k+1}) \not\subset \bigcup_{\mathfrak{p} \in \mathrm{Ass}(M)} \mathfrak{p}$ for some $k \in [0, n]$. If $k = n$, then $d(I, M) = 0$ and there is nothing to prove. If $k < n$, we shall show that there is a non-zerodivisor b of M such that $(a_1, \ldots, a_{k+1}) = (b, a_1, \ldots, a_k)$. Passing over to M/bM and $I/(b)$, the assertion of the proposition follows by induction.

Let $\{\mathfrak{p}_1, \ldots, \mathfrak{p}_s\}$ be the set of maximal elements of $\mathrm{Ass}(M)$ (with respect to inclusion). By hypothesis there is an element of the form $a + ra_{k+1}$ with $a \in (a_1, \ldots, a_k), r \in R$, which is not contained in $\mathfrak{p}_1 \cup \cdots \cup \mathfrak{p}_s$. If $a_{k+1} \in \mathfrak{p}_i$ for $i = 1, \ldots, \sigma$ and $a_{k+1} \notin \mathfrak{p}_j$ for $j = \sigma + 1, \ldots, s$, then we choose $t \in \bigcap_{j=\sigma+1}^{s} \mathfrak{p}_j$, $t \notin \bigcup_{i=1}^{\sigma} \mathfrak{p}_i$ and put $b := ta + a_{k+1}$. Then $b \notin \mathfrak{p}_i$ ($i = 1, \ldots, s$); that is, b is not a zero divisor of M. Further, $(a_1, \ldots, a_{k+1}) = (b, a_1, \ldots, a_k)$, q. e. d.

Now we can also sharpen V.5.12a).

Proposition 3.6. Let R be a Noetherian ring, $I \neq R$ an ideal generated by elements $a_1, \ldots, a_n$ for which

$$\alpha : R/I[X_1, \ldots, X_n] \to \mathrm{gr}_I(R) \qquad (\alpha(X_i) = a_i + I^2)$$

is an isomorphism (hence $\{a_1, \ldots, a_n\}$ is a quasi-regular sequence). Then I is also generated by an R-regular sequence of length n.

Proof. We shall show that if $n > 0$, then I contains a non-zerodivisor of R. Were this not so, by 2.8 there would be an $r \in R \setminus \{0\}$ with $Ir = (0)$. From $ra_1 = 0$ follows $L_I(r) = 0$ with the aid of V.5.2; therefore $r \in \bigcap_{\nu \in \mathbb{N}} I^\nu =: \tilde{I}$. By Krull's Intersection Theorem $\tilde{I} = I \cdot \tilde{I}$. If $(b_1, \ldots, b_s)$ is a generating system of $\tilde{I}$ then we have equations $b_i = \sum_{k=1}^{s} r_{ik} b_k$ ($i = 1, \ldots, s$) with $r_{ik} \in I$ for all i, k; and by Cramer's Rule we get an element of the form $i + 1$ ($i \in I$) with $(1 + i)\tilde{I} = (0)$ (compare with the proof of V.5.16). Then $r = -ri = 0$, since $I \cdot r = (0)$, contradicting $r \neq 0$.

Therefore, I contains a non-zerodivisor c, and by 3.5 we can assume that $I = (c, a_2, \ldots, a_n)$. Since α is an isomorphism, so is

$$R/I[X_1, \ldots, X_n] \to gr_I(R), \quad X_1 \mapsto c + I^2, \quad X_i \mapsto a_i + I^2 \quad (i = 2, \ldots, n).$$

If we put $R' := R/(c), a_i' := a_i + (c), I' := I/(c)$, then by V.5.3 and 5.4

$$\alpha' : R'/I'[X_2, \ldots, X_n] \to gr_{I'}(R') \quad (X_i \mapsto a_i' + I'^2)$$

is also an isomorphism, and the assertion of the proposition follows by induction.

From the considerations of Chapter V, §4 and §5 we now have the following criterion for complete intersections.

Corollary 3.7. Let R be a Noetherian ring, $I \neq R$ an ideal with $\mathrm{Rad}(I) = I$. I is a complete intersection in R if and only if I is generated by an R-regular sequence. In particular, an affine variety is an ideal-theoretic complete intersection if and only if its ideal in the polynomial ring is generated by a regular sequence.

We next want to discuss the connection between depth and Krull dimension.

Definition 3.8. The *dimension of a module* M over a ring R is the Krull dimension of $R/\mathrm{Ann}(M)$.

Of course, for $M = R$ we get nothing new. If M is a finitely generated module over a ring R, then by 2.9 the minimal prime divisors of $\mathrm{Ann}(M)$ are also the minimal elements of $\mathrm{Ass}(M)$ and of $\mathrm{Supp}(M)$. Hence we have the formula

$$\dim M = \operatorname*{Sup}_{\mathfrak{p} \in \mathrm{Ass}(M)} \{\dim R/\mathfrak{p}\} = \operatorname*{Sup}_{\mathfrak{p} \in \mathrm{Supp}(M)} \{\dim R/\mathfrak{p}\}. \tag{1}$$

Proposition 3.9. Let $(R, \mathfrak{m})$ be a Noetherian local ring, $M \neq \langle 0 \rangle$ a finitely generated R-module. Then

$$d(M) \leq \operatorname*{Min}_{\mathfrak{p} \in \mathrm{Ass}(M)} \{\dim R/\mathfrak{p}\} \leq \dim M.$$

Proof. Only the left inequality has to be proved. This is done by induction on $n := d(M)$. Since for $n = 0$ there is nothing to show, we assume that $\mathfrak{m}$ contains a non-zerodivisor a of M and that the proposition has already been verified for modules of depth $n - 1$.

Since $d(M/aM) = d(M) - 1$ by 3.4, we have

$$d(M/aM) \leq \operatorname*{Min}_{\mathfrak{p} \in \mathrm{Ass}(M/aM)} \{\dim R/\mathfrak{p}\}.$$

Hence it suffices to show that for any $\mathfrak{p} \in \mathrm{Ass}(M)$ there exists a $\mathfrak{p}' \in \mathrm{Ass}(M/aM)$ with $\mathfrak{p} \subset \mathfrak{p}'$, $\mathfrak{p} \neq \mathfrak{p}'$. We then can conclude that

$$\operatorname*{Min}_{\mathfrak{p}' \in \mathrm{Ass}(M/aM)} \{\dim R/\mathfrak{p}'\} < \operatorname*{Min}_{\mathfrak{p} \in \mathrm{Ass}(M)} \{\dim R/\mathfrak{p}\}$$

and the assertion follows.

For $\mathfrak{p} \in \mathrm{Ass}(M)$ we have $a \notin \mathfrak{p}$. We shall show that M/aM has a submodule $U \neq \langle 0 \rangle$ with $\mathfrak{p}U = 0$. Then $\mathfrak{p}' \in \mathrm{Ass}(U)$ implies $\mathfrak{p} \subset \mathfrak{p}'$ and $\mathfrak{p}' \neq \mathfrak{p}$ since $a \in \mathfrak{p}'$. Further, $\mathfrak{p}' \in \mathrm{Ass}(M/aM)$, and then the proposition is proved.

Since $\mathfrak{p} \in \mathrm{Ass}(M)$, $N' := \{ m \mid \mathfrak{p}m = \langle 0 \rangle \}$ is a submodule $\neq \langle 0 \rangle$ of M. Further, $N' \subset N := \{ m \mid \mathfrak{p}m \in aM \}$. If we had $N = aM$, then for all $n' \in N'$ we would have a representation $n' = am$ with $m \in M$ and from $\mathfrak{p}n' = \langle 0 \rangle$ would follow $m \in N'$, since a is not a zero divisor of M. Then $N' = aN'$, and by Nakayama's Lemma $N' = \langle 0 \rangle$. Thus $N \neq aM$ and $U := N/aM$ is the desired submodule $\neq \langle 0 \rangle$ of M with $\mathfrak{p}U = \langle 0 \rangle$.

If, under the assumptions of 3.9, $d(M) = \dim M$, this has far-reaching consequences for the properties of M. This will be pointed out in what follows.

Definition 3.10. Let M be a finitely generated module over a Noetherian ring R. If R is local we call M a *Cohen–Macaulay module* if $M = \langle 0 \rangle$ or if $d(m) = \dim M$. In the general case M is a Cohen–Macaulay module if $M_\mathfrak{m}$ (considered as an $R_\mathfrak{m}$-module) is Cohen–Macaulay for all $\mathfrak{m} \in \mathrm{Max}(R)$. R is called a *Cohen–Macaulay ring* if as an R-module R is Cohen–Macaulay.

Corollary 3.11. If $(R, \mathfrak{m})$ is a Noetherian local ring, M a finitely generated R-module and $\{a_1, \ldots, a_m\}$ an M-regular sequence in $\mathfrak{m}$, then M is a Cohen–Macaulay module if and only if $M/(a_1, \ldots, a_m)M$ as an $R/(a_1, \ldots, a_m)$-module is Cohen–Macaulay. In particular, if $\{a_1, \ldots, a_m\} \subset \mathfrak{m}$ is an R-regular sequence, then R is a Cohen–Macaulay ring if and only if $R/(a_1, \ldots, a_m)$ is.

Proof. It suffices to prove the statement for modules, and we have to consider only the case $m = 1$. Since $d(M/a_1M) = d(M) - 1$, we have only to show that $\dim M/a_1M = \dim M - 1$. Here, by the definition of the depth and the dimension of a module it makes no difference whether we consider M/a_1M as an R-module or an $R/(a_1)$-module.

By (1) we have $\dim M/a_1M = \mathrm{Max}_{\mathfrak{p} \in \mathrm{Supp}(M/a_1M)} \{\dim R/\mathfrak{p}\}$. Further, $\mathrm{Supp}(M/a_1M) = \mathrm{Supp}(M) \cap \mathfrak{V}(a_1)$. Since $a_1 \notin \mathfrak{p}$ for all $\mathfrak{p} \in \mathrm{Ass}(M)$ and since $\dim M = \mathrm{Max}_{\mathfrak{p} \in \mathrm{Ass}(M)} \{\dim R/\mathfrak{p}\}$ by (1), it follows first that $\dim M/a_1M < \dim M$. We now choose $\mathfrak{p} \in \mathrm{Ass}(M)$ with $\dim M = \dim R/\mathfrak{p}$ and a prime divisor $\mathfrak{P}$ of $\mathfrak{p} + (a_1)$ with $\dim R/\mathfrak{P} = \dim R/\mathfrak{p} + (a_1)$. Then $\mathfrak{P} \in \mathrm{Ass}(M/a_1M)$ and by V.4.12 we have $\dim R/\mathfrak{P} = \dim R/\mathfrak{p} - 1$. It follows that $\dim M/a_1M \geq \dim M - 1$, and so $\dim M/a_1M = \dim M - 1$, q. e. d.

Corollary 3.12. Under the hypotheses of 3.11 let M be a Cohen–Macaulay module. Then

$$\dim R/\mathfrak{p} = \dim M - m \qquad \text{for all} \qquad \mathfrak{p} \in \mathrm{Ass}(M/(a_1, \ldots, a_m)M).$$

$(a_1, \ldots, a_m)M$ (and in particular the zero module of M) has no embedded primary components.

Proof. By 3.9

$$d(M) - m = d(M/(a_1,\ldots,a_m)M) \leq \min_{\mathfrak{p}\in\text{Ass}(M/(a_1,\ldots,a_m)M)} \{\dim R/\mathfrak{p}\}$$

$$\leq \max_{\mathfrak{p}\in\text{Ass}(M/(a_1,\ldots,a_m)M)} \{\dim R/\mathfrak{p}\}$$

$$= \dim M/(a_1,\ldots,a_m)M = \dim M - m;$$

and from $d(M) = \dim M$ it follows that the equality sign holds throughout. Hence the assertion follows.

Corollary 3.13. Let $(R, \mathfrak{m})$ be a local Cohen–Macaulay ring and $\{a_1,\ldots,a_m\}$ a system of elements in $\mathfrak{m}$. Then $\{a_1,\ldots,a_m\}$ is an R-regular sequence if and only if $\{a_1,\ldots,a_m\}$ can be extended to a parameter system of R. In particular, the parameter systems of R are just the maximal R-regular sequences in $\mathfrak{m}$.

Proof. Since by 3.12 Ass(R) consists only of minimal prime ideals of R, for $a \in \mathfrak{m}$ we have $\dim R/(a) = \dim R - 1$ if and only if a is not a zero divisor of R. Since $R/(a)$ is then a Cohen–Macaulay ring, then assertion now follows by induction on m from the characterization V.4.12 of parameter systems.

Examples.

1. Any finitely generated module of dimension 0 over a Noetherian ring is Cohen–Macaulay; in particular, any 0-dimensional Noetherian ring is Cohen–Macaulay.

2. Any reduced Noetherian ring of dimension 1 is Cohen–Macaulay. For any $\mathfrak{m} \in \text{Max}(R), R_\mathfrak{m}$ is a field or $\dim R_\mathfrak{m} = 1$. In the second case $\mathfrak{m}R_\mathfrak{m} \notin \text{Ass}(R_\mathfrak{m})$, since in a reduced ring only the minimal prime ideals are associated. Therefore, $\mathfrak{m}R_\mathfrak{m}$ contains a non-zerodivisor, so $d(R_\mathfrak{m}) = 1$.

3. Any regular Noetherian ring is Cohen–Macaulay. Indeed, for any $\mathfrak{m} \in \text{Max}(R), \mathfrak{m}R_\mathfrak{m}$ is generated by an $R_\mathfrak{m}$-regular sequence of length $\dim R_\mathfrak{m}$; therefore $d(R_\mathfrak{m}) = \dim R_\mathfrak{m}$.

4. If K is a field, then $R = K[X_1, X_2]/(X_1^2, X_1 X_2)$ is not Cohen–Macaulay. Indeed, if $\mathfrak{m}$ is the ideal in R generated by the images of X_1 and X_2, then $\dim R_\mathfrak{m} = 1$, but $\mathfrak{m}R_\mathfrak{m}$ consists only of zero divisors. (Another example is contained in Exercise 1.)

The following theorem, proved by Macaulay [55] in the case of polynomial rings, was the starting point of the whole theory described here.

Theorem 3.14. Let R be a (not necessarily local) Cohen–Macaulay ring, $I \neq R$ an ideal of R with $h(I) =: n$. Then

a) $d(I, R) = n$

b) (Macaulay's Unmixedness Theorem) I is a complete intersection in R if and only if I is generated by an R-regular sequence. In this case all the prime ideals in Ass(R/I) have the same height n; in particular, I has no embedded components.

Proof. Let $\{a_1, \ldots, a_m\}$ be an R-regular sequence and $\mathfrak{p} \in \mathrm{Ass}(R/(a_1, \ldots, a_m))$. Choose a maximal ideal $\mathfrak{m}$ of R containing $\mathfrak{p}$. The images of the a_i in $R_\mathfrak{m}$ then form an $R_\mathfrak{m}$-regular sequence in $\mathfrak{m}R_\mathfrak{m}$ and (by 2.10)

$$\mathfrak{p}R_\mathfrak{m} \in \mathrm{Ass}(R_\mathfrak{m}/(a_1, \ldots, a_m)R_\mathfrak{m}).$$

From 3.12 it follows that $\mathfrak{p}R_\mathfrak{m}$ is a minimal prime divisor of $(a_1, \ldots, a_m)R_\mathfrak{m}$, and therefore $\mathfrak{p}$ is a minimal prime divisor of $(a_1, \ldots, a_m)$. Since regular sequences always generate complete intersections, it follows that $h(\mathfrak{p}) = m$. This proves the second statement in b).

Now let $\{a_1, \ldots, a_v\}$ be a maximal R-regular sequence in I. Then I consists only of zero divisors of the R-module $R/(a_1, \ldots, a_v)$; therefore, it is contained in an associated prime ideal $\mathfrak{p}$ of this module. As already shown, $h(\mathfrak{p}) = v$ and so $n \leq v$. On the other hand, if $\mathfrak{q}$ is a minimal prime divisor of I with $h(\mathfrak{q}) = h(I)$, then it follows from $(a_1, \ldots, a_v) \subset \mathfrak{q}$ that $h(\mathfrak{q}) \geq v$ and hence $v = n$, so $d(I, R) = n$.

If I is a complete intersection, then I is generated by n elements, and by 3.5 we can choose them to form an R-regular sequence. This proves the theorem.

Corollary 3.15. In a Cohen–Macaulay ring R the following holds:

a) For any $\mathfrak{p} \in \mathrm{Spec}(R)$, $R_\mathfrak{p}$ is also a Cohen–Macaulay ring.

b) For all $\mathfrak{p}, \mathfrak{q} \in \mathrm{Spec}(R)$ with $\mathfrak{p} \subset \mathfrak{q}$, we have

$$\dim R_\mathfrak{q} = \dim R_\mathfrak{p} + \dim R_\mathfrak{q}/\mathfrak{p}R_\mathfrak{q}$$

(R is a "chain ring").

In a local Cohen–Macaulay ring R, $\dim R = \dim R/I + h(I)$ for any ideal $I \neq R$.

Proof.

a) If $h(\mathfrak{p}) =: n$, then $\mathfrak{p}$ contains an R-regular sequence of length n. This is also an $R_\mathfrak{p}$-regular sequence in $\mathfrak{p}R_\mathfrak{p}$ and so $d(R_\mathfrak{p}) \geq n = h(\mathfrak{p}) = \dim R_\mathfrak{p}$. Since we always have $d(R_\mathfrak{p}) \leq \dim R_\mathfrak{p}$, $R_\mathfrak{p}$ is a Cohen–Macaulay ring.

b) Choose an R-regular sequence $\{a_1, \ldots, a_n\}$ in $\mathfrak{p}$ with $n = \dim R_\mathfrak{p}$ elements. This is also an $R_\mathfrak{q}$-regular sequence in $\mathfrak{p}R_\mathfrak{q}$, and $\mathfrak{p}R_\mathfrak{q}$ is a minimal prime divisor of $(a_1, \ldots, a_n)R_\mathfrak{q}$; therefore $\mathfrak{p}R_\mathfrak{q} \in \mathrm{Ass}(R_\mathfrak{q}/(a_1, \ldots, a_n)R_\mathfrak{q})$. Since $R_\mathfrak{q}$ is also a Cohen–Macaulay ring, 3.12 now shows that

$$\dim R_\mathfrak{q}/\mathfrak{p}R_\mathfrak{q} = \dim R_\mathfrak{q} - n = \dim R_\mathfrak{q} - \dim R_\mathfrak{p}.$$

If R is a local Cohen–Macaulay ring, it follows from 3.12 that $\dim R = \dim R/\mathfrak{p}$ for all minimal prime ideals $\mathfrak{p}$ of R. But then all the maximal prime ideal chains of R have length $\dim R$, and from the definitions of dimension and height it follows that $\dim R = \dim R/I + h(I)$ for any ideal $I \neq R$ of R.

Corollary 3.16. For any multiplicatively closed subset S of a Cohen–Macaulay ring R, R_S is also a Cohen–Macaulay ring.

If $(R, \mathfrak{m})$ is a Noetherian local ring and $\mathfrak{q}$ is an $\mathfrak{m}$-primary ideal, then the R-module $R/\mathfrak{q}$ is of finite length by V.2.6, for $\mathfrak{m}$ is the unique prime ideal $\neq R$ that contains $\mathfrak{q}$. The *socle* of $R/\mathfrak{q}$ is the set of all residue classes $\bar{r} \in R/\mathfrak{q}$ that are annihilated by $\mathfrak{m}$:

$$\mathfrak{S}(R/\mathfrak{q}) := \{\bar{r} \in R/\mathfrak{q} \mid \mathfrak{m} \cdot \bar{r} = 0\}.$$

It is a finite-dimensional vector space. Moreover, for any submodule $U \subset R/\mathfrak{q}$ with $U \neq 0$, we also have $U \cap \mathfrak{S}(R/\mathfrak{q}) \neq (0)$, for U, as a module of finite length, certainly has an element $\neq 0$ that is annihilated by $\mathfrak{m}$, since $\mathrm{Ass}(U) = \{\mathfrak{m}\}$ by 2.3 and V.2.6.

Proposition 3.17. Let $(R, \mathfrak{m})$ be a local Cohen–Macaulay ring, $\{a_1, \ldots, a_d\}$ a parameter system of R. The number

$$r := \dim_{R/\mathfrak{m}} (\mathfrak{S}(R/(a_1, \ldots, a_d)))$$

is independent of the choice of the system $\{a_1, \ldots, a_d\}$.

Proof. This goes according to the same pattern as the proof of 3.1. For $d = 0$ there is nothing to prove. We first consider the case $d = 1$.

Along with a_1 let there be given another non-zerodivisor $b_1 \in \mathfrak{m}$. Then $a_1 b_1$ is not a zero divisor either, and it suffices to show $\dim_{R/\mathfrak{m}}(\mathfrak{S}(R/(a_1))) = \dim_{R/\mathfrak{m}}(\mathfrak{S}(R/(a_1 b_1)))$.

If $r \in R \setminus (a_1)$ with $\mathfrak{m}r \subset (a_1)$ is given, then $rb_1 \subset R \setminus (a_1 b_1)$ and $\mathfrak{m}rb_1 \subset (a_1 b_1)$. Therefore, the multiplication mapping μ_{b_1} defines an injection $\varphi : \mathfrak{S}(R/(a_1)) \to \mathfrak{S}(R/(a_1 b_1))$. If $r \in R$ with $\mathfrak{m}r \subset (a_1 b_1)$ is given, then in particular $a_1 r \in (a_1 b_1)$ and so $r \in (b_1)$. Therefore φ is an isomorphism.

Now let $d > 1$ and suppose the assertion has been proved for rings of lower dimension. If $\{b_1, \ldots, b_d\}$ is another parameter system of R, then as in the proof of 3.1 we find a $c \in \mathfrak{m}$ such that $\{c, a_1, \ldots, a_{d-1}\}$ and $\{c, b_1, \ldots, b_{d-1}\}$ are also parameter systems of R. Since by 3.11 $R/(a_1, \ldots, a_{d-1})$ and $R/(b_1, \ldots, b_{d-1})$ are Cohen–Macaulay rings, it follows from the induction hypothesis that $\dim \mathfrak{S}(R/(a_1, \ldots, a_d)) = \dim \mathfrak{S}(R/(c, a_1, \ldots, a_{d-1})) = \dim \mathfrak{S}(R/(c, b_1, \ldots, b_{d-1})) = \dim \mathfrak{S}(R/(b_1, \ldots, b_d))$, q. e. d.

Definition 3.18. The number r in Proposition 3.17 is called the *type* of the Cohen–Macaulay ring $(R, \mathfrak{m})$. Cohen–Macaulay rings of type 1 are called *Gorenstein* rings. An arbitrary Noetherian ring R is called a Gorenstein ring if $R_\mathfrak{m}$ is a local Gorenstein ring for all $\mathfrak{m} \in \mathrm{Max}(R)$.

Of the many special ideal-theoretic properties of Gorenstein rings we mention only some that follow from the following lemma.

Lemma 3.19. Let $(R, \mathfrak{m})$ be a Noetherian local ring and $\mathfrak{q}$ an $\mathfrak{m}$-primary ideal. $\mathfrak{q}$ is an irreducible ideal if and only if $\dim_{R/\mathfrak{m}} \mathfrak{S}(R/\mathfrak{q}) = 1$.

Proof. $\mathfrak{q}$ is irreducible if and only if in $R/\mathfrak{q}$ the zero module is irreducible. If the zero module is reducible, then there are submodules $U_i \subset R/\mathfrak{q}$, $U_i \neq \langle 0 \rangle$ $(i = 1, 2)$ with $U_1 \cap U_2 = \langle 0 \rangle$. Since $U_i \cap \mathfrak{S}(R/\mathfrak{q}) \neq \langle 0 \rangle$, it follows that $\dim \mathfrak{S}(R/\mathfrak{q}) \geq 2$. Conversely, if this condition is fulfilled, then we can choose 1-dimensional subvector spaces $U_1, U_2 \subset \mathfrak{S}(R/\mathfrak{q})$ with $U_1 \cap U_2 = \langle 0 \rangle$. But this means that $\mathfrak{q}$ is reducible.

Corollary 3.20. A local Cohen–Macaulay ring $(R, \mathfrak{m})$ is a Gorenstein ring if and only if one (any) ideal generated by a parameter system of R is irreducible.

Definition 3.21. A Noetherian local ring $(R, \mathfrak{m})$ is called a *complete intersection* if there is a regular local ring $(A, \mathfrak{M})$ and an ideal $I \subset A$ that is a complete intersection in A such that $R = A/I$. An arbitrary Noetherian ring R is *locally a complete intersection* if $R_\mathfrak{m}$ is a complete intersection for all $\mathfrak{m} \in \mathrm{Max}(R)$.

One can show: If $(R, \mathfrak{m})$ is a complete intersection, then for any representation $R = A/I$ with a regular local ring A the ideal I is a complete intersection in A, so is generated by an A-regular sequence. (For the proof in a special case see Ch. V, §5, Exercise 14b).) An algebraic variety V is locally a complete intersection at $x \in V$ if and only if $\mathcal{O}_{V,x}$ is a complete intersection in the sense of 3.21.

Proposition 3.22. A Noetherian local ring that is a complete intersection (in particular, a regular local ring) is also a Gorenstein ring.

Proof. If $R = A/I$ with a regular local ring A and an ideal I that is generated by an A-regular sequence $\{a_1, \ldots, a_m\}$ in the maximal ideal $\mathfrak{M}$ of A, then R is a Cohen–Macaulay ring (3.11). We can enlarge $\{a_1, \ldots, a_m\}$ to a parameter system $\{a_1, \ldots, a_m, b_1, \ldots, b_n\}$ of A. Since $\mathfrak{M}$ is also generated by a parameter system,

$$\dim \mathfrak{S}(A/(a_1, \ldots, a_m, b_1, \ldots, b_n)) = \dim \mathfrak{S}(A/\mathfrak{M})$$

by 3.17. The images $\bar{b}_1, \ldots, \bar{b}_n$ of the b_i in R form a parameter system of this ring. Since $\dim \mathfrak{S}(R/(\bar{b}_1, \ldots, \bar{b}_n)) = 1$, R is a Gorenstein ring.

The diagram in Fig. 19 makes clear the *hierarchy of Noetherian rings*. Here we call a Noetherian ring *normal* if it is integrally closed in its full ring of quotients.

For the local properties of a variety V at a point x it is important to know to which of the classes in Fig. 19 the local ring $\mathcal{O}_{V,x}$ belongs. We thus get a rough classification of singularities (Gorenstein singularities, normal singularities, etc.). Varieties all of whose local rings belong to one of the classes shown (smooth varieties, Cohen–Macaulay varieties, etc.) are distinguished by special properties. But here we cannot go into this more deeply.

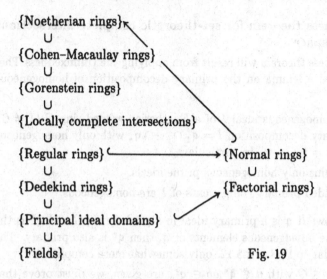

{Noetherian rings}

∪

{Cohen–Macaulay rings}

∪

{Gorenstein rings}

∪

{Locally complete intersections}

∪

{Regular rings} ⸦————————————→ {Normal rings}

∪   ∪

{Dedekind rings}   {Factorial rings}

∪

{Principal ideal domains}

∪

{Fields}   Fig. 19

Exercises

In the following K always denotes a field.

1. $R = K[X_1, X_2, X_3, X_4]/(X_1, X_2) \cap (X_3, X_4)$ is a reduced ring of dimension 2, but not a Cohen–Macaulay ring.

2. For $P := K[X, Y]_{(X,Y)}$, $R := P[Z]/(XZ, YZ, Z^2)$ is a local ring of depth 0. $\mathfrak{p} := (X, \xi)$, where ξ is the residue class of Z in R, is a prime ideal with $d(R_\mathfrak{p}) = 1$.

3. A finitely generated module over a Noetherian integral domain of dimension 1 is a Cohen–Macaulay module if and only if it is torsion-free.

4. Let $(R, \mathfrak{m})$ be a local Cohen–Macaulay ring of dimension 1. Then for any non-zerodivisor $a \in \mathfrak{m}$ we have an isomorphism of $R/\mathfrak{m}$-vector spaces $\mathfrak{S}(R/(a)) \cong \mathfrak{m}^{-1}/R$, where $\mathfrak{m}^{-1} := \{x \in Q(R) \mid \mathfrak{m}x \subset R\}$. R is a Gorenstein ring if and only if $\dim_{R/\mathfrak{m}}(\mathfrak{m}^{-1}/R) = 1$.

5. Let H be a numerical semigroup (V.§3 Exercise 3). The (completed) semigroup ring $K[\|H\|]$ is the subring of the ring $K[\|t\|]$ of formal power series in t consisting of all the series $\sum_{h \in H} a_h t^h$ $(a_h \in K)$.

   a) $K[\|H\|]$ is a Noetherian local ring of dimension 1; in particular it is a Cohen–Macaulay ring.

   b) $K[\|H\|]$ is a Gorenstein ring if and only if H is a symmetric semigroup (V.§3, Exercise 3).

   c) For $H = \mathbf{N} \cdot 5 + \mathbf{N} \cdot 6 + \mathbf{N} \cdot 7 + \mathbf{N} \cdot 8$ we get a Gorenstein ring that is not a complete intersection.

   d) $K[\|H\|]$ is regular if and only if $H = \mathbf{N}$.

   (A generalization of b) is given in [37].)

4. A connectedness theorem for set-theoretic complete intersections in projective space

The connectedness theorem will result from applying the Unmixedness Theorem 3.14. We need a lemma on the primary decomposition of homogeneous ideals.

Lemma 4.1. A homogeneous ideal I of a Noetherian graded ring G $(I \neq G)$ has a reduced primary decomposition $I = q_1 \cap \cdots \cap q_s$ with only homogeneous primary ideals q_i $(i = 1, \ldots, s)$. In particular:

a) $\mathrm{Ass}(G/I)$ contains only homogeneous prime ideals.

b) The non-embedded primary components of I are homogeneous.

Proof. We first show: If q is a primary ideal in a graded ring G and q^* is the ideal spanned by the homogeneous elements of q, then q^* is also primary. The proof of this is similar to that of I.5.12, only somewhat more complicated.

If elements $a, b \in G$ with $a \notin q^*, ab \in q^*$, are given, we must prove that q^* contains a power of b. Write $a = a_\mu + \cdots + a_m$, $b = b_\nu + \cdots + b_n$ with homogeneous elements a_i, b_i of degree i and $\mu \leq m, \nu \leq n$. If suffices to show that a power of each summand b_j of b lies in q^*, since then a high enough power of b also lies in q^*. We may assume that $a_m \notin q^*$; otherwise, subtract a_m from a, which changes nothing essential. Since q^* is homogeneous, $a_m b_n \in q^*$ and $a_m \notin q$, since $a_m \notin q^*$. Since q is primary, $b_n^\rho \in q$ and therefore $b_n^\rho \in q^*$ for suitable $\rho \in \mathbb{N}$.

Now suppose it has been shown that $b_n^\rho, \ldots, b_{n-j}^\rho \in q^*$ for some $j \geq 0$. Then $a \cdot (b - b_n - \cdots - b_{n-j})^{\rho \cdot (j+1)} \in q^*$, and the same argument as above shows that $b_{n-j-1}^\sigma \in q^*$ for some $\sigma \in \mathbb{N}$.

Now if $I = q_1 \cap \cdots \cap q_s$ is any reduced primary decomposition of I, then so is $I = q_1^* \cap \cdots \cap q_s^*$, since $I \subset q_i^* \subset q_i$ for each $i = 1, \ldots, s$; and we get again a reduced primary decomposition. The other statements of the lemma result from the two uniqueness theorems on primary decompositions.

Theorem 4.2. (Hartshorne [31], [32]) Let $V \subset \mathbf{P}^n(L)$ be a K-variety of dimension $d > 0$. If V is a set-theoretic complete intersection (with respect to K), then, for any K-subvariety $W \subset V$ of codimension ≥ 2, $V \setminus W$ is connected in the K-topology (in particular, V itself is connected).

Proof.

a) V is connected.

Let $r := n - d$ and let $F_1, \ldots, F_r$ be homogeneous polynomials in $K[X_0, \ldots, X_n]$ with $\mathfrak{V}(F_1, \ldots, F_r) = V$. $I = (F_1, \ldots, F_r)$ is then a homogeneous ideal and a complete intersection in $K[X_0, \ldots, X_n]$. By 4.1 and the Unmixedness Theorem, only homogeneous ideals q_i occur in the primary decomposition $I = q_1 \cap \ldots \cap q_s$ of I, and the corresponding prime ideals p_i are also homogeneous and all of height r. Let V_i be the irreducible component of V belonging to p_i $(i = 1, \ldots, s)$.

Suppose V is not connected. Say $V_1 \cup \cdots \cup V_t$ $(t < s)$ is a connected component of V; then put $\mathfrak{a} := \bigcap_{i=1}^{t} \mathfrak{q}_i$ and $\mathfrak{b} := \bigcap_{j=t+1}^{s} \mathfrak{q}_j$. We have $\mathfrak{a} \cdot \mathfrak{b} \subset \mathfrak{a} \cap \mathfrak{b} = I$ and $\mathrm{Rad}(\mathfrak{a} + \mathfrak{b}) = \mathfrak{m}$, the homogeneous maximal ideal of $K[X_0, \ldots, X_n]$, since $(V_1 \cup \cdots \cup V_t) \cap (V_{t+1} \cup \cdots \cup V_s) = \emptyset$.

We choose homogeneous elements $a \in \mathfrak{a} \setminus \bigcup_{j=t+1}^{s} \mathfrak{p}_j$ and $b \in \mathfrak{b} \setminus \bigcup_{i=1}^{t} \mathfrak{p}_i$, of which we may also assume that they have the same degree. Then $H := a + b \notin \bigcap_{i=1}^{s} \mathfrak{p}_i$, and hence $J := (F_1, \ldots, F_r, H)$ has height $r + 1$ and is a complete intersection in $K[X_0, \ldots, X_n]$. By the Unmixedness Theorem J has no embedded primary components, and all minimal prime divisors of J have height $r + 1 = n - d + 1 < n + 1$, since $d > 0$.

We shall show that $\mathfrak{m} \in \mathrm{Ass}(K[X_0, \ldots, X_n]/J)$ and thus get a contradiction, because $h(\mathfrak{m}) = n + 1$.

We have $a \notin J$, since from $a = \lambda_0(a + b) + \lambda_1 F_1 + \cdots + \lambda_r F_r$ with homogeneous polynomials λ_i would follow $\lambda_0 \in K$ (by degree considerations); from $(1 - \lambda_0)a = \lambda_0 b + \lambda_1 F_1 + \cdots + \lambda_r F_r$ would follow $b \in I \subset \mathfrak{a}$ (in case $\lambda_0 = 1$) or $a \in (b, I) \subset \mathfrak{b}$ (in case $\lambda_0 \neq 1$), contradicting the construction of a and b.

For any $z \in \mathfrak{m}$, $z^\rho \in \mathfrak{a} + \mathfrak{b}$ for some $\rho \in \mathbb{N}$. If, say, $z^\rho = a_1 + b_1$ ($a_1 \in \mathfrak{a}$, $b_1 \in \mathfrak{b}$), then $z^\rho a = (a_1 + b_1)a \equiv a_1 a \equiv a_1(a + b) \equiv a_1 H \bmod I$, and so $z^\rho a \equiv 0 \bmod J$. Let ρ be the smallest number with $z^\rho a \equiv 0 \bmod J$ ($\rho > 0$, since $a \not\equiv 0 \bmod J$). Then $z(z^{\rho-1}a) \equiv 0 \bmod J$, $z^{\rho-1}a \not\equiv 0 \bmod J$; and it has been shown that any $z \in \mathfrak{m}$ is a zero divisor of $K[X_0, \ldots, X_n]/J$. Therefore $\mathfrak{m} \in \mathrm{Ass}(K[X_0, \ldots, X_n]/J)$.

The assumption that V is not connected has led to a contradiction. This proves a).

b) Suppose there is a subvariety $W \subset V$ of codimension ≥ 2 such that $V \setminus W$ is not connected in the K-topology. Then neither is it connected in the L-topology, so we may assume that $K = L$ is algebraically closed (in particular, infinite).

Using the Noether Normalization Theorem (II.3.1d)) one shows that there is a linear variety $\Lambda \subset \mathbf{P}^n$ with $W \cap \Lambda = \emptyset$ that intersects any irreducible component of V in a curve. Namely, if I is the ideal of W in the homogeneous coordinate ring $K[V]$ of V, then there are algebraically independent elements $Y_0, \ldots, Y_d \in K[V]$ over K that are homogeneous of degree 1 and there is $i \in [0, d]$ such that $K[V]$ is a finite module over $K[Y_0, \ldots, Y_d]$ and $I \cap K[Y_0, \ldots, Y_d] = (Y_0, \ldots, Y_i)$. Here $i + 1$ is the height of I, so $i \geq 1$, since $\mathrm{codim}_V W = h(I) \geq 2$.

Now if Λ is the linear variety given by the system of equations $Y_2 = \cdots = Y_d = 0$, then $\dim \Lambda = n - d + 1$. Further, $\dim K[V]/(Y_2, \ldots, Y_d) = 2$, so $\dim(V \cap \Lambda) = 1$; and this also holds if V is replaced by any of its irreducible components, since they all have dimension d. Since $V \cap \Lambda$ can be described by $n - d + d - 1 = n - 1$ equations, $V \cap \Lambda$ is a set-theoretic complete intersection.

From $(Y_0, \ldots, Y_d) \subset I + (Y_2, \ldots, Y_d)$ it follows that $\mathrm{Rad}(I + (Y_2, \ldots, Y_d))$ is the irrelevant maximal ideal of $K[V]$; therefore $W \cap \Lambda = \emptyset$.

But $V \setminus W$ was not connected. From $W \cap \Lambda = \emptyset$ it follows that $V \cap \Lambda$ is a non-connected curve. Since $V \cap \Lambda$ is a set-theoretic complete intersection, we have contradicted a). This proves the theorem.

Corollary 4.3. If a K-variety $V \subset \mathbf{P}^n(L)$ $(n \geq 2)$ is representable as the intersection of $r \leq (n+1)/2$ K-hypersurfaces, then it is connected.

Proof. Let $V = V_1 \cup \cdots \cup V_s$ be the decomposition of V into irreducible components. Then $\dim V_i \geq n - r \geq (n-1)/2 > 0$ $(i = 1, \ldots, s)$ by V.3.6. If all the components have dimension $(n-1)/2$, then V is a complete intersection and hence by 4.2 is connected. On the other hand, if, say, $\dim V_1 > (n-1)/2$, then for $i = 2, \ldots, s$ we have $\dim V_1 + \dim V_i \geq n$, so $V_1 \cap V_i \neq \emptyset$ by V.3.10, and it again follows that V is connected.

Examples 4.4.

a) Let $C_1, C_2 \subset \mathbf{P}^3$ be two algebraic curves that do not intersect (say two skew lines). Then $C_1 \cup C_2$ is not representable as the intersection of two algebraic surfaces.

b) Let $F_1, F_2 \subset \mathbf{P}^4$ be two irreducible algebraic surfaces with only one point in common (say two planes with only one point of intersection). Then $F_1 \cup F_2$ is connected but is not a set-theoretic complete intersection.

For if we remove the point of intersection from $F_1 \cup F_2$ (a subvariety of codimension 2 in $F_1 \cup F_2$), we get a non-connected space.

There is a local analogue (see Exercise 1) of Theorem 4.2 that is also due to Hartshorne [31]. Similar, but somewhat more general connectedness theorems than 4.2 may be found in Rung [67].

Exercises

1. Apply the argument from the proof of Theorem 4.2 to prove: If $(R, \mathfrak{m})$ is a Noetherian local ring for which $\mathrm{Spec}(R) \setminus \{\mathfrak{m}\}$ is not connected, then $d(R) \leq 1$.

2. From this and III.§1, Exercise 3, deduce: Let R be a Noetherian ring. If $X := \mathrm{Spec}(R)$ is connected and $Y \subset X$ is a closed subset such that $d(R_{\mathfrak{p}}) \geq 2$ for all $\mathfrak{p} \in Y$, then $X \setminus Y$ is connected.

References

The theory presented in §1 of the regular points of varieties and their relations to local rings goes back to Zariski [84]. An important theorem on regular local rings says that they are factorial. There is a simple proof of this in Kaplansky [D]. More generally, the question of under what conditions Noetherian local rings are factorial has been the subject of numerous investigations. A survey of the results on this theme is given in Lipman's lecture [54].

Some very important invariants of Noetherian local rings that have not appeared in the text are the multiplicity and the Hilbert function. They are treated in detail in most textbooks (see for example [E], [F], [G]). Additional properties of Gorenstein rings and Cohen–Macaulay rings may be found, for example, in Bass [7] or [36]. The ideal theory of Gorenstein rings was first studied by Gröbner [28]. On the generation of ideals more precise statements can often be made if they are ideals in Noetherian rings of a special type. A synopsis of results of this kind can be found in Judith Sally's work [69].

Important connectedness theorems for projective varieties have been discovered in recent years, see: G. Faltings, Some theorems about formal functions, *Publ. RIMS. Kyoto Univ.* **16**, 721–737 (1980); W. Fulton and J. Hansen, A connectedness theorem for projective varieties, with applications to intersections and singularities of mappings, *Ann. of Math.* **110**, 159–166 (1980); W. Fulton and R. Lazarsfeld, Connectivity and its applications in algebraic geometry. In: Algebraic Geometry, *Springer Lecture Notes in Math.* **862**, 26–92 (1981).

The theory of singularities is a lively, extensive part of algebraic geometry, with close points of contact to many other branches of mathematics. For a survey of this subject see the lectures in [1].

Chapter VII
Projective resolutions

This chapter deals with theorems on regular rings and complete intersections whose proofs use methods of homological algebra. For the results developed here we need very little homology theory; in fact, we use only projective resolutions and make repeated application of the "Snake Lemma." The main results of this chapter are Hilbert's Syzygy Theorem, its generalization to regular rings due to Auslander–Buchsbaum and Serre, a characterization of local complete intersections by their conormal module, and Szpiro's result on space curves mentioned in Chapter V.3.13d.

1. The projective dimension of modules

For any module M over a ring R there is an exact sequence

$$0 \to K_1 \to F_0 \overset{\epsilon}{\to} M \to 0, \tag{1}$$

where F_0 is a free R-module and $K_1 := \mathrm{Ker}(\epsilon)$. By iterating this construction we get the concept of a free resolution of a module. If an exact seqence

$$0 \to K_i \to F_{i-1} \overset{\alpha_{i-2}}{\to} F_{i-2} \to \cdots \overset{\alpha_0}{\to} F_0 \to M \to 0 \tag{2}$$

with free R-modules F_i has been constructed, consider a sequence (1) for K_i:

$$0 \to K_{i+1} \to F_i \to K_i \to 0 \qquad (F_i \text{ free}). \tag{3}$$

(2) and (3) can be put together into an exact sequence

$$0 \to K_{i+1} \to F_i \longrightarrow F_{i-1} \to \cdots \to F_0 \to M \to 0.$$
$$\searrow \quad \nearrow$$
$$K_i$$

K_i is called an i-th *relation module* or *syzygy module*. One hopes to get information about the structure of M from studying the syzygy modules of M.

If R is a Noetherian ring and M is finitely generated, then the exact sequences above can be constructed so that all the F_i are finitely generated R-modules.

Definition 1.1. An exact sequence

$$\cdots \to F_{i+1} \xrightarrow{\alpha_i} F_i \to \cdots \to F_1 \xrightarrow{\alpha_0} F_0 \xrightarrow{\epsilon} M \to 0 \tag{4}$$

with only free (resp. projective) modules F_i $(i = 0, 1, \ldots)$ is called a *free* (resp. *projective*) *resolution* of M.

As will be shown, it is useful to investigate, instead of the class of free resolutions, the more comprehensive class of projective resolutions.

Rules 1.2.

a) If $S \subset R$ is a multiplicatively closed subset and (4) is a free (resp. projective) resolution of M, then

$$\cdots \to (F_{i+1})_S \xrightarrow{(\alpha_i)_S} (F_i)_S \to \cdots \to (F_1)_S \xrightarrow{(\alpha_0)_S} (F_0)_S \xrightarrow{\epsilon_S} M_S \to 0$$

is a free (resp. projective) resolution of the R_S-module M_S.

b) Let P/R be a ring extension, where P is free as an R-module; then, along with (4),

$$\cdots \to P \underset{R}{\otimes} F_{i+1} \xrightarrow{P \otimes \alpha_i} P \underset{R}{\otimes} F_i \to \cdots \to P \underset{R}{\otimes} F_1 \xrightarrow{P \otimes \alpha_0} P \underset{R}{\otimes} F_0 \xrightarrow{P \otimes \epsilon} P \underset{R}{\otimes} M \to 0 \tag{4'}$$

is a free (resp. projective) resolution of the P-module $P \otimes_R M$.

Proof. a) follows from III.4.17. To prove b) note first that the $P \otimes_R F_i$ are free (resp. projective) P-modules. If $P \cong R^\Lambda$, then (4') is just a "direct sum" of as many sequences (4) as Λ has elements, so it too is exact.

In particular, Rule 1.2b) can be applied when $P = R[X_1, \ldots, X_n]$ is a polynomial ring over R.

The case is especially important where the zero module occurs in a projective resolution.

Definition 1.3. M is said to be of *finite projective dimension* if there is a projective resolution of the form

$$0 \to F_n \to F_{n-1} \to \cdots \to F_1 \to F_0 \to M \to 0.$$

The minimum of the lengths n of such resolutions is called the *projective dimension* of M $(\mathrm{pd}(M))$. If there is no such resolution, then we put $\mathrm{pd}(M) = \infty$. (We often denote the projective dimension of M by $\mathrm{pd}_R(M)$ when it is useful to emphasize that M is being considered an R-module.)

A module M is projective if and only if $\mathrm{pd}(M) = 0$. In general, $\mathrm{pd}(M)$ can be considered a measure of how far the module is from being projective. Of course, a module has infinitely many projective resolutions. Nevertheless, we have:

Proposition 1.4. Let there be given two exact sequences of R-modules

$$0 \to K_n \xrightarrow{\alpha_{n-1}} F_{n-1} \xrightarrow{\alpha_{n-2}} \cdots \xrightarrow{\alpha_0} F_0 \xrightarrow{\epsilon} M \to 0$$

$$0 \to K'_n \xrightarrow{\alpha'_{n-1}} F'_{n-1} \xrightarrow{\alpha'_{n-2}} \cdots \xrightarrow{\alpha'_0} F'_0 \xrightarrow{\epsilon'} M \to 0$$

where $n \geq 1$ and the F_i, F'_i $(i = 0, \ldots, n - 1)$ are projective R-modules. Then:

a) $K_n \oplus F'_{n-1} \oplus F_{n-2} \oplus \cdots \cong K'_n \oplus F_{n-1} \oplus F'_{n-2} \oplus \cdots$.

b) K_n is projective if and only if K'_n is.

Proof. b) follows from a) because a direct summand of a projective module is projective. (Note that this is not always true for free modules.)

a) has been proved in IV.1.15 in the case $n = 1$ for free modules F_0, F'_0. For projective modules the proof is similar. The general case results from this by induction on n. If $n > 1$ and the proposition has been proved for sequences of length $n - 1$, let $K_{n-1} := \mathrm{Im}(\alpha_{n-2})$, $K'_{n-1} := \mathrm{Im}(\alpha'_{n-2})$. Then

$$K'_{n-1} \oplus F_{n-2} \oplus F'_{n-3} \oplus \cdots \cong K_{n-1} \oplus F'_{n-2} \oplus F_{n-3} \oplus \cdots,$$

and we have exact sequences (coming from the exact sequences $0 \to K_n \to F_{n-1} \to K_{n-1} \to 0$ and $0 \to K'_n \to F'_{n-1} \to K'_{n-1} \to 0$)

$$0 \to K_n \to (F_{n-1} \oplus F'_{n-2} \oplus F_{n-3} \oplus \cdots) \to (K_{n-1} \oplus F'_{n-2} \oplus F_{n-3} \oplus \cdots) \to 0$$

$$0 \to K'_n \to (F'_{n-1} \oplus F_{n-2} \oplus F'_{n-3} \oplus \cdots) \to (K'_{n-1} \oplus F_{n-2} \oplus F'_{n-3} \oplus \cdots) \to 0$$

Another application of IV.1.15 now yields the assertion.

Corollary 1.5. Let $0 \to K_n \to F_{n-1} \to \cdots \to F_0 \to M \to 0$ be an exact sequence with projective R-modules F_i $(i = 0, \ldots, n - 1)$. Then we have

a) $\mathrm{pd}(M) \leq n$ if and only if K_n is projective.

b) If $\mathrm{pd}(M) \geq n$, then $\mathrm{pd}(K_n) = \mathrm{pd}(M) - n$.

Proof.

a) If K_n is projective, then $\mathrm{pd}(M) \leq n$. Conversely, if $\mathrm{pd}(M) =: m \leq n$, then we have a projective resolution

$$0 \to F'_m \to F'_{m-1} \to \cdots \to F'_0 \to M \to 0$$

and by 1.4 K_n is a direct summand of a projective module, so is itself projective.

b) If $\mathrm{pd}(M) = \infty$, then $\mathrm{pd}(K_n) = \infty$. On the other hand, if $\mathrm{pd}(M) =: m$ with $n \leq m < \infty$, then consider an exact sequence with projective modules $F_i : 0 \to K_m \to F_{m-1} \to \cdots \to F_n \to K_n \to 0$. This can be spliced with the given sequence:

$$0 \to K_m \to F_{m-1} \to \cdots \to F_n \overset{\searrow \qquad \nearrow}{\underset{K_n}{\longrightarrow}} F_{n-1} \to \cdots \to F_0 \to M \to 0$$

and by a) it follows that K_m is projective. Then $\mathrm{pd}(K_n) \leq m - n$. But it cannot be that $\mathrm{pd}(K_n) < m - n$, since otherwise $\mathrm{pd}(M) < m$.

Corollary 1.6. If R is a Noetherian ring, M a finitely generated R-module, then

$$\mathrm{pd}_R(M) = \mathrm{Sup}_{\mathfrak{m} \in \mathrm{Max}(R)} \{\mathrm{pd}_{R_\mathfrak{m}}(M_\mathfrak{m})\}.$$

Proof. Denote the supremum by d. By 1.2a) $\mathrm{pd}_R(M) \geq d$. Hence only for $d < \infty$ is there something to be shown. In this case we consider an exact sequence

$$0 \to K_d \to F_{d-1} \to \cdots \to F_0 \to M \to 0$$

with free R-modules F_i of finite rank. Then K_d is a finitely generated R-module and by 1.5a) $(K_d)_\mathfrak{m}$ is a projective (therefore free) $R_\mathfrak{m}$-module for all $\mathfrak{m} \in \mathrm{Max}(R)$. By IV.3.6 K_d is itself projective, and so $\mathrm{pd}(M) = d$.

Corollary 1.7. For arbitrary R-modules M_1, M_2 we have

$$\mathrm{pd}(M_1 \oplus M_2) = \mathrm{Sup}\{\mathrm{pd}(M_1), \mathrm{pd}(M_2)\}.$$

Proof. If projective resolutions

$$\cdots \to F_n \overset{\alpha_{n-1}}{\to} F_{n-1} \to \cdots \to F_0 \overset{\epsilon}{\to} M_1 \to 0$$

$$\cdots \to F_n' \overset{\alpha_{n-1}'}{\to} F_{n-1}' \to \cdots \to F_0' \overset{\epsilon'}{\to} M_2 \to 0$$

are given, then

$$\cdots \to F_n \oplus F_n' \overset{\alpha_{n-1} \oplus \alpha_{n-1}'}{\longrightarrow} F_{n-1} \oplus F_{n-1}' \to \cdots \to F_0 \oplus F_0' \overset{\epsilon \oplus \epsilon'}{\longrightarrow} M_1 \oplus M_2 \to 0$$

is a projective resolution of $M_1 \oplus M_2$. Let $K_{n+1} := \mathrm{Im}(\alpha_n), K_{n+1}' := \mathrm{Im}(\alpha_n')$. $K_{n+1} \oplus K_{n+1}'$ is projective if and only if K_{n+1} and K_{n+1}' are. The assertion now follows from 1.5.

1.7 is a special case of the following Comparison Theorem.

Proposition 1.8. Let $0 \to M_1 \overset{\beta_1}{\to} M_2 \overset{\beta_2}{\to} M_3 \to 0$ be an exact sequence of R-modules. Then:

a) If two modules in the sequence have finite projective dimension, so does the third.

b) If this is the case, then $\mathrm{pd}(M_2) \leq \mathrm{Max}\{\mathrm{pd}(M_1), \mathrm{pd}(M_3)\}$. And if $\mathrm{pd}(M_2) < \mathrm{Max}\{\mathrm{pd}(M_1), \mathrm{pd}(M_3)\}$, then $\mathrm{pd}(M_3) = \mathrm{pd}(M_1) + 1$.

In the proof we use the widely useful

Lemma 1.9. (Snake Lemma) Let there be given a commutative diagram of R-modules with exact rows and columns

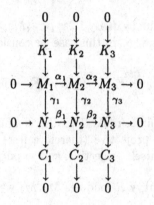

Let $\alpha_i' : K_i \rightarrow K_{i+1}$ and $\beta_i' : C_i \rightarrow C_{i+1}$ $(i = 1, 2)$ be the homomorphisms induced by α_i and β_i. Then there is a linear mapping $\delta : K_3 \rightarrow C_1$ (called *connecting homomorphism*) such that the sequence

$$0 \rightarrow K_1 \xrightarrow{\alpha_1'} K_2 \xrightarrow{\alpha_2'} K_3 \xrightarrow{\delta} C_1 \xrightarrow{\beta_1'} C_2 \xrightarrow{\beta_2'} C_3 \rightarrow 0 \tag{5}$$

is exact.

Proof. We consider the injections in the diagram above as inclusion mappings.

a) Construction of δ. For $x \in K_3$ we choose $x' \in M_2$ with $\alpha_2(x') = x$ and put $y' := \gamma_2(x')$. Then $\beta_2(y') = \gamma_3(\alpha_2(x')) = \gamma_3(x) = 0$, and so $y' \in N_1$. Let $\delta(x)$ be the image of y' in C_1. $\delta(x)$ does not depend on the choice of x', for if $x'' \in M_2$ with $\alpha_2(x'') = x$ and if $y'' := \gamma_2(x'')$, then $x' - x'' \in \mathrm{Ker}(\alpha_2) = \mathrm{Im}(\alpha_1)$ and $y' - y'' \in \mathrm{Im}(\gamma_1)$. But then y' and y'' have the same image in C_1.

From the definition of δ it immediately follows that δ is an R-linear mapping.

b) Exactness of the sequence (5) at K_3. If $\delta(x) = 0$, then $y' \in \mathrm{Im}(\gamma_1)$. If we choose $y'' \in M_1$ with $\gamma_1(y'') = y'$ and put $x'' := \alpha_1(y'')$, then $\alpha_2(x' - x'') = x$ and $\gamma_2(x' - x'') = 0$, so $x' - x'' \in K_2$. This shows that $\mathrm{Im}(\alpha_2') \supset \mathrm{Ker}(\delta)$. That $\mathrm{Im}(\alpha_2') \subset \mathrm{Ker}(\delta)$ is clear by the construction of δ.

c) Exactness of the sequence (5) at C_1. From the definition of δ it immediately follows that $Im(\delta) \subset \mathrm{Ker}(\beta_1')$. Conversely, if $z \in \mathrm{Ker}(\beta_1')$ is given and $y \in N_1$ is a representative of z, then $\beta_1(y) \in \mathrm{Im}(\gamma_2)$. Then there is $x' \in M_2$ with $\gamma_2(x') = \beta_1(y)$. If we put $x := \alpha_2(x')$, then $\delta(x) = z$ by the construction of δ, which shows $\mathrm{Ker}(\beta_1') \subset \mathrm{Im}(\delta)$.

The exactness of the sequence (5) at the other places is very easily verified.

Proof of 1.8.

1. We choose exact sequences

$$0 \to K_1 \to F_0 \xrightarrow{\gamma_1} M_1 \to 0$$

$$0 \to K_3 \to F_0' \xrightarrow{\gamma_3} M_3 \to 0$$

with projective R-modules F_0, F_0', and we construct a commutative diagram with exact rows

$$
\begin{array}{ccccccccc}
0 & \to & F_0 & \xrightarrow{\alpha_1} & F_0 \oplus F_0' & \xrightarrow{\alpha_2} & F_0' & \to & 0 \\
 & & \downarrow{\gamma_1} & & \downarrow{\gamma_2} & & \downarrow{\gamma_3} & & \\
0 & \to & M_1 & \xrightarrow{\beta_1} & M_2 & \xrightarrow{\beta_2} & M_3 & \to & 0.
\end{array}
\tag{6}
$$

Here α_1 is the canonical injection and α_2 the canonical projection. γ_2 is defined as follows. There is a linear mapping $\tilde{\gamma} : F_0' \to M_2$ with $\beta_2 \circ \tilde{\gamma} = \gamma_3$ since F_0' is projective. For $(y, y') \in F_0 \oplus F_0'$ put $\gamma_2(y, y') := \beta_1(\gamma_1(y)) + \tilde{\gamma}(y')$. One immediately checks that the diagram is commutative and γ_2 is a surjective linear mapping.

If $K_2 := \mathrm{Ker}(\gamma_2)$, then by 1.9 we have an exact sequence

$$0 \to K_1 \xrightarrow{\alpha_1'} K_2 \xrightarrow{\alpha_3'} K_3 \to 0. \tag{7}$$

2a) Suppose two of the modules M_i $(i = 1, 2, 3)$ have finite projective dimension, and let m be the maximum of these dimensions. If $m = 0$, then both modules are projective. If M_3 is among them, then the exact sequence $0 \to M_1 \to M_2 \to M_3 \to 0$ splits and all three modules are projective. If M_3 is not among them, then the sequence is a projective resolution of M_3, so $\mathrm{pd}(M_3) \leq 1$. Now let $m > 0$. Then by 1.5 two of the modules in the exact sequence (7) have finite projective dimension $< m$. By the induction hypothesis the third module in (7) also has finite projective dimension, and a) is proved.

b) follows by induction on $d := \mathrm{pd}(M_2)$. If $d = 0$, then $\mathrm{pd}(M_3) = \mathrm{pd}(M_1) + 1$ or $\mathrm{pd}(M_3) = 0$ by 1.5. In the second case, $\mathrm{pd}(M_1) = 0$. This proves the assertion for $d = 0$. Now let $d > 0$ and suppose the assertion proved for all exact sequences in which the middle module has projective dimension $< d$. Then

$$\mathrm{pd}(K_2) \leq \mathrm{Max}\{\mathrm{pd}(K_1), \mathrm{pd}(K_3)\},$$

and if $\mathrm{pd}(K_2) < \mathrm{Max}\{\mathrm{pd}(K_1), \mathrm{pd}(K_3)\}$, we have $\mathrm{pd}(K_3) = \mathrm{pd}(K_1) + 1$. Then $\mathrm{pd}(K_i) = \mathrm{pd}(M_i) - 1$ $(i = 1, 2, 3)$ if M_1 and M_3 both are not projective. Assertion b) follows from the above formulas for the modules K_i. If M_3 is projective, then $\mathrm{pd}(M_2) = \mathrm{pd}(M_1) = \mathrm{Max}\{\mathrm{pd}(M_1), \mathrm{pd}(M_3)\}$ by 1.7. If M_1 is projective then in diagram (6) we can take F_0 to be the module M_1. Then $K_1 = \langle 0 \rangle$ and $\mathrm{pd}(M_2) - 1 = \mathrm{pd}(K_2) = \mathrm{pd}(K_3) = \mathrm{pd}(M_3) - 1$, which proves b) in this case. The proof of Proposition 1.8 is now complete.

In what follows let $(R, \mathfrak{m})$ be a Noetherian local ring and M a finitely generated R-module.

A free resolution of M

$$\cdots \xrightarrow{\alpha_n} F_n \xrightarrow{\alpha_{n-1}} F_{n-1} \to \cdots \to F_1 \xrightarrow{\alpha_0} F_0 \xrightarrow{\epsilon} M \to 0 \tag{8}$$

is called *minimal* if $\mathrm{Im}(\alpha_n) \subset \mathfrak{m}F_n$ for all $n \in \mathbf{N}$. Then if $K_n := \mathrm{Im}(\alpha_{n-1})$ $(n \geq 1)$, it follows from Nakayama's Lemma that $\mu(F_0) = \mu(M)$ and $\mu(F_n) = \mu(K_n)$ $(n > 0)$.

It is clear that M has a minimal free resolution: Choose F_0 so that $\mu(F_0) = \mu(M)$. Then $K_1 := \mathrm{Ker}(\epsilon) \subset \mathfrak{m}F_0$. Now choose F_1 so that $\mu(F_1) = \mu(K_1)$ etc.

Proposition 1.10. If two minimal free resolutions of M are given

$$\cdots \to F_n \xrightarrow{\alpha_{n-1}} F_{n-1} \to \cdots \xrightarrow{\alpha_0} F_0 \xrightarrow{\epsilon} M \to 0$$

$$\cdots \to F_n' \xrightarrow{\alpha_{n-1}'} F_{n-1}' \to \cdots \xrightarrow{\alpha_0'} F_0' \xrightarrow{\epsilon'} M \to 0$$

then $\mu(F_n) = \mu(F_n')$ for all $n \in \mathbf{N}$.

Proof. We have $\mu(F_0) = \mu(F_0') = \mu(M)$. Let $K_n := \mathrm{Im}(\alpha_{n-1}), K_n' := \mathrm{Im}(\alpha_{n-1}')$ $(n \geq 1)$. By 1.4

$$K_n \oplus F_{n-1}' \oplus F_{n-2} \oplus \cdots \cong K_n' \oplus F_{n-1} \oplus F_{n-2}' \oplus \cdots.$$

If it has been proved that $\mu(F_i) = \mu(F_i')$ for $i < n$, then $\mu(F_n) = \mu(K_n) = \mu(K_n') = \mu(F_n')$.

The invariants $\beta_i := \mu(F_i)$ are called the *Betti numbers* of the module M. By definition, the Betti numbers of R are the Betti numbers of the R-module $R/\mathfrak{m}$.

Corollary 1.11. If $\mathrm{pd}(M) =: n < \infty$ and if (8) is a minimal free resolution of M, then $F_m = \langle 0 \rangle$ for all $m > n$ (and of course $F_m \neq \langle 0 \rangle$ for $m \leq n$).

By 1.5 $K_n := \mathrm{Im}(\alpha_{n-1})$ is a free R-module and hence $0 \to K_n \to F_{n-1} \to \cdots \to F_0 \to M \to 0$ is a minimal free resolution of M. The assertion results at once from 1.10.

There is a close connection between the projective dimension and the depth of M.

Proposition 1.12. (Auslander-Buchsbaum [6]) Let M be a finitely generated module over a Noetherian local ring $(R, \mathfrak{m})$. If $\mathrm{pd}(M) < \infty$ then

$$\mathrm{pd}(M) + d(M) = d(R).$$

The proof requires some preparation.

We choose an exact sequence $0 \to K \to F \to M \to 0$ with a free R-module of finite rank F and for $x \in \mathfrak{m}$ consider the commutative diagram with exact rows and columns

$$
\begin{array}{ccccccccc}
 & & 0 & & 0 & & 0 & & \\
 & & \downarrow & & \downarrow & & \downarrow & & \\
 & & K' & & F' & & M' & & \\
 & & \downarrow & & \downarrow & & \downarrow & & \\
0 \to & & K & \to & F & \to & M & \to & 0 \\
 & & \downarrow{\scriptstyle \mu_x} & & \downarrow{\scriptstyle \mu_x} & & \downarrow{\scriptstyle \mu_x} & & \\
0 \to & & K & \to & F & \to & M & \to & 0 \\
 & & \downarrow & & \downarrow & & \downarrow & & \\
 & & K/xK & & F/xF & & M/xM & & \\
 & & \downarrow & & \downarrow & & \downarrow & & \\
 & & 0 & & 0 & & 0 & &
\end{array}
$$

where $M' := \{m \in M \mid xm = 0\} = \operatorname{Ker}\mu_x$, likewise F', K'. The Snake Lemma provides us with an exact sequence

$$0 \to K' \to F' \to M' \to K/xK \to F/xF \to M/xM \to 0. \tag{9}$$

From this we read off:

Lemma 1.13. If x is not a zero divisor of M, then M is free if and only if M/xM is a free $R/(x)$-module.

Proof. Let M/xM be a free $R/(x)$-module. We may assume that $\mu(F) = \mu(M)$. Then $F/xF \to M/xM$ is an isomorphism. Since x is not a zero divisor of $M, M' = \langle 0 \rangle$ and therefore $K/xK = \langle 0 \rangle$. From Nakayama's Lemma it follows that $K = \langle 0 \rangle$ and $M \cong F$.

More generally:

Lemma 1.14. If x is not a zero divisor of R and M, then

$$\operatorname{pd}_R(M) = \operatorname{pd}_{R/(x)}(M/xM).$$

Proof. From (9) we get an exact sequence $0 \to K/xK \to F/xF \to M/xM \to 0$; moreover, $K' = \langle 0 \rangle$ since $F' = \langle 0 \rangle$, and therefore x is not a zero divisor of K. If both projective dimensions in the formula are ∞, there is nothing to show; by 1.13 there is also nothing to show if one is 0.

Let $\operatorname{pd}(M) =: m$ and $0 < m < \infty$. Then $\operatorname{pd}(K) = m-1$, and by induction we may assume that $\operatorname{pd}_{R/(x)}(K/xK) = \operatorname{pd}_R(K)$. Since also $\operatorname{pd}_{R/(x)}(M/xM) \neq 0$, it follows that

$$\operatorname{pd}_{R/(x)}(M/xM) = \operatorname{pd}_{R/(x)}(K/xK) + 1 = \operatorname{pd}_R(K) + 1 = \operatorname{pd}_R(M).$$

One argues similarly if $\operatorname{pd}_{R/(x)}(M/xM) =: m$ with $0 < m < \infty$.

Lemma 1.15. If $d(R) > 0, d(M) = 0$, then $d(K) = 1$.

Proof. Let $x \in \mathfrak{m}$ be a non-zerodivisor of R. In this case (9) yields an exact sequence

$$0 \to M' \to K/xK \to F/xF \to M/xM \to 0.$$

Since $d(M) = 0$, there is $m \in M \setminus \langle 0 \rangle$ with $\mathfrak{m} \cdot m = 0$ (VI.2.8). Then $m \in M'$, so $\mathfrak{m} \in \mathrm{Ass}(M')$ and hence also $\mathfrak{m} \in \mathrm{Ass}(K/xK)$, that is $d(K/xK) = 0$. Since x is not a zero divisor of K ($K' = \langle 0 \rangle$!), we have $d(K) = 1$.

Proof of 1.12.

Let $n := \mathrm{pd}(M)$ and

$$0 \to F_n \overset{\alpha_{n-1}}{\to} F_{n-1} \to \cdots \overset{\alpha_0}{\to} F_0 \to M \to 0$$

be a minimal free resolution of M, $K_i := \mathrm{Im}(\alpha_{i-1})$ $(i = 1, \ldots, n)$.

We argue by induction on $d := d(R)$. For $d = 0$ there is an $x \in R$ with $\mathfrak{m} \cdot x = 0, x \neq 0$. If we had $n > 0$, then from $F_n \subset \mathfrak{m}F_{n-1}$ would follow $xF_n \subset x\mathfrak{m}F_{n-1} = \langle 0 \rangle$, so $F_n = \langle 0 \rangle$, a contradiction. Therefore $n = 0, M$ is free, and $d(M) = d(R) = 0$.

Now let $d > 0$ and suppose the proposition proved for local rings of lower depth. If also $d(M) > 0$, there is $x \in \mathfrak{m}$ that is not a zero divisor of R or M, since $\mathfrak{m} \not\subset \bigcup \mathfrak{p}$ if $\mathfrak{p}$ varies over all of $\mathrm{Ass}(R) \cup \mathrm{Ass}(M)$. Then $d(R/(x)) = d - 1$, $d(M/xM) = d(M) - 1$, and by 1.14 $\mathrm{pd}_{R/(x)}(M/xM) = \mathrm{pd}(M)$. The assertion now follows from the induction hypothesis.

If $d > 0$ and $d(M) = 0$, then $d(K_1) = 1$ by 1.15, $\mathrm{pd}(K_1) = \mathrm{pd}(M) - 1$, and as already proved

$$\mathrm{pd}(K_1) + d(K_1) = d(R),$$

whence follows $\mathrm{pd}(M) = d(R)$, q. e. d.

Exercises

1. Give an example of a module of infinite projective dimension.

2. Let $\alpha : R \to S$ be a ring homomorphism, M an S-module. Consider S and M as R-modules via α. Then

$$\mathrm{pd}_R(M) \leq \mathrm{pd}_S(M) + \mathrm{pd}_R(S).$$

3. Let $P := R[X_1, \ldots, X_n]$ be a polynomial ring over a ring R, M an R-module. We have $\mathrm{pd}_P(P \underset{R}{\otimes} M) = \mathrm{pd}_R(M)$.

4. Let x be a non-zerodivisor of a ring $R, S := R/(x)$, and $M \neq \langle 0 \rangle$ an S-module of finite projective dimension. Consider M as an R-module via the canonical epimorphism $R \to S$. Then

$$\text{pd}_R(M) = \text{pd}_S(M) + 1.$$

5. Let R be a ring, $\{x_1, \ldots, x_n\}$ an R-regular sequence. In the seqence

$$0 \to F_n \xrightarrow{d_{n-1}} F_{n-1} \to \cdots \to F_1 \xrightarrow{d_0} R \xrightarrow{\epsilon} R/(x_1, \ldots, x_n) \to 0$$

let ϵ be the canonical epimorphism, F_p for $p = 1, \ldots, n$ a free R-module with basis $\{e_{i_1 \cdots i_p} \mid 1 \leq i_1 < \cdots < i_p \leq n\}, d_0$ the linear mapping with $d_0(e_i) = x_i$ $(i = 1, \ldots, n)$, and in general d_p $(p > 0)$ the linear mapping with

$$d_p(e_{i_1 \ldots i_{p+1}}) = \sum_{k=1}^{p+1} (-1)^{k+1} x_k e_{i_1 \ldots \hat{i}_k \ldots i_{p+1}}.$$

a) In this way we get a free resolution of the R-module $R/(x_1, \ldots, x_n)$.

b) If $(R, \mathfrak{m})$ is a regular local ring and $\{x_1, \ldots, x_n\}$ is a regular parameter system of R, then we get a minimal free resolution of $R/\mathfrak{m}$.

6. The *Betti series* (or *Poincaré series*) of a Noetherian local ring $(R, \mathfrak{m})$ is the formal power series $P_R(T) = \sum_{i=0}^{\infty} \beta_i(R) T^i$, where the $\beta_i(R)$ are the Betti numbers of R. If R is a regular local ring of dimension d, then

$$P_R(T) = (1 + T)^d.$$

7. Let M be a finitely generated module over a Noetherian local ring R and $0 \to K \to F \to M \to 0$ an exact sequence, where F is a free R-module of finite rank.

a) If $d(R) > d(M)$, then $d(K) = d(M) + 1$.

b) If $d(R) = d(M)$, then $d(K) = d(M)$.

2. Homological characterizations of regular rings and local complete intersections

Most of the results of this section can be deduced rather quickly from the following theorem.

Theorem 2.1. (Ferrand [21], Vasconcelos [81]) Let R be a Noetherian ring, $I \neq R$ an ideal of R.

a) If I is generated by an R-regular sequence, then I/I^2 is a free R/I-module and $\text{pd}_R(R/I) < \infty$.

b) If R is local, the converse of this statement also holds.

Proof.

a) Let $I = (x_1, \ldots, x_t)$ with an R-regular sequence $\{x_1, \ldots, x_t\}$. That I/I^2 is a free R/I-module has been shown in V.5.11. We prove

$$\text{pd}_R(R/I) = t \tag{1}$$

by induction on t. For $t = 0$ there is nothing to prove. If $t > 0$, then from the exact sequence

$$0 \to R/(x_1, \ldots, x_{t-1}) \xrightarrow{\mu_{x_t}} R/(x_1, \ldots, x_{t-1}) \to R/(x_1, \ldots, x_t) \to 0,$$

if $\text{pd}_R(R/(x_1, \ldots, x_{t-1})) = t - 1$ has been shown, $\text{pd}_R(R/(x_1, \ldots, x_t)) \leq t$ must hold (1.8). If $\mathfrak{p} \in \text{Spec}(R)$, $\mathfrak{p} \supset \mathfrak{P}$ is given, then $\{x_1, \ldots, x_t\}$ is also an $R_\mathfrak{p}$-regular sequence, so $d(R_\mathfrak{p}) - d(R_\mathfrak{p}/(x_1, \ldots, x_t)R_\mathfrak{p}) = t$ by VI.3.4. From 1.12 and 1.6 it now follows that

$$t = d(R_\mathfrak{p}) - d(R_\mathfrak{p}/(x_1, \ldots, x_t)R_\mathfrak{p}) = \text{pd}_{R_\mathfrak{p}}(R_\mathfrak{p}/(x_1, \ldots, x_t)R_\mathfrak{p})$$
$$\leq \text{pd}_R(R/(x_1, \ldots, x_t))$$

and hence we obtain (1).

b) Now let $(R, \mathfrak{m})$ be a Noetherian local ring. For the ideal $I \neq R$ suppose I/I^2 is a free R/I-module of rank t and $\text{pd}_R(R/I) < \infty$. We shall deduce that I is generated by an R-regular sequence of length t. For $t = 0$ we have $I = I^2$, and by Nakayama $I = (0)$. Hence we may assume $t > 0$ and that the assertion has been proved for smaller rank. We first show that I contains a non-zerodivisor of R. This results from applying the following lemma to $M := R/I$.

Lemma 2.2. Let M be a finitely generated module over a Noetherian ring $R \neq \{0\}$. Let M have a free resolution

$$0 \to F_n \to F_{n-1} \to \cdots \to F_0 \to M \to 0,$$

where F_i have finite rank $r(F_i)$ $(i = 0, \ldots, n)$. Then the following statements are equivalent.

a) $\text{Ann}(M) \neq (0)$.

b) $\sum_{i=0}^n (-1)^i r(F_i) = 0$.

c) $\text{Ann}(M)$ contains a non-zerodivisor of R.

Proof. For $\mathfrak{p} \in \text{Ass}(R)$ we have $\text{pd}_{R_\mathfrak{p}}(M_\mathfrak{p}) < \infty$ and $d(R_\mathfrak{p}) = 0$; therefore $\text{pd}_{R_\mathfrak{p}}(M_\mathfrak{p}) = 0$ by 1.12, so $M_\mathfrak{p}$ is a free $R_\mathfrak{p}$-module. From the exact sequence

$$0 \to (F_n)_\mathfrak{p} \to (F_{n-1})_\mathfrak{p} \to \cdots \to (F_0)_\mathfrak{p} \to M_\mathfrak{p} \to 0$$

it follows by a simple induction argument (as with vector spaces) that

$$r(M_{\mathfrak{p}}) = \sum_{i=0}^{n}(-1)^i r(F_i). \tag{2}$$

a)→b). If $\mathfrak{a} := \mathrm{Ann}(M) \neq \langle 0 \rangle$, we shall show that there is a $\mathfrak{p} \in \mathrm{Ass}(R)$ with $M_{\mathfrak{p}} = \langle 0 \rangle$. Were this not so, because $M_{\mathfrak{p}}$ is free we would have $\mathfrak{a}R_{\mathfrak{p}} = (0)$ for all $\mathfrak{p} \in \mathrm{Ass}(R)$, so by III.4.6 $\mathrm{Ann}(\mathfrak{a}) \not\subset \mathfrak{p}$ for all $\mathfrak{p} \in \mathrm{Ass}(R)$. Then $\mathrm{Ann}(\mathfrak{a})$ would contain a non-zerodivisor x of R. From $x \cdot \mathfrak{a} = (0)$ would follow $\mathfrak{a} = (0)$, contradicting our assumption. Now if we choose $\mathfrak{p} \in \mathrm{Ass}(R)$ with $M_{\mathfrak{p}} = \langle 0 \rangle$, then (2) yields the formula in b).

b)→c). From b) and formula (2) follows $M_{\mathfrak{p}} = \langle 0 \rangle$, and so $\mathfrak{a}R_{\mathfrak{p}} = R_{\mathfrak{p}}$ for all $\mathfrak{p} \in \mathrm{Ass}(R)$. Then $\mathfrak{a} \not\subset \mathfrak{p}$ for all $\mathfrak{p} \in \mathrm{Ass}(R)$; that is, $\mathfrak{a}$ contains a non-zerodivisor of R, q. e. d.

Now let $x_1, \ldots, x_t \in I$ be elements whose images $\overline{x}_1, \ldots, \overline{x}_t$ in I/I^2 form a basis of this R/I-module.

We put $J := (x_2, \ldots, x_t) + I^2$. Then $\mathfrak{V}(J) = \mathfrak{V}(I)$ (zero set in $\mathrm{Spec}(R)$) and $\mu(I/J) = 1$. Let $\{\mathfrak{p}_1, \ldots, \mathfrak{p}_s\}$ be the set of maximal elements of $\mathrm{Ass}(R)$. Then $I \not\subset \bigcup_{j=1}^{s} \mathfrak{p}_j$, since I contains a non-zerodivisor of R because $t > 0$. Therefore, the hypotheses of V.4.7 are fulfilled, and there is an $a \in I$, $a \notin \bigcup_{j=1}^{s} \mathfrak{p}_j$ with $I/J = R \cdot \overline{a}$ if $\overline{a}$ is the image of a. This means that, without loss of generality, we may assume that x_1 is not a zero divisor of R (if necessary, replace x_1 by a).

Let $S := R/(x_1)$, $I' := I/(x_1)$. We shall show that I' satisfies analogous hypotheses as I. We have

$$I'/I'^2 \cong I/(x_1) + I^2 \cong I/I^2/(x_1) + I^2/I^2 \cong R/I \cdot \overline{x_2} \oplus \cdots \oplus R/I \cdot \overline{x_t}$$
$$= S/I' \cdot \overline{x_2} \oplus \cdots \oplus S/I' \cdot \overline{x_t}$$

and this is a free S/I'-module of rank $t - 1$ (if we identify R/I with S/I').

Further, it follows from $I = Rx_1 + J$ that $I/x_1 I = Rx_1/Ix_1 + J/x_1 I$. Here $Rx_1/Ix_1 \cong R/I$, since x_1 is not a zero divisor of R and $J \cap Rx_1 = x_1 I$, since x_1 is a basis element of the R/I-module I/J. It follows that $J/x_1 I \cong J/J \cap Rx_1 \cong I/Rx_1 = I'$ and $Rx_1/Ix_1 \cap J/x_1 I = \langle 0 \rangle$. Therefore $I/x_1 I \cong R/I \oplus I'$.

By 1.14 (and because $\mathrm{pd}_R(R/I) < \infty$), we have $\mathrm{pd}_S(I/x_1 I) = \mathrm{pd}_R(I) < \infty$. And since I' is a direct summand of $I/x_1 I$, we also have $\mathrm{pd}_S(I') < \infty$. From the exact sequence $0 \to I' \to S \to S/I' \to 0$ it follows that also $\mathrm{pd}_S(S/I') < \infty$. By the induction hypothesis I' is generated by an S-regular sequence of length $t - 1$; therefore, I is generated by an R-regular sequence of length t, q. e. d.

Corollary 2.3. For a Noetherian local ring $(R, \mathfrak{m})$ the following statements are equivalent.

a) R is regular.

b) $\mathrm{pd}_R(R/\mathfrak{m}) < \infty$.

If a) or b) is satisfied then $\mathrm{pd}_R(R/\mathfrak{m}) = \dim R$.

Proof. Since $\mathfrak{m}/\mathfrak{m}^2$ is a free $R/\mathfrak{m}$-module, $\mathrm{pd}_R(R/\mathfrak{m}) < \infty$ if and only if $\mathfrak{m}$ is generated by an R-regular sequence, which is equivalent to regularity of R. If $\dim R =: d$ and R is regular, then $\mathfrak{m}$ is generated by a regular sequence of length d, and it follows that $\mathrm{pd}_R(R/\mathfrak{m}) = d$ by formula (1) in the proof of 2.1.

From the homological characterization of regular local rings given by 2.3 a characterization of global regular rings now follows.

Proposition 2.4. Let R be a Noetherian ring of Krull dimension $d < \infty$. Then the following statements are equivalent.

a) R is regular.

b) Any finitely generated R-module M has projective dimension $\leq d$.

c) Any finitely generated R-module has finite projective dimension.

Proof.

a)$\to$b). Let M be a finitely generated R-module. By 1.6 it suffices to show that $\mathrm{pd}_{R_\mathfrak{p}}(M_\mathfrak{p}) \leq d$ for all $\mathfrak{p} \in \mathrm{Max}(R)$. If $\dim R_\mathfrak{p} = 0$, then $R_\mathfrak{p}$ is a field, and the statement is correct. Hence let $\dim R_\mathfrak{p} > 0$ and $x \in \mathfrak{p}R_\mathfrak{p} \setminus \mathfrak{p}^2 R_\mathfrak{p}$. Then $R_\mathfrak{p}/xR_\mathfrak{p}$ is also a regular local ring (VI.1.10) and $\dim R_\mathfrak{p}/xR_\mathfrak{p} = \dim R_\mathfrak{p} - 1$. We consider an exact sequence of $R_\mathfrak{p}$-modules

$$0 \to K \to F \to M_\mathfrak{p} \to 0$$

with a free $R_\mathfrak{p}$-module of finite rank F. Since x is not a zero divisor of K (or $K = \langle 0 \rangle$), 1.14 and the induction hypothesis show that $\mathrm{pd}_{R_\mathfrak{p}}(K) = \mathrm{pd}_{R_\mathfrak{p}/xR_\mathfrak{p}}(K/xK) \leq \dim R_\mathfrak{p} - 1$ and hence $\mathrm{pd}_{R_\mathfrak{p}}(M_\mathfrak{p}) \leq \dim R_\mathfrak{p}$.

c)$\to$a). For any $\mathfrak{p} \in \mathrm{Max}(R)$ we have $\mathrm{pd}_R(R/\mathfrak{p}) < \infty$ and consequently also $\mathrm{pd}_{R_\mathfrak{p}}(R_\mathfrak{p}/\mathfrak{p}R_\mathfrak{p}) < \infty$.

From 2.3 it follows that $R_\mathfrak{p}$ is regular.

Remark. It can be proved that for a regular Noetherian ring R with $\dim R = d < \infty$ we always have $\mathrm{pd}_R(M) \leq d$, even if M is not finitely generated. But we shall not need this. If $\mathfrak{m} \in \mathrm{Max}(R)$ with $h(\mathfrak{m}) = d$ is given, then we have shown that $\mathrm{pd}_R(R/\mathfrak{m}) = d$. The bound d for the projective dimension will therefore always be attained. We say that R has *homological dimension* d.

Corollary 2.5.

a) (Hilbert's Syzygy Theorem) If K is a field, then any finitely generated module M over the polynomial ring $K[X_1, \ldots, X_n]$ has a free resolution of length $\leq n$.

b) If K is a principal ideal domain, then any finitely generated $K[X_1, \ldots, X_n]$-module has a free resolution of length $\leq n + 1$.

Proof. $K[X_1, \ldots, X_n]$ is a regular Noetherian ring of dimension n in case a), of dimension $n + 1$ in case b). Since finitely generated projective $K[X_1, \ldots, X_n]$-modules are free, the assertion follows from 2.4b).

Corollary 2.6. If a Noetherian ring R is regular (resp. locally a complete intersection), then so is any ring of fractions R_S.

Proof. If suffices to show that $R_{\mathfrak{p}}$ is regular (resp. is a complete intersection) for all $\mathfrak{p} \in \operatorname{Spec}(R)$. Let $\mathfrak{m}$ be a maximal ideal of R containing $\mathfrak{p}$. If R is regular, then by 1.6 and 2.4

$$\operatorname{pd}_{R_{\mathfrak{p}}}(R_{\mathfrak{p}}/\mathfrak{p}R_{\mathfrak{p}}) \leq \operatorname{pd}_{R_{\mathfrak{m}}}(R_{\mathfrak{m}}/\mathfrak{p}R_{\mathfrak{m}}) < \infty,$$

and therefore $R_{\mathfrak{p}}$ is regular by 2.3.

If R is locally a complete intersection, then $R_{\mathfrak{m}} \cong A/I$, where A is a regular local ring and the ideal $I \subset A$ is generated by an A-regular sequence. Further, $R_{\mathfrak{p}} \cong (R_{\mathfrak{m}})_{\mathfrak{p}R_{\mathfrak{m}}} \cong A_{\mathfrak{P}}/IA_{\mathfrak{P}}$, if $\mathfrak{P}$ is the inverse image of $\mathfrak{p}R_{\mathfrak{m}}$ in A. As already shown, $A_{\mathfrak{P}}$ is regular. Since $\mathfrak{P} \supset I$, $IA_{\mathfrak{P}}$ is generated by an $A_{\mathfrak{P}}$-regular sequence. This shows that $R_{\mathfrak{p}}$ is a complete intersection.

Corollary 2.7. If a Noetherian ring R is regular (resp. locally a complete intersection), then so is the polynomial ring $R[X_1, \ldots, X_n]$.

Proof. Let $\mathfrak{P} \in \operatorname{Spec}(R[X_1, \ldots, X_n])$ and $\mathfrak{p} := \mathfrak{P} \cap R$. Then $R[X_1, \ldots, X_n]_{\mathfrak{P}} = R_{\mathfrak{p}}[X_1, \ldots, X_n]_{\mathfrak{P}}$, and the maximal ideal of this ring intersects $R_{\mathfrak{p}}$ in the maximal ideal $\mathfrak{p}R_{\mathfrak{p}}$. If $R_{\mathfrak{p}}$ is regular, then so is $R[X_1, \ldots, X_n]_{\mathfrak{P}}$ by VI.1.7 and induction on n.

If $R_{\mathfrak{p}} \cong A/I$ with a regular local ring A and an ideal I generated by an A-regular sequence, then $R_{\mathfrak{p}}[X_1, \ldots, X_n]_{\mathfrak{P}} \cong A[X_1, \ldots, X_n]_{\Omega}/IA[X_1, \ldots, X_n]_{\Omega}$, where Ω is the pre-image of $\mathfrak{P}$ in $A[X_1, \ldots, X_n]$. As already shown, the ring $A[X_1, \ldots, X_n]_{\Omega}$ is regular. Further, it is clear that $IA[X_1, \ldots, X_n]_{\Omega}$ is generated by an $A[X_1, \ldots, X_n]_{\Omega}$-regular sequence; in fact, the A-regular sequence generating I is also $A[X_1, \ldots, X_n]_{\Omega}$-regular. Therefore, $R[X_1, \ldots, X_n]_{\mathfrak{P}}$ is a complete intersection.

On the basis of 2.4, Theorem 2.1 yields the following characterization of complete intersections in regular rings.

Corollary 2.8. For an ideal $I \neq R$ of a regular Noetherian ring R the following statements are equivalent.

a) I is locally a complete intersection in R.

b) The conormal module I/I^2 is a projective R/I-module.

If the conormal module I/I^2 is a free R/I-module, then we can often conclude that I is globally a complete intersection (Mohan Kumar [56]):

Corollary 2.9. Let I be an ideal of the polynomial ring $R = K[X_1, \ldots, X_n]$ over a field. If I/I^2 is a free R/I-module and $2 \cdot \dim R/I + 2 \leq n$, then I is a complete intersection in R.

Proof. Let r be the rank of I/I^2. By 2.8 I is locally a complete intersection in R, and so $r = h(I) = n - \dim R/I$. From $\mu(I/I^2) = n - \dim R/I \geq \dim R/I + 2$ follows $\mu(I) = \mu(I/I^2) = r$ by V.5.20. Therefore I is also globally a complete intersection.

Without the assumption that $2 \cdot \dim R/I + 2 \leq n$, it can be shown that I is a set-theoretic complete intersection (see: M. Boratyński, A note on set-theoretic complete intersection ideals, *J. Alg.* **54**, 1–5 (1978)).

Exercises

1. Let $R \subset S$ be Noetherian rings of finite Krull dimension, where S is a free R-module. If S is regular, then so is R. (Hint: Apply the characterization of regularity in 2.4 and the formula in §1, Exercise 2.)

2. Let $V \subset A^n(L)$ be a smooth algebraic variety, I its ideal in $L[X_1, \ldots, X_n]$. If $K \subset L$ is a field of definition of I (Ch. I, §2, Exercise 9), then any $x \in V$ is also a K-regular point of V.

3. For a module M over a ring R denote by $[M]$ the isomorphism class of M. Let $\mathcal{F}_R$ be the free Abelian group on the isomorphism classes of finitely generated R-modules, and $U \subset \mathcal{F}_R$ the subgroup generated by the elements $[M_2] - [M_1] - [M_3]$ for which there is an exact sequence

$$0 \to M_1 \to M_2 \to M_3 \to 0. \qquad (*)$$

$\mathbf{K}(R) := \mathcal{F}_R/U$ is called the *Grothendieck group* of R.

   It has the following universal property. If χ assigns to each $[M]$ (M finitely generated) an element of an Abelian group G in such a way that

$$\chi([M_2]) = \chi([M_1]) + \chi([M_3])$$

for modules in an exact sequence $(*)$, then there is a unique group homomorphism $\epsilon : \mathbf{K}(R) \to G$ with $\epsilon([M] + U) = \chi([M])$ for all $[M]$.

4. We now consider the isomorphism classes $[P]$ of finitely generated projective R-modules and the group $P(R)$ constructed from them in the same way as $\mathbf{K}(R)$. We have a canonical homomorphism $\alpha : P(R) \to \mathbf{K}(R)$ that assigns to the residue class of $[P]$ in $P(R)$ the corresponding residue class in $\mathbf{K}(R)$.

   If R is a regular Noetherian ring of finite Krull dimension, then α is an isomorphism. (Hint: For a finitely generated R-module M choose a projective resolution $0 \to P_n \to P_{n-1} \to \cdots \to P_0 \to M \to 0$ and assign to the class of $[M]$ in $\mathbf{K}(R)$ the class of $\sum_{i=0}^{n}(-1)^i[P_i]$ in $P(R)$.)

5. If K is a principal ideal domain, then

$$\mathbf{K}(K[X_1, \ldots, X_n]) \cong P(K[X_1, \ldots, X_n]) \cong \mathbf{Z}.$$

3. Modules of projective dimension ≤ 1

About these we can make more precise statements than about arbitrary modules of finite projective dimension. We begin with an example in which such modules occur.

Example 3.1. In a Cohen–Macaulay ring R let there be given an ideal $I \neq R$ that is locally a complete intersection, where $h(I_\mathfrak{p}) = 2$ for all $\mathfrak{p} \in \mathfrak{V}(I)$. Then $I_\mathfrak{p}$ is generated by an $R_\mathfrak{p}$-regular sequence of length 2 (VI.3.14). By §2, (1) we have $\mathrm{pd}_{R_\mathfrak{p}}(R_\mathfrak{p}/I_\mathfrak{p}) = 2$ and so by 1.5b) $\mathrm{pd}_{R_\mathfrak{p}}(I_\mathfrak{p}) = 1$. Then also $\mathrm{pd}_R(I) = 1$ by 1.6.

In particular, this holds for an ideal $I \subset K[X_1, X_2, X_3]$ (K a field) that defines a space curve and is locally a complete intersection, say the vanishing ideal of a space curve that is locally a complete intersection.

The results of the present section will be applied to this case in §4.

In what follows let R be a Noetherian ring $\neq \{0\}$ and M a finitely generated R-module with $\mathrm{pd}(M) \leq 1$. We consider a projective resolution

$$0 \to P_1 \xrightarrow{\alpha} P_0 \xrightarrow{\epsilon} M \to 0, \tag{1}$$

where the P_i ($i = 0, 1$) are finitely generated. By passing to dual modules and transposed mappings ($M^* := \mathrm{Hom}_R(M, R), \alpha^* = \mathrm{Hom}_R(\alpha, R)$) we get an exact sequence

$$0 \to M^* \xrightarrow{\epsilon^*} P_0^* \xrightarrow{\alpha^*} P_1^* \to E \to 0, \tag{1*}$$

where $E := \mathrm{Coker}(\alpha^*)$.

If along with (1) we are given a similar resolution $0 \to \overline{P}_1 \xrightarrow{\overline{\alpha}} \overline{P}_0 \xrightarrow{\overline{\epsilon}} M \to 0$ and $\overline{E} := \mathrm{Coker}(\overline{\alpha}^*)$, then:
Lemma 3.2. $\overline{E} \cong E$.

Proof. Consider the exact sequence

$$0 \to P \xrightarrow{\gamma} P_0 \oplus \overline{P}_0 \xrightarrow{\delta} M \to 0,$$

where $\delta(x, y) = \epsilon(x) + \overline{\epsilon}(y)$ and $P := \mathrm{Ker}(\delta)$. By 1.5 this too is a projective resolution of M. It suffices to show that E and $\overline{E}$ are isomorphic to $\mathrm{Coker}(\gamma^*)$. Therefore, we may assume that there is an epimorphism

$$\varphi : \overline{P}_0 \to P_0$$

with $\epsilon \cdot \varphi = \overline{\epsilon}$. If $K := \mathrm{Ker}(\varphi)$, then we get the following diagrams with exact rows and columns:

$$
\begin{array}{cc}
& \begin{array}{cc} 0 & 0 \\ \downarrow & \downarrow \end{array} \\
0 \to K \to \overline{P}_1 \to P_1 \to 0 \\
\| \quad \downarrow \quad \downarrow \\
0 \to K \to \overline{P}_0 \xrightarrow{\varphi} P_0 \to 0 \\
\downarrow \overline{\epsilon} \quad \downarrow \epsilon \\
M = M \\
\downarrow \quad \downarrow \\
0 \quad 0
\end{array}
\qquad\qquad
\begin{array}{cc}
0 \quad 0 \\
\downarrow \quad \downarrow \\
M^* = M^* \\
\downarrow \quad \downarrow \\
0 \to P_0^* \to \overline{P}_0^* \to K^* \to 0 \\
\downarrow \quad \downarrow \quad \| \\
0 \to P_1^* \to \overline{P}_1^* \to K^* \to 0 \\
\downarrow \quad \downarrow \\
E \quad \overline{E} \\
\downarrow \quad \downarrow \\
0 \quad 0
\end{array}
$$

in which the rows of the second diagram are exact because the rows split in the first diagram. From the Snake Lemma it now follows that $E \cong \overline{E}$.

In the future we shall denote by $E(M)$ the unique (up to isomorphism) module E assigned to M. From the definition of $E(M)$ and the compatibility of localization with exact sequences and dualization, it follows at once that

$$E(M_{\mathfrak{p}}) = E(M)_{\mathfrak{p}}, \qquad \text{for all} \qquad \mathfrak{p} \in \mathrm{Spec}(R). \tag{2}$$

Lemma 3.3. M is projective if and only if $E(M) = \langle 0 \rangle$.

Proof. If M is projective, then (1) splits, and it follows that $E(M) = \langle 0 \rangle$. Conversely, if $E(M) = \langle 0 \rangle$, then the exact sequence (1^*) $0 \to M^* \to P_0^* \to P_1^* \to 0$ splits, since if P_1 is projective, so is the dual module P_1^* (IV.3.17a)). This shows that M^* is projective.

From the commutative diagram with exact rows

$$
\begin{array}{ccccccccc}
0 & \to & P_0 & \to & P_1 & \to & M & \to & 0 \\
& & \downarrow & & \downarrow & & \downarrow & & \\
0 & \to & P_0^{**} & \to & P_1^{**} & \to & M^{**} & \to & 0
\end{array}
$$

in which the lower row arises by dualizing (1^*) and the vertical arrows are the canonical mappings into the bidual modules, it follows that $M \cong M^{**}$, since the $P_i \to P_i^{**}$ $(i = 0, 1)$ are isomorphisms (IV.3.17b)). Since M^* is projective also M^{**} and hence M is projective.

Corollary 3.4. If $M = I$ is an ideal of R with $\mathrm{pd}(I) \leq 1$, then $\mathrm{Supp}(E(I)) \subset \mathfrak{V}(I)$.

Indeed, if $\mathfrak{p} \in \mathrm{Spec}(R) \setminus \mathfrak{V}(I)$, then $I_{\mathfrak{p}} = R_{\mathfrak{p}}$ and $E(I)_{\mathfrak{p}} \cong E(I_{\mathfrak{p}}) = \langle 0 \rangle$ by 3.3, so $\mathfrak{p} \notin \mathrm{Supp}(E(I))$.

The following proposition (essentially going back to Serre [73]) shows that $E(M)$ often also contains information on the number of generators of M. Note that for any minimal prime ideal $\mathfrak{p}$ of R from the formula 1.12 of Auslander–Buchsbaum

$$\mathrm{pd}_{R_\mathfrak{p}}(M_\mathfrak{p}) + d(M_\mathfrak{p}) = d(R_\mathfrak{p}) = 0,$$

it follows that $\mathrm{pd}_{R_\mathfrak{p}}(M_\mathfrak{p}) = 0$, so $M_\mathfrak{p}$ is a free $R_\mathfrak{p}$-module. If all finitely generated projective R-modules are free, then §2, (2), shows that the rank r of $M_\mathfrak{p}$ is independent of the minimal prime ideal $\mathfrak{p}$ chosen. We then call r the rank of M.

Proposition 3.5. Let R be a Noetherian ring over which all finitely generated projective modules are free. Let M be a finitely generated R-module with $\mathrm{pd}(M) \leq 1$ that has rank r. Then the following statements are equivalent.

a) $E(M)$ is generated by s elements.

b) M is generated by $r + s$ elements.

Proof.

b)→a). If M is generated by $r + s$ elements, then there is a free resolution (1), where P_0 has rank $r + s$. Let $\mathfrak{p}$ be a minimal prime ideal of R. Then

$$0 \to (P_1)_\mathfrak{p} \to (P_0)_\mathfrak{p} \to M_\mathfrak{p} \to 0$$

is a split exact sequence, and it follows that rank $(P_1)_\mathfrak{p} = \mathrm{rank}(P_1) = s$. Since P_1^* is also free of rank s, (1^*) shows that $E(M)$ is generated by s elements.

a)→b). Suppose $E(M)$ is generated by s elements. We choose a free R-module $\overline{P}_1$ of rank s and a surjective R-linear mapping $\gamma : \overline{P}_1^* \to E(M)$. Further, let there be given an arbitrary sequence (1). Then there is also a linear mapping $\beta : P_1 \to \overline{P}_1$ such that the diagram

$$\overline{P}_1^* \xrightarrow{\ \beta^*\ } P_1^*$$
$$\gamma \searrow \quad \swarrow \tau$$
$$E(M)$$

is commutative, where τ is the mapping occurring in (1^*).

If $\overline{P}_0 := P_0 \amalg_{P_1} \overline{P}_1$ is the fiber sum formed with respect to $\alpha : P_1 \to P_0$ and $\beta : P_1 \to \overline{P}_1$, then we have a commutative diagram with exact rows and columns

$$
\begin{array}{ccc}
0 & & 0 \\
\downarrow & & \downarrow \\
P_1 & = & P_1 \\
\downarrow \eta & & \downarrow \alpha \\
0 \to \overline{P}_1 \to P_0 \oplus \overline{P}_1 \to P_0 \to 0 \\
\| \qquad\qquad \downarrow \qquad\qquad \downarrow \epsilon \\
0 \to \overline{P}_1 \to \overline{P}_0 \to M \to 0 \\
\downarrow \qquad\qquad \downarrow \\
0 \qquad\qquad 0
\end{array}
$$

where $\eta(x) = (\alpha(x), -\beta(x))$ for all $x \in P_1$. Obviously $\mathrm{pd}(\overline{P}_0) \leq 1$. We shall see that $E(\overline{P}_0) = \langle 0 \rangle$, so $\overline{P}_0$ is projective (and therefore free) (3.3). If $\mathfrak{p}$ is a minimal prime ideal of R, because rank $(\overline{P}_1) = s$ the exact sequence $0 \rightarrow (\overline{P}_1)_\mathfrak{p} \rightarrow (\overline{P}_0)_\mathfrak{p} \rightarrow M_\mathfrak{p} \rightarrow 0$ then shows that $\mathrm{rank}(\overline{P}_0) = r + s$. Therefore, M is generated by $r + s$ elements.

We pass to the dual of the diagram above:

$$
\begin{array}{ccccc}
& & 0 & & 0 \\
& & \downarrow & & \downarrow \\
0 \rightarrow & M^* & \rightarrow & \overline{P}_0^* & \rightarrow & \overline{P}_1^* \\
& & \downarrow & & \downarrow \\
0 \rightarrow & P_0^* & \rightarrow & P_0^* \oplus \overline{P}_1^* & \rightarrow \overline{P}_1^* \rightarrow 0 \\
& & \downarrow & & \downarrow \eta^* \\
& P_1^* & = & P_1^* \\
& & \downarrow \tau & & \downarrow \\
& E(M) & & E(\overline{P}_0) \\
& & \downarrow & & \downarrow \\
& & 0 & & 0
\end{array}
$$

The Snake Lemma yields an exact sequence

$$ 0 \rightarrow M^* \rightarrow \overline{P}_0^* \rightarrow \overline{P}_1^* \xrightarrow{\delta} E(M) \rightarrow E(\overline{P}_0) \rightarrow 0. $$

Here, if $z \in \overline{P}_1^*$, by the construction of the connecting homomorphism in 1.9 we have $\delta(z) = \tau(\eta^*(0, z)) = \tau(-\beta^*(z)) = -\gamma(z)$. Therefore $\delta = -\gamma$ is surjective, and so $E(\overline{P}_0) = \langle 0 \rangle$, q. e. d.

As a first application we get a sharpening of the Forster–Swan Theorem (IV.2.14) in a special case. Observe that if under the hypotheses of 3.5 the module M is an ideal $\neq (0)$ in R, then M has rank 1.

Corollary 3.6. Let R be a Noetherian ring over which any finitely generated projective module is free. If an ideal $I \neq R$ of R with $\mathrm{pd}(I) \leq 1$ and $\dim R/I =: d$ is everywhere locally generated by s elements, then it is globally generated by $s + d$ elements.

Proof. We need only consider the case $I \neq (0)$. By 3.5 for all $\mathfrak{p} \in \mathrm{Spec}(R), E(I_\mathfrak{p})$ is generated by $s - 1$ elements. Since $\mathrm{Supp}(E(I)) \subset \mathfrak{V}(I)$ by 3.4, by the Forster–Swan Theorem $E(I)$ is globally generated by $s - 1 + d$ elements. Another application of 3.5 shows that I is globally generated by $s + d$ elements.

The corollary contains the statement that the ideal of an affine curve in $\mathbb{A}^3$ that is locally a complete intersection is generated by 3 elements; that has been shown more generally in V.5.22.

Using the module $E(I)$ one can decide in certain cases whether I is locally or globally a complete intersection.

Corollary 3.7. Let R be a Cohen–Macaulay ring. For an ideal $I \neq R$ with $h(I_{\mathfrak{p}}) = 2$ for all $\mathfrak{p} \in \mathfrak{V}(I)$ the following statements are equivalent.

a) I is locally a complete intersection.

b) $\mathrm{pd}(I) \leq 1$ and $E(I)$ is locally generated by one element.

c) $\mathrm{pd}(I) \leq 1$ and $E(I)$ is a projective R/I-module of rank 1.

Proof. b) follows trivially from c); using 3.5 it follows from b) that I is locally generated by two elements, so it is locally a complete intersection. Hence it only remains to prove a)$\rightarrow$c).

We start from a) and assume that R is local. Then $I = (x_1, x_2)$ for some R-regular sequence $\{x_1, x_2\}$ (VI.3.14). The sequence of R-modules

$$0 \to R \xrightarrow{\alpha} R \oplus R \xrightarrow{\epsilon} I \to 0$$

with $\epsilon(r_1, r_2) = r_1 x_1 + r_2 x_2, \alpha(1) = (-x_2, x_1)$ is exact: Obviously $\mathrm{Im}(\alpha) \subset \mathrm{Ker}(\epsilon)$. Further, if $\epsilon(r_1, r_2) = 0$, then there is an $r \in R$ with $r_1 = x_2 r$ and $r_2 = -x_1 r$ since x_1 is not a zero divisor and x_2 is not a zero divisor mod (x_1). Hence also $\mathrm{Ker}(\epsilon) \subset \mathrm{Im}(\alpha)$.

For the transposed mapping $\alpha^* : R \oplus R \to R$ we have $\alpha^*(r_1, r_2) = -r_1 x_2 + r_2 x_1$, whence $\mathrm{Im}(\alpha^*) = I$ and $E(I) = \mathrm{Coker}(\alpha^*) = R/I$.

Now if R is a global ring, then for all $\mathfrak{m} \in \mathrm{Max}(R)$

$$(I \cdot E(I))_{\mathfrak{m}} = I_{\mathfrak{m}} \cdot E(I_{\mathfrak{m}}) \cong I_{\mathfrak{m}} \cdot R_{\mathfrak{m}}/I_{\mathfrak{m}} = \langle 0 \rangle$$

and therefore $IE(I) = \langle 0 \rangle$. $E(I)$ is an R/I-module and as such, as shown, is locally free of rank 1, q. e. d.

Corollary 3.8. Let R be a Cohen–Macaulay ring over which all finitely generated projective modules are free. For an ideal $I \neq R$ of height 2 in R the following statements are equivalent.

a) I is (globally) a complete intersection.

b) $\mathrm{pd}(I) \leq 1$ and $E(I) \cong R/I$.

c) $\mathrm{pd}(I) < \infty$ and I/I^2 is a free R/I module.

Proof. b) follows from a) as in the proof of 3.7; here $E(I)$ is a free R/I module of rank 1 because it is projective of rank 1 and (by 3.5) is generated by one element. By 3.5, a) follows from b); and c) follows from a) by 2.1. Hence it only remains to deduce b) from c).

By 2.1 c) implies that I is locally a complete intersection; thus $\mathrm{pd}(I) \leq 1$ and $E(I)$ is defined. Further, the R/I-module I/I^2 has rank 2 and $I \cdot E(I) = \langle 0 \rangle$ as was shown in the proof of 3.7.

We choose $x_1, x_2 \in I$ such that their images in I/I^2 form a basis of this R/I-module. There is then an $x_0 \in I^2$ such that $I = (x_0, x_1, x_2)$ (V.5.16). We have a free resolution

$$0 \to R^2 \xrightarrow{\alpha} R^3 \xrightarrow{\epsilon} I \to 0 \qquad (3)$$

with $\epsilon(r_0, r_1, r_2) = r_0 x_0 + r_1 x_1 + r_2 x_2$ for all $(r_0, r_1, r_2) \in R^3$. Here $\alpha(1,0) = (a_0, a_1, a_2)$, $\alpha(0,1) = (a'_0, a'_1, a'_2)$ with $a_1, a_2, a'_1, a'_2 \in I$ according to the choice of x_1, x_2.

(3) yields an exact sequence (modulo I)

$$(R/I)^2 \xrightarrow{\overline{\alpha}} (R/I)^3 \xrightarrow{\overline{\epsilon}} I/I^2 \to 0.$$

Here $\mathrm{Im}(\overline{\alpha}) = \mathrm{Ker}(\overline{\epsilon})$ is the R/I-module spanned by $(1,0,0)$. If we denote by $\overline{a}_0, \overline{a}'_0$ the images of a_0, a'_0 in R/I we see that there are elements $\overline{s}_1, \overline{s}_2 \in R/I$ with $-\overline{s}_1 \overline{a}_0 + \overline{s}_2 \overline{a}'_0 = 1$.

From the dual sequence (3^*) of (3) we get, because $IE(I) = \langle 0 \rangle$, an exact sequence (modulo I):

$$(R/I)^3 \xrightarrow{\overline{\alpha}^*} (R/I)^2 \to E(I) \to 0.$$

Here $\overline{\alpha}^*(1,0,0) = (\overline{a}_0, \overline{a}'_0)$ and $\overline{\alpha}^*(0,1,0) = \overline{\alpha}^*(0,0,1) = 0$. Since

$$\det \begin{pmatrix} \overline{a}_0 & \overline{a}'_0 \\ \overline{s}_1 & \overline{s}_2 \end{pmatrix} = 1$$

we can extend $(\overline{a}_0, \overline{a}'_0)$ to a basis of $(R/I)^2$. In fact, it now follows that $E(I) \cong R/I$.

In particular, the corollary can be applied when $R = K[X_1, \ldots, X_n]$ is the polynomial ring over a field. If, for an ideal I of height 2 in R, the R/I-module I/I^2 is free, then I is (globally) a complete intersection in R. For $n \leq 4$ this statement holds for ideals of arbitrary height, since for $h(I) \leq 1$ the statement is easy to prove, otherwise 2.9 or 3.8 can be applied.

We shall prove now a duality statement, which will play an important role in the next section.

Proposition 3.9. Let R be a Cohen-Macaulay ring, $I \neq R$ an ideal in R that is locally a complete intersection with $h(I_\mathfrak{p}) = 2$ for all $\mathfrak{p} \in \mathfrak{V}(I)$. Let

$$0 \to P_1 \xrightarrow{\alpha} P_0 \xrightarrow{\epsilon} I \to 0$$

be a projective resolution of I with finitely generated modules P_i $(i = 0, 1)$. In the corresponding dual sequence

$$0 \to I^* \xrightarrow{\epsilon^*} P_0^* \xrightarrow{\alpha^*} P_1^* \to E(I) \to 0$$

let $K := \mathrm{Im}(\alpha^*)$. Then $\mathrm{pd}(K) \leq 1$ and $E(K) \cong R/I$.

Proof.

a) We first show that the transposed mapping $i^* : R^* \to I^*$ corresponding to the inclusion $i : I \to R$ is an isomorphism. It suffices to prove this locally for all $\mathfrak{m} \in \mathrm{Max}(R)$. Hence we may assume that R is a local ring and $I = (x_1, x_2)$ is generated by an R-regular sequence $\{x_1, x_2\}$. It is clear that i^* is injective, since I contains a non-zerodivisor of R. The assertion results if we can show that any linear form $l : I \to R$ is multiplication by some $r \in R$. But it follows from $x_1 l(x_2) - x_2 l(x_1) = 0$ and because $\{x_1, x_2\}$ is a regular sequence, that x_i is a divisor of $l(x_i)$ ($i = 1, 2$) and indeed $l(x_i) = r x_i$ for some $r \in R$ independent of i. Therefore l is in fact multiplication by r. Since the diagram

$$
\begin{array}{ccc}
I & \xrightarrow{\ i\ } & R \\
\downarrow & & \downarrow{\wr} \\
I^{**} & \xrightarrow[i^{**}]{\ \sim\ } & R^{**}
\end{array}
$$

in which the vertical arrows denote the canonical homomorphism into the bidual modules, is commutative, we may identify $I \to I^{**}$ with the inclusion $i : I \to R$.

b) Because $I^* \cong R$, the exact sequence $0 \to I^* \xrightarrow{\varepsilon} P_0^* \to K \to 0$ shows that $\mathrm{pd}(K) \leq 1$. Therefore $E(K)$ is defined. In the commutative diagram with exact rows

$$
\begin{array}{ccccccccc}
0 & \to & P_1 & \to & P_0 & \to & I & \to & 0 \\
& & & & \downarrow & & \downarrow & & \\
0 & \to & K^* & \to & P_0^{**} & \to & I^{**} & \to & E(K) \to 0
\end{array}
$$

$P_0 \to P_0^{**}$ is an isomorphism and $I \to I^{**}$ is identified with $i : I \to R$. It follows that $E(K) \cong R/I$.

Exercises

1. Let R be a 2-dimensional regular ring over which all finitely generated projective modules are free. Any maximal ideal of R is generated by 2 elements.

2. In the polynomial ring $R := K[X_1, \ldots, X_n]$ in $n \geq 2$ variables over a field K let there be given an ideal $I = (f_1, \ldots, f_m)$ for which R/I is a Cohen–Macaulay ring of dimension $n - 2$. Let $0 \to F \to R^m \to I \to 0$ be the presentation corresponding to $\{f_1, \ldots, f_m\}$.

    a) $F \subset R^m$ is a free R-module of rank $m - 1$.

    b) For a basis $\{v_1, \ldots, v_{m-1}\}$ of F put $\tilde{f}_i := \det(v_1, \ldots, v_{m-1}, e_i)$ ($i = 1, \ldots, m$) where e_i is the i-th canonical basis element of R^m. There is an $r \in K(X_1, \ldots, X_n)$ with $\tilde{f}_i = r f_i$ ($i = 1, \ldots, m$).

    c) In fact, $r \in K \setminus \{0\}$. (Hint: Write $r = p/q$ in lowest terms with $p, q \in R$. Then $q \in K \setminus \{0\}$, since $h(I) = 2$. And $p \in K \setminus \{0\}$, since otherwise $v_1, \ldots, v_{m-1}$ would be linearly dependent over R.)

The structure theorem given by this exercise says, in other words, that the ideals I of $K[X_1,\ldots,X_n]$ considered have the following property. For any generating system $\{f_1,\ldots,f_m\}$ of I there is an $(m-1) \times m$-matrix with coefficients in $K[X_1,\ldots,X_n]$ such that the f_i are the maximal minors of this matrix.

3. Write the generators of the ideals in V.3.13, Example f) that are not generated by 2 elements as the 2×2-subdeterminants of a 2×3-matrix.

4. (R. Waldi) In this exercise a smooth curve $C \subset \mathsf{A}^3(L)$ is to be constructed that is not an ideal-theoretic complete intersection.

Let K be a subfield of L. Give the polynomial ring $K[X,Y]$ the grading in which $\deg X = -3$, $\deg Y = -4$. The leading form $L(G)$ of a polynomial $G \neq 0$ in $K[X,Y]$ is its homogeneous component of lowest degree with respect to this grading. $\deg G$ is the degree of $L(G)$.

Let $A := K[X,Y]/(F)$ with $F := XY + Y^3 - X^4$ and $x := X + (F)$, $y := Y + (F)$. For $g \in A \setminus \{0\}$ let $\deg(g)$ be defined as the maximum of the degrees of all representatives of the residue class g in $K[X,Y]$. And put $\deg(0) = \infty$.

a) A representative G of $g \in A \setminus \{0\}$ has the same degree as g if and only if $L(G)$ is not divisible by $L(F) = Y^3 - X^4$.

b) For $g, h \in A$ we have $\deg(gh) = \deg(g) + \deg(h)$; in particular, A is an integral domain (and so F is irreducible).

c) If $\deg(g) < -4$, then $\{1, x, y\}$ is K-linearly independent modulo gA. In general, for any $g \in A \setminus K$

$$\dim_K(A/gA) \geq 3.$$

d) In $Q(A)$ consider $z := y^2/x$ and $\overline{A} := A[z]$. Then $\overline{A} = A \oplus Kz$ and $(x,y)A$ is the conductor $f_{\overline{A}/A}$ of $\overline{A}$ along A. (Ch. IV, §1, Exercise 6.)

e) The kernel I of the K-epimorphism $\varphi : K[X,Y,Z] \to \overline{A}$ with $\varphi(X) = x$, $\varphi(Y) = y$, $\varphi(Z) = z$ is generated by the polynomials

$$\Delta_1 := X^3 - Y(Z+1), \quad \Delta_2 := Y^2 - ZX, \quad \Delta_3 := Z(Z+1) - X^2Y.$$

f) The zero set C of I in $\mathsf{A}^3(L)$ is an irreducible K-regular curve. (Hint: Show this in the case $K = L$ by using the Jacobian criterion and applying §2, Exercise 2.)

g) We have a free resolution (with $R := K[X,Y,Z]$)

$$0 \to R^2 \xrightarrow{\alpha} R^3 \xrightarrow{\epsilon} I \to 0,$$

where α is given by the matrix $\left(\begin{smallmatrix} Z & X^2 & Y \\ Y & Z+1 & X \end{smallmatrix}\right)$ and $\epsilon(e_i) = \Delta_i$ $(i = 1, 2, 3)$ (cf. Exercise 2).

h) By dualizing the sequence in g) and passing to $\overline{A}$ we get an exact sequence of $\overline{A}$-modules (see formula (4) in the proof of 3.8)

$$\overline{A}^3 \xrightarrow{\overline{\alpha}^*} \overline{A}^2 \to E(I) \to 0.$$

The $\overline{A}$-linear mapping $\overline{A}^2 \to \overline{A}$ ($e_1 \mapsto -x, e_2 \mapsto y$) induces an isomorphism $E(I) \xrightarrow{\sim} f_{\overline{A}/A}$ of $\overline{A}$-modules.

i) $f_{\overline{A}/A}$ is not a principal ideal in $\overline{A}$, hence $E(I)$ is not generated by one element and C is not an ideal-theoretic complete intersection. (Apply c) and note that $\dim_K(\overline{A}/A) < \infty$ so that $\dim_K(\overline{A}/g\overline{A}) = \dim_K(A/gA)$ for all $g \in A \setminus \{0\}$.)

k) We have $I^2 \subset (\Delta_2, X^2\Delta_1 + (Z+1)\Delta_3) \subset I$; therefore, C is the intersection of the surfaces with equations

$$\Delta_2 = 0, X^2\Delta_1 + (Z+1)\Delta_3 = 0.$$

5. In the case $L = \mathbb{C}$ start with $F := X^2 + Y^3 - 2X^3$ and as in Exercise 4, construct a curve $C \subset \mathsf{A}^3(\mathbb{C})$ whose coordinate ring is the integral closure $\overline{A}$ of $A := \mathbb{Q}[X,Y]/(F)$ in its field of fractions. Just as there show:

a) Over any subfield $K \subset \mathbb{C}, C$ is an irreducible K-regular curve.

b) C is not an ideal-theoretic complete intersection over $\mathbb{Q}$, but it is over $\mathbb{C}$.

c) C is the intersection of two surfaces defined over $\mathbb{Q}$.

4. Algebraic curves in A^3 that are locally complete intersections can be represented as the intersection of two algebraic surfaces

In the proof of this theorem we use many previous results: the theorem of Quillen–Suslin, Serre's Splitting-off Theorem, and the statements in §3 on modules of projective dimension ≤ 1. Moreover, the following construction (taken from algebraic deformation theory) plays a decisive role.

Let R be a Cohen–Macaulay ring, $I \neq R$ an ideal of R, P a projective R/I-module of rank 1, and $\pi : I/I^2 \to P$ an epimorphism of R/I-modules. We can form the fiber sum $S := P \amalg_{I/I^2} R/I^2$ with respect to π and the inclusion $i : I/I^2 \to R/I^2$. Using the rules III.5.3 we get a commutative diagram with exact rows

$$
\begin{array}{ccccccccc}
0 \to & I/I^2 & \xrightarrow{i} & R/I^2 & \to & R/I & \to 0 \\
 & \downarrow{\pi} & & \downarrow & & \| & \\
0 \to & P & \to & S & \to & R/I & \to 0,
\end{array}
\tag{1}
$$

where $R/I^2 \to S$ is also an epimorphism. Hence $S = R/J$ for some ideal J for which $I^2 \subset J \subset I$; so, in particular, $\mathfrak{V}(I) = \mathfrak{V}(J)$.

Proposition 4.1. If, under the above hypotheses, I is locally a complete intersection, so is J.

Proof. The formation of the fiber sum commutes with localization (III.5.3e)). Hence we may assume that R is a local ring. I/I^2 is then a free R/I-module, and P is free of rank 1. Hence we can find a generating system $\{x_1, \ldots, x_n\}$ of I such that the images $\bar{x}_i \in I/I^2$ form a basis of I/I^2 and $\{\bar{x}_2, \ldots, \bar{x}_n\}$ is a basis of $\mathrm{Ker}(\pi)$.

By III.5.3 i maps the kernel of π isomorphically onto the kernel J/I^2 of $R/I^2 \to S$. Therefore $J = (x_2, \ldots, x_n) + I^2 = (x_1^2, x_2, \ldots, x_n)$. Since I and J have the same height, if I is a complete intersection, so is J.

We now apply the construction in the case where R is a 3-dimensional Cohen–Macaulay ring over which every finitely generated projective module is free, and where $I_{\mathfrak{p}}$ is of height 2 and is a complete intersection for all $\mathfrak{p} \in \mathfrak{V}(I)$.

Then I/I^2 is a projective R/I-module of rank 2, and by 3.7 $E(I)$ is projective of rank 1. Since $\dim R/I = 1$, by Serre's Splitting-off Theorem (IV.3.18) there is an epimorphism $\pi : I/I^2 \to E(I)$. We apply this to the construction of the diagram (1) and of the ideal J.

From the lower row of (1) we get the exact sequence

$$0 \to E(I) \to R/J \to R/I \to 0,$$

whence we get an isomophism $E(I) \cong I/J$. We show:

Proposition 4.2. The ideal J thus constructed is generated by 2 elements. In particular, $\mathrm{Rad}(I) = \mathrm{Rad}(f_1, f_2)$ with $f_1, f_2 \in I$ for any ideal I with the properties indicated above.

Proof. By 4.1 J is locally generated by 2 elements. Then $\mathrm{pd}(J) \leq 1$, and the results of §3 can be applied to J just as to I. In particular, the module $E(J)$ is defined. By 3.5 it suffices to show that $E(J)$ is generated by one element.

For this we construct a commutative diagram with exact rows and columns:

$$
\begin{array}{ccccccccc}
& & 0 & & 0 & & 0 & & \\
& & \downarrow & & \downarrow & & \downarrow & & \\
0 & \to & G_1 & \xrightarrow{\alpha} & G_0 & \to & J & \to & 0 \\
& & \downarrow{\scriptstyle\beta} & & \downarrow & & \downarrow & & \\
0 & \to & F_1 & \to & F_0 \oplus G_0 & \xrightarrow{\epsilon} & I & \to & 0 \\
& & \downarrow & & \downarrow & & \downarrow & & \\
0 & \to & K & \to & F_0 & \to & E(I) & \to & 0 \\
& & \downarrow & & \downarrow & & \downarrow & & \\
& & 0 & & 0 & & 0 & &
\end{array}
$$

Here $0 \to G_1 \to G_0 \to J \to 0$ is a free resolution of J and

$$0 \to J \to I \to E(I) \to 0$$

is the sequence corresponding to the isomorphism $E(I) \cong I/J$. K is the module defined in 3.9 using a projective resolution $0 \to P_1 \to P_0 \to I \to 0$, F_0 is the P_1^* there, and $0 \to K \to F_0 \to E(I) \to 0$ is the sequence $0 \to K \to P_1^* \to E(I) \to 0$ there. The epimorphism $\epsilon : F_0 \oplus G_0 \to I$ is constructed as in the proof of 1.8; F_1 is its kernel. The exact sequence $0 \to G_1 \to F_1 \to K \to 0$ results from the Snake Lemma.

By dualizing we get the following commutative diagram with exact rows and columns.

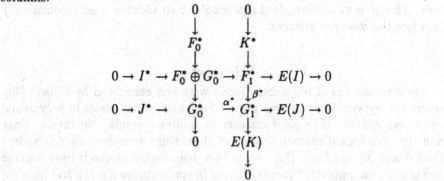

Since $\mathrm{Im}(\alpha^*) \subset \mathrm{Im}(\beta^*), G_1^* \to E(K)$ induces an epimorphism $\psi_0 : E(J) \to E(K)$. By 3.9, $E(K) \cong R/I$; therefore, we also have an epimorphism $\psi : E(J) \to R/I$ and a commutative diagram (since $E(J)$ and R/I are both R/J-modules)

$$
\begin{array}{ccc}
 & R/J & \\
{\scriptstyle \chi}\swarrow & & \downarrow{\scriptstyle c} \\
E(J) \underset{\psi}{\to} & & R/I \to 0,
\end{array}
$$

where c is the canonical epimorphism.

We shall show that χ is bijective. It suffices to show this locally; hence we may assume that R is a local ring. In this case $E(J) \cong R/J$ by 4.1 and 3.7, and we have a commutative diagram

$$
\begin{array}{ccc}
 & R/J & \\
{\scriptstyle \chi}\swarrow & & \downarrow{\scriptstyle c} \\
R/J \underset{\psi}{\to} & & R/I.
\end{array}
$$

If $\chi(1)$ were not a unit in R/J, then, because R is local, neither would $\psi(\chi(1)) = c(1)$ be a unit in R/I. But $c(1) = 1$. Therefore, since $\chi(1)$ is a unit in R/J, χ is an isomorphism, q. e. d.

The hypotheses above are fulfilled if $R = K[X_1, X_2, X_3]$ is the polynomial ring in 3 variables over a field K, and I is an ideal that is locally a complete intersection and defines an algebraic curve in A^3. In particular, we have proved:

Theorem 4.3. (Szpiro [78]) Let $C \subset \mathsf{A}^3(L)$ be a curve defined over K that is locally a complete intersection (with respect to K). Then C is the intersection of two surfaces in $\mathsf{A}^3(L)$ defined over K.

In particular, the theorem can be applied for $K = L$ to smooth curves in A^3. Its proof shows that we actually got a more general result, since the ideal I in the discussion above need not be the complete vanishing ideal of the curve: We actually proved a theorem on 1-dimensional subschemes of A^3. In fact, it was indispensable to the proof that arbitrary ideals defining the curve be kept in view. This gives us an example of how proofs about varieties often automatically lead into the theory of schemes.

References

Free resolutions of polynomial ideals were first considered by Hilbert [38], where the Syzygy Theorem was proved for homogeneous ideals in polynomial rings over a field. The generalization to arbitrary regular Noetherian rings and the homological characterization of these rings were done by Auslander–Buchsbaum [6] and Serre [72]. Since then homological methods have become one of the most important means of proof in commutative algebra and have led to many remarkable results.

The idea of studying modules of projective dimension ≤ 1 to get results on affine varieties in codimension 2 goes back to Serre [73]. The connections with the problem of complete intersection brought to light by this method were a strong motivation for deciding Serre's problem on projective modules.

In the language of homological algebra the module $E(M)$ considered in §3 is the module $\mathrm{Ext}_R^1(M, R)$; in the case of an ideal I, $E(I) = \mathrm{Ext}_R^1(I, R) \cong \mathrm{Ext}_R^2(R/I, R)$. If R is a polynomial ring in 3 variables over a field and I is the ideal of a curve $C \subset \mathsf{A}^3$ that is locally a complete intersection, then $\mathrm{Ext}_R^2(R/I, R)$ is also called the *canonical* (or *dualizing*) module of C. In this case Proposition 3.9 says that $\mathrm{Ext}_R^2(\mathrm{Ext}_R^2(R/I, R), R) \cong R/I$, a (very) special case of Grothendieck's Local Duality Theorem [29] (cf. also [36]). A characterization of the ideals I with $\mathrm{pd}(I) \leq 1$ is given by Ohm [63].

Theorem 4.3, due to Szpiro, was inspired by a result of Ferrand's [22] on projective space curves, which yields however in the projective case only a weaker result. As mentioned in Chapter V, Theorem 4.3 was extended by Mohan Kumar [56] to the case of curves in A^n ($n \geq 3$). But the proof requires somewhat more than the elementary methods used here. By skillfully applying the Frobenius homomorphism Cowsik–Nori [11] proved that if L has prime characteristic, then arbitrary curves in $\mathsf{A}^n(L)$ are set-theoretic complete intersections. Whether this is true for characteristic zero also seems to be still an open problem presently.

The question of under what conditions smooth affine curves are ideal-theoretic complete intersections is studied by, among others, Abhyankar [2], Abhyankar-Sathaye [4], Geyer [25], and Ohm [63].

For many years it was an open question whether the Betti-series (Poincaré-series) $P_R(T)$ of a Noetherian local ring R (see §1 Exercise 6) is always rational, that is, if there exist polynomials $P(T), Q(T) \in \mathbf{Z}(T), Q(T) \neq 0$ with $Q(T) \cdot P_R(T) = P(T)$. This is known to be true for many R (e.g. if R is a complete intersection) but recently D. Anick (Construction d'espace de lacets et d'anneaux locaux à série Poincaré-Betti non rationelles, *Comptes rendus Acad. Sci. Paris* **290**, 729–732 (1980)) has also given a counter-example. J. E. Roos (Homology of loop space and local rings, *Proc. 18th Scandinavian Congress of Mathematicians*, 336–346 (1981)) gives a report on the whole subject.

Bibliography

A. Textbooks

I. On Commutative Algebra

[A] *Atiyah, M. F.* and *I. G. Macdonald*, Introduction to Commutative Algebra. Addison-Wesley, Reading, Mass. (1969).

[B] *Bourbaki, N.*, Algèbre Commutative. Hermann, Paris (1961–1965).

[C] *Kaplansky, I.*, Fields and Rings. Univ. of Chicago Press (1969).

[D] *Kaplansky, I.*, Commutative Rings. Allyn and Bacon, Boston (1970).

[E] *Matsumura, H.*, Commutative Algebra, Second edition. Benjamin, New York (1980).

[F] *Nagata, M.*, Local Rings. Interscience Tracts in Pure and Appl. Math. Wiley, New York (1962).

[G] *Serre, J. P.*, Algèbre locale. Multiplicités. Springer Lecture Notes in Math. **11** (1965).

[H] *Zariski, O.* and *P. Samuel*, Commutative Algebra. Vol. I–II. Van Nostrand, Princeton (1958, 1960).

II. On Algebraic Geometry

[I] *Borel, A.*, Linear Algebraic Groups. Benjamin, New York (1969).

[K] *Dieudonné, J.*, Cours de géométrie algébrique. Presses Univ. France (1974).

[L] *Fulton, W.*, Algebraic Curves. Benjamin, New York (1969).

[M] *Grothendieck, A.* and *J. Dieudonné*, Eléments de géométrie algébrique. Publ. Math. IHES **4** (1960), **8** (1961), **11** (1961), **17** (1963), **20** (1964), **24** (1965), **28** (1966), **32** (1967).

[N] *Hartshorne, R.*, Algebraic Geometry. Springer, Heidelberg (1977).

[O] *Hodge, W. V. D.* and *D. Pedoe*, Methods of Algebraic Geometry. Vol. I–II. Cambridge University Press (1968).

[P] *Mumford, D.*, Algebraic Geometry I. Complex Projective Varieties. Springer, Heidelberg (1976).

[Q] *Lang, S.*, Introduction to Algebraic Geometry. Addison-Wesley, Reading, Mass. (1968).

[R] *Lang, S.*, Diophantine Geometry. Interscience Publishers (1962).

[S] *Seidenberg, A.*, Elements of the Theory of Algebraic Curves. Addison-Wesley, Reading, Mass. (1969).

[T] *Semple, J. G.* and *G. T. Kneebone*, Algebraic Curves. At the Clarendon Press, Oxford (1959).

[U] *Shafarevich, I. R.*, Basic Algebraic Geometry. Springer, Heidelberg (1974).

[V] *Van der Waerden, B.*, Einführung in die algebraische Geometrie. Springer, Heidelberg (1973).

[W] *Walker, R. J.*, Algebraic Curves. Dover (1950).

[X] *Weil, A.*, Foundations of Algebraic Geometry. Am. Math. Soc. Coll. Publications **29** (1962).

[Y] *Griffiths, P.* and *J. Harris*, Principles of Algebraic Geometry. Wiley, New York (1978).

[Z] *Brieskorn, E.* and *H. Knörrer*, Ebene algebraische Kurven. Birkhäuser, Basel-Boston-Stuttgart (1981).

AA] *Fulton, W.*, Intersection Theory. Springer, Berlin-Heidelberg-New York-Tokyo (1984).

B. Research Papers

[1] Algebraic Geometry. Arcata 1974. Proc. Symp. Pure Math. **29** (1975)

[2] *Abhyankar, S. S.*, Algebraic Space Curves. Sém. Math. Sup. **43**, Les presses de l'université de Montréal (1971).

[3] *Abhyankar, S. S.*, On Macaulay's Example. In: Conf. Comm. Algebra; Lawrence 1972. Springer Lecture Notes in Math. **311** (1973), 1–16.

[4] *Abhyankar, S. S.* and *A. M. Sathaye*, Geometric Theory of Algebraic Space Curves. Springer Lecture Notes in Math. **423** (1974).

[5] *Artin, E.* and *J. Tate*, A Note on Finite Ring Extensions. J. Math. Soc. Japan **3** (1951), 74–77.

[6] *Auslander, M.* and *D. A. Buchsbaum*, Homological Dimension in Local Rings. Trans. Am. Math. Soc. **85** (1957), 390–405.

[7] *Bass, H.*, On the Ubiquity of Gorenstein Rings. Math. Z. **82** (1963), 8–28.

[8] *Bass, H.*, Libération des modules projectifs sur certains anneaux de polynômes. Sém. Bourbaki 1973/74, n° 448.

[9] *Chevalley, C.*, On the Notion of the Ring of Quotients of a Prime Ideal. Bull. Am. Math. Soc. **50** (1944), 93–97.

[10] *Cohen, I. S.* and *A. Seidenberg*, Prime Ideals and Integral Dependence. Bull. Am. Math. Soc. **52** (1946), 252–261.

[11] *Cowsik, R. C.* and *M. V. Nori*, Curves in Characteristic p are Set Theoretic Complete Intersections. Inv. Math. **45** (1978), 111–114.

[12] *Davis, E. D.*, Ideals of the Principal Class, R-Sequences and a Certain Monoidal Transformation. Pac. J. Math. **20** (1967), 197–205.

[13] *Davis, E. D.*, Further Remarks on Ideals of the Principal Class. Pac. J. Math. **27** (1968), 49–51.

[14] *Davis, E. D.*, Prime Elements and Prime Sequences in Polynomial Rings. Proc. Am. Math. Soc. **27** (1978), 33–38.

[15] *Davis, E. D,* and *A. V. Geramita*, Efficient Generation of Maximal Ideals in Polynomial Rings. Trans. Am. Math. Soc. **231** (1977), 497–505.

[16] *Eagon, J. A.* and *D. G. Northcott*, Ideals Defined by Matrices and a Certain Complex Associated with them. Proc. Royal Soc. England **A269** (1962), 188–204.

[17] *Eisenbud, D.* and *E. G. Evans, Jr.*, Every Algebraic Set in n-Space is the Intersection of n Hypersurfaces. Inv. Math. **19** (1973), 107–112.

[18] *Eisenbud, D.* and *E. G. Evans, Jr.*, A Generalized Principal Ideal Theorem. Nagoya Math. J. **62** (1976), 41–53.

[19] *Eisenbud, D.* and *E. G. Evans, Jr.*, Generating Modules Efficiently: Theorems from Algebraic K-Theory. J. Alg. **27** (1973), 278–315.

[20] *Eisenbud, D.* and *E. G. Evans, Jr.*, Three Conjectures about Modules over Polynomial Rings. In: Conf. Comm. Algebra; Lawrence 1972. Springer Lecture Notes in Math. **311** (1973), 78–89.

[21] *Ferrand, D.*, Suite régulière et intersection complète. C. R. Acad. Sci. Paris **264** (1967), 427–428.

[22] *Ferrand, D.*, Courbes gauches et fibré de rang deux. C. R. Acad. Sci. Paris **281** (1975), 345–347.

[23] *Ferrand, D.*, Les modules projectifs de type fini sur un anneau de polynômes sur un corps sont libres. Sém. Bourbaki 1975/1976, n° 484.

[24] *Forster, O.*, Über die Anzahl der Erzeugenden eines Ideals in einem noetherschen Ring. Math. Z. **84** (1964), 80–87.

[25] *Geyer, W. D.*, On the Number of Equations which are Necessary to Describe an Algebraic Set in n-Space. Atas 3^a Escola de Algebra, Brasilia (1976), 183–317.

[26] *Goldman, O.*, Hilbert Rings and the Hilbert Nullstellensatz. Math. Z. **54** (1952), 136–140.

[27] *Grell, H.*, Beziehung zwischen den Idealen verschiedener Ringe. Math. Ann. **97** (1927), 490–523.

[28] *Gröbner, W.*, Über irreduzibel Ideale in kommutativen Ringen. Math. Ann. **110** (1934), 197–222.

[29] *Grothendieck, A.*, Local Cohomology, Springer Lecture Notes in Math. **41** (1967).

[30] *Grothendieck, A.*, et al., Séminaire de Géometrie Algébrique. Springer Lecture Notes in Math. **151, 152, 153, 224, 225, 269, 270, 288, 305, 340, 569**; and (SGA 2) North-Holland, Amsterdam (1968).

[31] *Hartshorne, R.*, Complete Intersections and Connectedness. Am. J. Math. **84** (1962), 497–508.

[32] *Hartshorne, R.*, Cohomological Dimension of Algebraic Varieties. Ann. of Math. **88** (1968), 403–450.

[33] *Heitmann, R. C.*, A Negative Answer to the Prime Sequence Question. Proc. Amer. Math. Soc. **77** (1979), 23–26.

[34] *Herzog, J.*, Generators and Relations of Abelian Semigroups and Semigroup-Rings. Manuscripta Math. **3** (1970), 153–193.

[35] *Herzog, J.*, Ein Cohen-Macaulay-Kriterium mit Anwendungen auf den Differentialmodul und den Konormalenmodul. Math. Z. **163** (1978), 149–162.

[36] *Herzog, J.* and *E. Kunz* (ed.), Der kanonische Modul eines Cohen-Macaulay-Rings. Springer Lecture Notes in Math. **238** (1971).

[37] *Herzog, J.* and *E. Kunz*, Die Wertehalbgruppe eines lokalen Rings der Dimension 1. S. B. Heidelberger Akad. Wiss. **2** (1971).

[38] *Hilbert, D.*, Über die Theorie der algebraischen Formen. Math. Ann. **36** (1890), 473–534.

[39] *Hilbert, D.*, Über die vollen Invariantensysteme. Math. Ann. **42** (1893), 313–373.

[40] *Horrocks, G.*, Projective Modules over an Extension of a Local Ring. Proc. London Math. Soc. **14** (1964), 714–718.

[41] *Kneser, M.*, Über die Darstellung algebraischer Raumkurven als Durchschnitte von Flächen. Arch. Math. **11** (1960), 157–158.

[42] *Kronecker, L.*, Grundzüge einer arithmetischen Theorie der algebraischen Größen. J. reine angew. Math. **92** (1882), 1–123.

[43] *Krull, W.*, Primidealketten in allgemeinen Ringbereichen. S. B. Heidelberger Akad. Wiss. **7** (1928).

[44] *Krull, W.*, Über einen Hauptsatz der allgemeinen Ringtheorie. S. B. Heidelberger Akad. Wiss. **2** (1929).

[45] *Krull, W.*, Idealtheorie. Ergebnisse d. Math. **4**, Nr. 3, Springer (1935).

[46] *Krull, W.*, Dimensionstheorie in Stellenringen. J. reine angew. Math. **179** (1938), 204–226.

[47] *Krull, W.*, Jacobsonsche Ringe, Hilbertscher Nullstellensatz, Dimensionstheorie. Math. Z. **54** (1951), 354–387.

[48] *Kunz, E.*, Characterizations of Regular Local Rings of Characteristic p. Am. J. Math. **91** (1969), 772–784.

[49] *Kunz, E.*, On Noetherian Rings of Characteristic p. Am. J. Math. **98** (1976), 999–1013.

[50] *Lam, T. Y.*, Serre's Conjecture. Springer Lecture Notes in Math. **635** (1978).

[51] *Lasker, E.*, Zur Theorie der Moduln und Ideale. Math. Ann. **60** (1905), 20–116.

[52] *Lindel, H.*, Eine Bemerkung zur Quillenschen Lösung des Serreschen Problems. Math. Ann. **230** (1977), 97–100.

[53] *Lindel, H.*, Projektive Moduln über Polynomringen $A[T_1, \ldots, T_m]$ mit einem regulären Grundring A. Manuscripta Math. **23** (1978), 143–154.

[54] *Lipman, J.*, Unique Factorization in Complete Local Rings. In: Algebraic Geometry. Arcata 1974. Proc. Symp. Pure Math. **29** (1975).

[55] *Macaulay, F. S.*, Algebraic Theory of Modular Systems. Cambridge Tracts **19** (1916).

[56] *Mohan Kumar, N.*, On two Conjectures about Polynomial Rings. Inv. Math. **46** (1978), 225–236.

[57] *Murthy, M. P.*, Generators for Certain Ideals in Regular Rings of Dimension Three. Comm. Math. Helv. **47** (1972), 179–184.

[58] *Murthy, M. P.*, Complete Intersections. In: Conf. Comm. Algebra; Kingston 1975. Queen's Papers Pure Appl. Math. **42**.

[59] *Noether, E.*, Idealtheorie in Ringbereichen. Math. Ann. **83** (1921), 24–66.

[60] *Noether, E.*, Der Endlichkeitssatz der Invarianten endlicher linearer Gruppen der Charakteristik p. Nachr. Ges. Wiss. Göttingen (1926), 28–35.

[61] *Northcott, D. G. and D. Rees*, Reduction of Ideals in Local Rings. Proc. Cambridge Phil. Soc. **50** (1954), 145–158.

[62] *Northcott, D. G.* and *D. Rees*, Extensions and Simplifications of the Theory of Regular Local Rings. Proc. Cambridge Phil. Soc. **57** (1961), 483–488.

[63] *Ohm, J.*, Space Curves as Ideal-theoretic Complete Intersections. In: Studies in Math. **20**. Math. Assoc. Amer. (1980), 47–115.

[64] *Perron, O.*, Über das Vahlensche Beispiel zu einem Satz von Kronecker. Math. Z. **47** (1942), 318–324.

[65] *Perron, O.*, Beweis und Verschärfung eines Satzes von Kronecker. Math. Ann. **118** (1941/43), 441–448.

[66] *Quillen, D.*, Projective Modules over Polynomial Rings. Inv. Math. **36** (1976), 167–171.

[67] *Rung, J.*, Mengentheoretische Durchschnitte und Zusammenhang. Regensburger Math. Schriften **3** (1978).

[68] *Sarges, H.*, Ein Beweis des Hilbertschen Basissatzes. J. reine angew. Math. **283/284** (1976), 436–437.

[69] *Sally, J.*, Number of Generators of Ideals in Local Rings. Lecture Notes Pure Appl. Math. Series 35. Dekker, New York (1978).

[70] *Schenzel, P.* and *W. Vogel*, On Set-theoretic Intersections. J. Alg. **48** (1977), 401–408.

[71] *Sathaye, A.*, On the Forster-Eisenbud-Evans Conjecture. Inv. Math. **46** (1978), 211–224.

[72] *Serre, J. P.*, Sur la dimension des anneaux et des modules noetheriens. Proc. Int. Symp. Alg. Number Theory, Tokyo and Nikko (1955).

[73] *Serre, J. P.*, Sur les modules projectifs. Sém. Dubreil-Pisot 1960/61.

[74] *Silhol, R.*, Géometrie algébrique sur un corps non algébriquement clos. Commun. Alg. **6** (1978), 1131–1155.

[75] *Suslin, A.*, Projective Modules over Polynomial Rings (Russian). Dokl. Akad. Nauk S. S. S. R. **26** (1976).

[76] *Swan R.*, The Number of Generators of a Module. Math. Z. **102** (1967), 318–322.

[77] *Swan, R.*, Serre's Problem. In: Conf. Comm. Algebra; Kingston 1975. Queen's Papers Pure Appl. Math. **42**.

[78] *Szpiro, L.*, Lectures on Equations Defining Space Curves (Notes by N. Mohan Kumar). Tata Inst. Lecture Notes in Math., Bombay (1979).

[79] *Storch, U.*, Bemerkung zu einem Satz von M. Kneser. Arch. Math. **23** (1972), 403–404.

[80] *Uzkov, A. I.*, On Quotient Rings of Commutative Rings (Russian). Mat. Sbornik N.S. **22** (64) (1948), 439–441.

[81] *Vasconcelos, W. V.*, Ideals Generated by R-Sequences. J. Alg. **6** (1967), 309–316.

[82] *Whitney, H.*, Elementary Structure of Real Algebraic Varieties. Ann. of Math. **66** (1957), 545–556.

[83] *Zariski, O.*, A New Proof of the Hilbert Nullstellensatz. Bull. Am. Math. Soc. **53** (1947), 363–368.

[84] *Zariski, O.*, The Concept of a Simple Point of an Abstract Algebraic Variety. Trans. Am. Math. Soc. **62** (1947), 1–52.

List of Symbols

:=	is defined as
a)→b)	a) implies b)
a)↔b)	a) and b) are equivalent
{ }	set consisting of
{ | }	set of all ... such that
∈	is an element of
∉	is not an element of
⊂	is a subset of
$M \cap N$	intersection of the sets M and N
$M \cup N$	union of the sets M and N
N	the set of natural numbers $0, 1, 2, \ldots$
Z	the set of integers
Q	the set of rational numbers
R	the set of real numbers
C	the set of complex numbers
∅	empty set
$M \times N$	cartesian product of the sets M and N
R^n	the set of all n-tuples of elements of R
R^Λ	the set of all mappings $\Lambda \to R$
$M \setminus N$	complementary set of N in M
$M \circ N$	composition of mappings M and N
$a \mapsto b$	a is mapped onto b
$\mathbb{A}^n(K)$	affine n-space over K
$\mathbb{P}^n(K)$	projective n-space over K
$K^\times$	set of all elements $\neq 0$ of a field K
$\mathfrak{I}(V)$	vanishing ideal of the variety V
$\mathfrak{I}_V(W)$	vanishing ideal of the subvariety W of V
$\mathfrak{p}_x$	prime ideal of the point x
$\mathfrak{V}(I)$	zero-set of the ideal I
$\mathfrak{V}_V(I)$	zero-set on V of the ideal I
$K[V]$	coordinate ring of the K-variety V

$R(V)$	ring of rational functions on a variety V
$\mathrm{Rad}(I)$	radical of an ideal I
IS	extension ideal of the ideal I in S
R_{red}	reduced ring associated with R
$\mathrm{Spec}(R)$	spectrum of the ring R
$J(R)$	J-spectrum of the ring R
$\mathrm{Max}(R)$	maximal spectrum of the ring R
$\mathrm{Proj}(G)$	homogeneous spectrum of a graded ring G
$\overline{A}$	closure of the set A
$\overline{A}$	integral closure of the ring A
$M \oplus N$	direct sum of M and N
trdeg	transcendence degree
$\dim$	Krull dimension
$\dim_x V$	dimension of V at x
$\dim_K$	dimension as a K-vector space
codim	codimension
edim	embedding dimension
J-$\dim$	J-dimension
g-$\dim$	g-dimension
$h(I)$	height of the ideal I
$D(f)$	see Ch. III, §1
$\mathcal{O}(U)$	algebra of regular functions on U
$\mathcal{O}_{V,W}$	local ring of V at W
$\mathcal{O}_{V,x}$	local ring of V at x
μ_r	multiplication by r
$\mu(M)$	length of a shortest system of generators of M
$\mu_\mathfrak{p}(M)$	length of a shortest system of generators of $M_\mathfrak{p}$
M_S	module of fractions with denominator set S
R_S	ring of fractions with denominator set S
M_g	module of fractions with respect to $\{1, g, g^2, \dots\}$
R_g	ring of fractions with respect to $\{1, g, g^2, \dots\}$
$M_\mathfrak{p}$	localization of M with respect to a prime $\mathfrak{p}$
$R_\mathfrak{p}$	localization of R with respect to a prime $\mathfrak{p}$
$Q(R)$	full ring of fractions of R
$\mathrm{Ann}(M)$	annihilator of the module M
$\mathrm{Ann}(m)$	annihilator of the element m

$T(M)$	torsion of M
$\mathrm{Supp}(M)$	support of M
$\mathrm{Aut}(M)$	group of automorphisms of M
M^*	dual module of M
M^{**}	bidual module of M
$F_i(M)$	i-th Fitting ideal of M
$l(M)$	length of the module M
$l(R)$	length of the ring R
$\mathrm{Ass}(M)$	set of associated primes of M
$d(I, M)$	I-depth of M
$d(M)$	depth of M
$\mathfrak{S}(M)$	socle of M
$\mathrm{pd}(M)$	projective dimension of M
$E(M)$	see Ch. VII, §3
$M_1 \amalg_N M_2$	fiber product of M_1 and M_2 with respect to N
$M_1 \amalg_N M_2$	fiber sum of M_1 and M_2 with respect to N
$S(U)$	S-component of a submodule, ideal U
$\mathfrak{p}^{(i)}$	i-th symbolic power of a prime ideal $\mathfrak{p}$
$M(r \times s, R)$	module of $r \times s$-matrices over R
$Gl(n, R)$	group of invertible $n \times n$-matrices over R
$A_1 \sim A_2$	A_1 and A_2 are equivalent matrices (Ch. IV, §1)
$N[X]$	extension module of N to $R[X]$
$\mathfrak{f}_{S/R}$	conductor of a ring extension S/R
$gr_I(R)$	graded ring of R with respect to I
$\mathfrak{R}_I(R)$	Rees ring of R with respect to I
$T_x(V)$	tangent space of a variety V at x
$\mathrm{Reg}(R)$	regular locus of the ring R
$\mathrm{Reg}(V)$	regular locus of the variety V
$\mathrm{Sing}(R)$	singular locus of the ring R
$\mathrm{Sing}(V)$	singular locus of the variety V

Index

affine algebra 19
  of a variety 19
affine
  algebraic variety 1
  coordinate transformation 1
  scheme 27, 87
  space 1
algebraic
  curve 57
  group 4
  point 24
  surface 57
algebraic variety 1
  affine 1
  irreducible 12
  linear 1, 30
  non-singular 166
  normal 175
  projective 30
  quasihomogeneous 3
  smooth 166
  unmixed 134
algebraic vector bundle 110
annihilator
  of an element 78
  of a module 78
Artin–Rees lemma 151
Artinian
  module 128
  ring 128
associated prime ideal 176
Auslander–Buchsbaum formula 202
avoiding prime ideals 64, 142

basic element 105
basis of a module 14
  canonical 14
basis theorem of Hilbert
  for modules 14
  for polynomial rings 11

for power series rings 160
Betti numbers
  of a module 202
  of a ring 202
Betti series 205

canonical module 222
chain ring 53
Chinese remainder theorem 41
closed immersion 73
closure
  integral (of a ring) 45
  projective (of an affine variety) 35
codimension 39
coheight 40
Cohen–Macaulay
  module 186
  ring 186
  singularity 190
  variety 190
Cohen–Seidenberg 46
comaximal ideals 41
comparison theorem
  for projective dimension 199
complete intersection 190
  ideal theoretic 134, 135
  local 135
  set theoretic 134, 135
complexification 15
component
  homogeneous (of an element) 32
  irreducible
    (of a top. space, variety) 12
composition series 127
conductor 103
cone 3
  affine (of a proj. variety) 33
conjugate points 9, 24
connectedness theorem
  of Hartshorne 192

238   INDEX

Zariski topology
  of an affine variety 12
  of a projective variety 33
  of the spectrum 23
  of the projective spectrum 35

zero divisor
  of a module 152
  of a ring 26
zero set 6, 33